Shaping the Vulcan

DESIGN AND DEVELOPMENT OF AVRO'S V-BOMBER
SHAPING THE VULCAN

STEPHEN LIDDLE

TEMPEST BOOKS

ACKNOWLEDGEMENTS

Although I am obviously indebted to Prof Herbert Pearcey (whom I did not meet) and Dr Robert Pleming (who is greatly missed) for providing the original seed from which this work has grown, the meat on the bones relied strongly on three sources. The first is the Aerade online archive, a repository of significant reports from the UK Aeronautical Research Council (ARC) and others. The second was extremely fortuitous: the RAeS Aeronautical Journal archive was made accessible to members just as I kicked off work on the original paper, and the sources I was able to find and incorporate made it come alive. The third was however the most critical, containing as it did Newby's original report on the Vulcan leading-edge work. This was the archive of the Farnborough Air Sciences Trust (FAST), which not only maintains an almost complete set of reports issued by the Royal Aircraft Establishment on site at Farnborough, but an excellent museum too. I am grateful to Geoff Butler, Mike Millar and Alan Brown there, and the support of Michael Trotter of VTST in acquiring these documents.

It would be remiss of me not to record the ongoing support and encouragement of my VTST colleagues. Our chairman John Sharman and Sir Donald Spiers were kind enough to read the original draft of PVN, while a general warmth has emanated from the remainder of the Board: Sir Gerald Howarth, Ken Smart, Phil Spiers, Ed Jarrod, Richard Clarke and our CEO Marc Walters. Bob and Isi Jackson of the Stratford and Events Team have (almost) always made me feel welcome, once they had established who I was at least. I cannot name everyone, but to feel a part of the volunteer family that looks after XH558 is very special and never less than a privilege. I will thank Adrian Sumner, the captain and left-hand seat occupant who kindly supervised me in the right-hand seat, when – like Stanley Hooker – I was able to throttle slam an Olympus. It would not have happened without Kevin 'Taff' Stone, Sam Scrimshaw and my partner in crime for the day, Simon Wray.

Many people have freely given of advice, encouragement and their time, including but very much not limited to: Robert Hopkins III, Matt Bone, Tony Buttler, Matthew Willis, Stephen McParlin, Richard Brown, Calum Douglas, David Lednicer, Joe Coles, Andy Foster, John Falk, Hugh Dibley and Guy Rennie. Dan Sharp of Tempest Books firmly falls into this category, and I must of course thank him for contacting me about writing this book in the first place.

One of my hopes for the book was that it would include previously unpublished images of the aircraft in service, to go along with the data from technical reports. I had felt that it would be rather dry otherwise; in the end, we're all here for the howls and our memories of those extraordinary shapes in the air! This naturally posed a problem for someone of my age, but I have been delighted and privileged to be able to use images from some generous individuals: James Alderman (and ARA), Frank Croom, David Francis Clarke, Doug Greaves, the late Dick Gilbert, David Henderson, Mike Grierson, Ian Perry, Jack Kell, Paul Vincent, Rob Mather and Joe Barr.

When I tentatively wrote to Neville Mangion with a request to use some of the superb images captured in the 1960s and 1970s by his father Godfrey Mangion in Malta, I could not have hoped for the positivity that I received from them both. In attempting to achieve this aim, these dramatic views against landscapes (and sunshine) that would be atypical of Lincolnshire, really have made an enormous difference, and I am very grateful.

As someone fascinated to the point of obsession with the history of these aircraft and how they were used, it has always been both an honour and privilege to have met those involved. The combination of 'good egg' with steel behind the eyes seems to me something to admire, evident none more so than with my friend Martin Withers, who was generous enough to write the Foreword to this book. For their contributions to the journey over the years, which they will likely not remember but which I do, thanks to Keith Mans, Andrew Brookes, Bill Ramsey, and Kev Rumans.

None of this would have happened without my parents having gone out of their way to show me amazing things, while letting me develop my passions. Thank you. These days I am just as supported by my wife Danielle and our eventual replacements Dylan and Charlotte. All my love; none of it would matter without you.

Steve Liddle
February 2025

Published in Great Britain by Tempest Books
an imprint of Mortons Books Ltd.
Media Centre, Morton Way
Horncastle LN9 6JR
www.mortonsbooks.co.uk

© 2025 by Tempest Books

All rights reserved. No part of this publication may be reproduced or transmitted in any form or by any means, electronic or mechanical including photocopying, recording, or any information storage retrieval system without prior permission in writing from the publisher.

ISBN 978-1-911704-07-2

The right of Stephen Liddle to be identified as the author of this work has been asserted in accordance with the Copyright, Designs and Patents Act 1988.
Typeset by Jayne Clements (jayne@hinoki.co.uk), Hinoki Design and Typesetting
Original cover artwork by Piotr Forkasiewicz

For Danielle, Dylan and Charlotte, who willingly put up with this nonsense, and Simon, who would otherwise have been ten books in by now.

CONTENTS

Foreword	9
Preface	10
1 Origins	15
2 Towards the UK Deterrent: With a Union Jack on Top	26
3 Defining the Bomber	36
4 The Building Blocks	54
5 Forged in Etna: The Start of the Vulcan	87
6 Emerging Knowledge: The Major Redesign of 1949	104
7 The Imperfect Delta: Making the Vulcan Work	116
8 The Victor	139
9 The Other Broad Delta: The Gloster Javelin	161
10 Second Generation: The Vulcan and Victor Mark 2	169
11 Bolt from the Blue	199
12 Apogee and Re-Entry	220
13 From Camelot to Nassau	230
14 Beyond the V-Force	249
15 Under the Radar	262
16 In the Reckoning	277
Bibliography	291
Endnotes	296
Index	300

FOREWORD

WITH TANKS nearly empty and contemplating an ejection into the unwelcoming Atlantic, we were delighted to see the Victor appear in a turn ahead of us, its refuelling basket trailing behind. It was the most beautiful sight in the world. We were very low on fuel!

Approximately 11 hours earlier, in the company of 11 Victor tankers, we had set off at night from Ascension Island (ASI) refuelling the aircraft five times, tasked with bombing the heavily defended runway on the airport at Stanley on the Falkland Islands, which had been invaded by Argentina three weeks earlier. We then had to return to ASI, a round-trip of 8,000 miles lasting 15:45 hours.

Our attack had been successful and the bomb which made a crater on the runway prevented any Argentinian fast jets from using the airfield as a Forward Operating Base for the rest of their occupation until the Falklands were liberated six weeks later.

The fuel consumption of the Vulcan, carrying 21 x 1,000lb bombs and constantly filling to full while flying at a speed and height to suit the heavy Victors, themselves refuelling one another, had been seriously underestimated and it was a close-run thing for all the aircraft to get back safely.

This was the first time since the Vulcan and the other V bombers had been used in anger, although fortunately not in the role of 'nuclear bomber', just one stick of 21 conventional 'thousand pounders'.

The author, who has written at length about the development of both military and where relevant post war civil aircraft, is very well qualified in aerodynamics and the use of wind tunnels and points out the unique features of the Vulcan. The design of the Vulcan was extremely advanced considering that it came off the Avro production line only 11 years after the last Lancaster was built.

Designed for high-altitude flight, with its large wing area the Vulcan could outmanoeuvre any fighter jet of its era at height by simply slowing down and turning, yet was still able to pull 2.5 G, and with its large elevons see the attacker or heat-seeking missile go sailing past. Without any flaps or slats, it was so easy to fly and when light, could fly happily at 150 knots (with a high nose attitude!) or catch up to join with the Red Arrows flying at 300 knots.

I was lucky enough to be invited by the Vulcan to the Sky Team to display XH558 at airshows. The only remaining flying Vulcan was always the star of the show. The crowds loved to see it, manoeuvring, slowly, quietly and gracefully, and then creating a huge noise with its four powerful Olympus engines howling as the aircraft was pulled up into a steep climb and rolled into a wingover just on the buffet.

I still remember the first time I saw a Vulcan as a boy in 1952, flying over the top of me when I was staying with family near Woodford where the aircraft was built. I was amazed by it and will never forget the experience. I never expected to be flying this superb aircraft in the future.

On October 15, 2015, I landed Vulcan 558 for the last time at Doncaster. No Vulcan would ever fly again.

This fascinating book is packed with information about the incredible advancements made in aviation over the years and answers questions as to why changes had to be made and the thought processes involved. A brilliant book for all aviation history fans.

Martin Withers
DFC FRAeS

PREFACE

In 2015 Dr Robert Pleming, in his capacity as chief executive of the Vulcan to the Sky Trust (VTST), made contact with Professor Herbert Pearcey, whom he had been advised had been involved in the design of the Avro Vulcan and particularly the leading-edge modifications, widely known to have been embodied to improve buffet performance. The resulting brief correspondence between them included Pearcey's own description of his actual involvement, which concluded with a remarkable implication — that the Vulcan's redesigned leading-edge was a point on the development path to the supercritical aerofoil.

> "Our back-room research received quite a boost from retrospective tests on two-dimensional aerofoils by comparing the flow for the section normal to the leading-edge of the Vulcan in its drooped form as designed by Ken Newby of the RAE with that for the original. The drooped leading-edge produced the rapid 'peaky' supersonic acceleration that, after reflection with change of sign at the boundary between locally supersonic and subsonic speeds in the outer flow, helps to reduce the strength of the ultimate shock wave. It was the understanding and exploitation of this phenomenon that was to lead us ultimately to the supercritical wing sections that are now commonplace on most civil airliners."[1]

Five years or so earlier, the author had happened to be invited to a dinner at the Royal Aeronautical Society, where Dr Pleming accepted an award on behalf of his charity, in recognition of the extraordinary effort to return Vulcan XH558 to flight. As a self-diagnosed Vulcan obsessive, it would have been remiss not to congratulate a man I was meeting for the first time, and one who was the cornerstone of a project that had brought excitement to myself and thousands of others. If we were to head back in time further still, then there would be ample evidence — photos against typically grey September skies of South Yorkshire, a battered Airfix model, a brochure for the final season of the RAF's Vulcan Display Flight — of the role of this mighty machine in inspiring my direction in life. Back in the Argyll Room at 4 Hamilton Place, I did not need to feign interest and this meeting would lead directly to my involvement as a trustee in VTST.

My professional career as an aerodynamicist meant that mine was a natural direction for Robert to refer Prof Pearcey's correspondence. Might we, he enquired, make use of this remarkable link, which was almost perfectly aligned with our aims for the (sadly inevitable) post-flight phase of XH558's life? To my regret, I could not immediately investigate this question in the depth required, and events soon overtook this request for the charity, as finding a new home the aircraft became an imperative.

More years passed. I had long held an ambition — which I did not seriously expect to fulfil — to

PREFACE

In July 2015, Vulcan XH558 made its final appearance at the Royal International Air Tattoo. After completing the display with a flypast with the Red Arrows, the last airworthy V-bomber was marshalled back on stand by their ground crew, the famous 'Blues'. The watchful eyes of Vulcan to the Sky's Taff Stone and his engineering team clearly remained attentive, however.

The V-bombers, Britain's airborne deterrent. Closest to the camera, the interim Vickers Valiant, leading the formation, the Avro Vulcan and farthest away, the last to enter service but perhaps offering the greatest potential in terms of long-range, high-altitude load carrying, the Handley Page Victor.

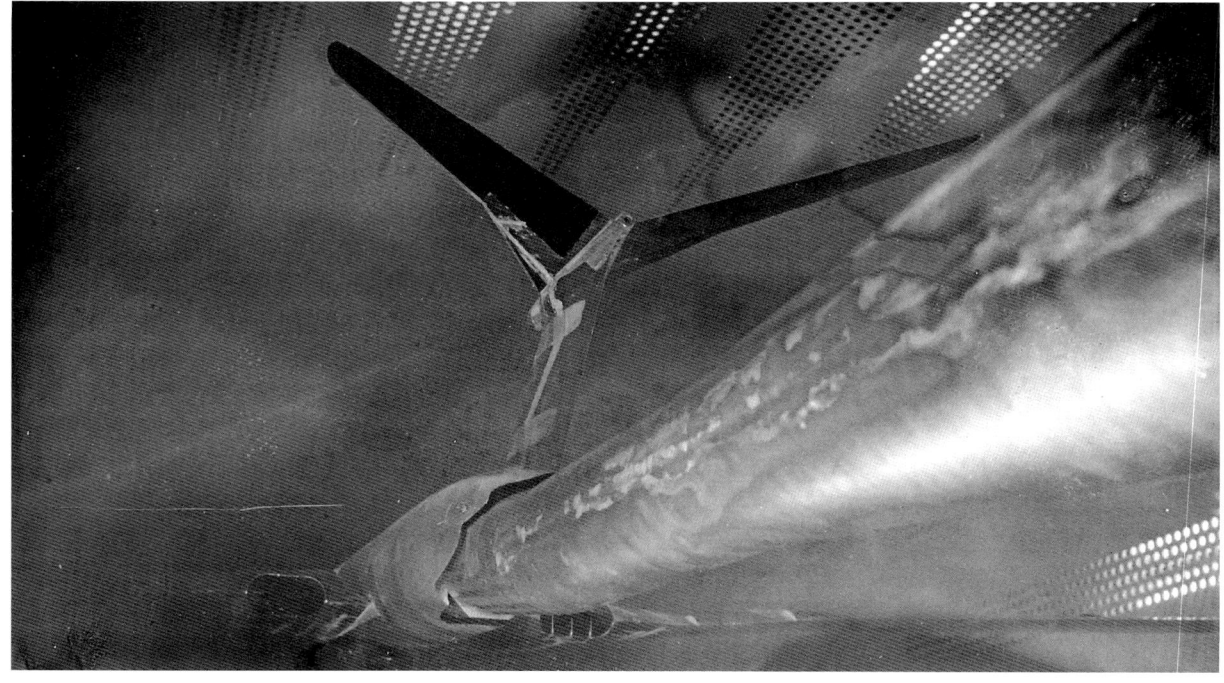

The last airworthy V-bomber, Avro Vulcan XH558 of the Vulcan to the Sky Trust, displaying at the Imperial War Museum Duxford, in September 2014.

Testing the Handley Page Victor in ARA transonic wind tunnel.

contribute something to the published record on the Vulcan, but of course after many decades the question was one of, what could I say that was new? I was however inspired by the work of one of my old lecturers from university days—Dr John Ackroyd, who as well as making cutting edge contributions to aerodynamics in his time, is a renowned technical historian. I was particularly impressed with two works he had produced for the *RAeS Journal of Aeronautical History*, one on the aerodynamics of the Spitfire, and a second, reassuringly vast work on the aerodynamic efficiency of British aircraft leading up to and during the Second World War. What could I do that was of a similar ilk in respect of the Vulcan; what new angle could there be? I had of course, been sitting on the answer to that for several years: aside from Robert, I was the only other person in possession of Herbert Pearcey's correspondence.

As can so often be the case, a simple description of an historical event would grow arms and legs, in this case on the drooped leading edges and a lengthy sojourn into the technical development of the UK's post-war airliners. I was able to emulate Dr Ackroyd in having the resulting paper, entitled *Pearcey, Newby and the Vulcan*, accepted for publication in the journal during the nominally post-COVID summer of 2021. The work showed that Pearcey's assertion was entirely sound. Along with his colleagues at the National Physical Laboratory (NPL), he had postulated, understood and exploited an entirely new aerodynamic mechanism which, when deployed by Ken Newby of the Royal Aircraft Establishment, had fundamentally saved the Vulcan.

Subsequently, it had formed one of the pillars of the advanced wing technology that, in the first instance, won the UK aerospace industry its place in the Airbus project, and then allowed that company to demonstrate technical parity if not dominance in the field. The technology would have happened in time, but would it have been soon enough for the A300B, Hawker Siddeley's involvement and the success that has followed? We will never know, but it would certainly have been a close-run thing.

This book has grown from that work, in order to try and answer some more of the questions about not just the aerodynamics of the Vulcan, but its great rival for the UK nuclear bomber contract, the Handley Page Victor. It is about something else too; these then-futuristic machines were created under difficult circumstances, in dreary, bomb-damaged surroundings, by people who must have felt keenly the presence of a threat. The received wisdom, or myth passed down, is often that the best ideas had been plundered from the ruins of Germany. The truth, as shall hopefully be seen, is much more subtle. It does however involve the presence of German nationals who had been the King's enemies just months previously. How remarkable is the magnanimity in which their fellow Europeans absorbed them. In the end, this is a story of high technology as it is moulded by people.

This is not a guide to serial numbers, squadrons and bases, nor eye-patches and Black Buck. You already have those on your shelf.[2] This is, I believe, a view of these charismatic, historically significant and much-admired aircraft through a different prism: the story of their aerodynamic development. It is an origin story, where the engineers were the heroes.

Vulcan XH558 encountering favourable lighting over Duxford in 2014, emphasising the camber and droop of the outboard leading edge.

1

ORIGINS

SEVEN DECADES after their first flights, the Avro Vulcan and Handley Page Victor remain iconic examples of British engineering and military capability, firmly rooted in the public consciousness. There is an underlying logic to this since, far from being merely the latest advanced aircraft to enter service, they formed the spearhead of a completely new defensive system, charged with saving Britain from the disaster of another all-encompassing total war.

An attack, should it ever come, would be met with an overwhelming retaliation, the shape of which had been demonstrated clearly in the fate of two devastated Japanese cities whose names would become tragically familiar. In a country where everyone had suffered for years and knew someone that had been lost, the importance of the nuclear deterrent was absolutely clear. And in 1947, there was only one plausible way in which it might be delivered.

The development of Britain's strategic nuclear bomber force required a step change in the complexity of the engineering brought to bear, compared to the previous generation of piston-engined bombers designed and manufactured by the same companies. While not a direct measure, the difference was reflected in the price paid by the Ministry of Supply. Whereas the unit price for an Avro Lancaster in 1943 was of the order of £17,000, the price per Avro Vulcan a decade later had risen to more than £400,000. Even using a pessimistic figure for the older aircraft—that achieved on the first order by a new subcontractor in 1945 to the tune of about £35,000—the order of magnitude increase remains readily apparent.[3]

The pace of technological advancement is perhaps even more impressive, at least superficially, if a two-decade time period is considered. In 1933, 20 years prior to the first flight of the Victor, Handley Page was building the first examples of the Heyford bomber for the RAF. This large biplane strategic bomber had a top speed at optimum altitude of well below 200mph. In terms of technology, it was scarcely better than the O/400 and V/1500 bombers—the 'Bloody Paralysers'—that the same company had supplied to the RAF around the time of its formation.

In a lecture in 1964, Handley Page's long term senior engineer Gustav Victor Lachmann graphically illustrated this progression in both cost and complexity, using the tube-and-fabric biplane baseline progressing to the 1940s equivalent, the Halifax, and finally the Victor.[4] The same sixfold increase in production cost that he cited for the final decade, the step from Second World War heavy bomber to Cold War nuclear bomber, had not been the rule during the preceding three.

In considering the development path that can be traced by the aircraft that physically existed however, there is the danger of excluding from the analysis those that were not built. This seems germane to the story of the Victor, particularly because of the philosophy that HP might have pursued had its focus not been almost entirely on the production, development and support of the Hampden

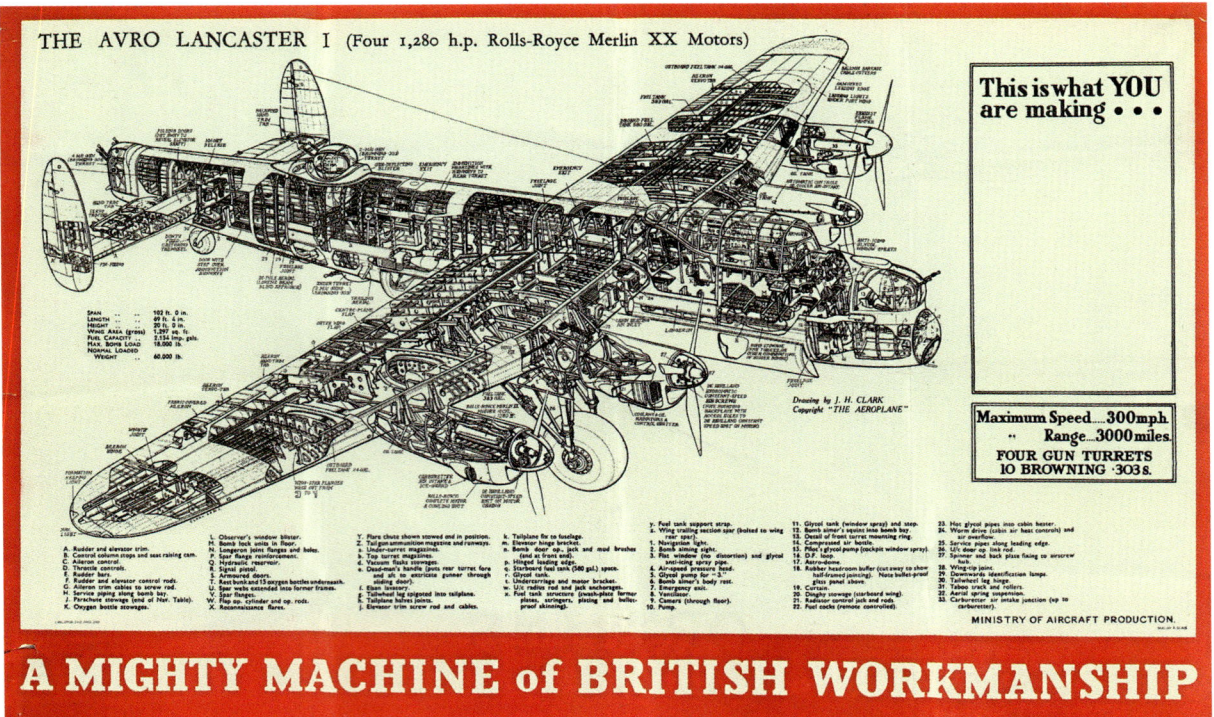

The ultimate expression, during wartime at least, of the British bomber: the Avro Lancaster. This poster from the Ministry of Aircraft Production illustrates the advance towards stressed skin construction in predominantly aluminium alloys, in the decade following Avro's tentative steps in that direction with Fokker's licenced technology.

and Halifax bombers that were sequentially perceived as essential to the war effort.

Two examples are particularly relevant to this. Even before the war, HP and specifically its chief designer George Volkert had proposed an unarmed, high-speed bomber as an alternative to the turret-armed (and performance compromised) aircraft that were being specified by the RAF. Contrary to the commonly accepted viewpoint, this was in line with the thinking of Air Marshal Edgar Ludlow-Hewitt, Air Officer Commanding-in-Chief of RAF Bomber Command, and was one of several proposals that culminated in the procurement of the de Havilland Mosquito.

Secondly, HP had for many years invested in research towards the production of a viable tailless aircraft, which was certainly perceived as an advantageous layout for a high-performance heavy bomber. This led to actual proposals during the war years and had one of these come to fruition the result might have been an aircraft equivalent in technology to the American Boeing B-29 Superfortress bomber.

As it was, the UK abortively pursued production under licence of the B-29 and Consolidated B-32 Dominator, eventually settling on the unconventional technological dead end that was the Vickers Windsor. What might actually have transpired, had Bomber Command been required to fully engage in trans-Pacific attacks on Japan, is open to conjecture.

In 1944, the south east of England came under sustained attack from first cruise missiles and then ballistic missiles. The potential of these delivery systems was frightening; while the V-1 pulsejet-powered flying bomb could be intercepted, the supersonic V-2 rocket could be tracked but not countered by the technology of the time. Despite this, it delivered its one-tonne payload over only about 200 miles. Development in the right conditions could be rapid, but the first nuclear weapons would tip the scales at over four times this payload; and by their very nature they were not something that could be metaphorically lobbed over the wall. The payload-range performance of a manned aircraft would be needed for many years into the future.

The genesis of the V-Bomber programme can be traced through two specific events: the first horrific, played out in front of the world and the source of a long shadow, the second a calm resolution behind closed doors.

The atomic blast of Little Boy over Hiroshima, on the morning of August 6, 1945, fundamentally changed the relationship between nation states. The extent of the destruction caused by Operation Meetinghouse a matter of months earlier, in which a quarter of the urban landscape of Tokyo was razed and around 100,000 lives lost, had showed the now terrifying effectiveness of conventional

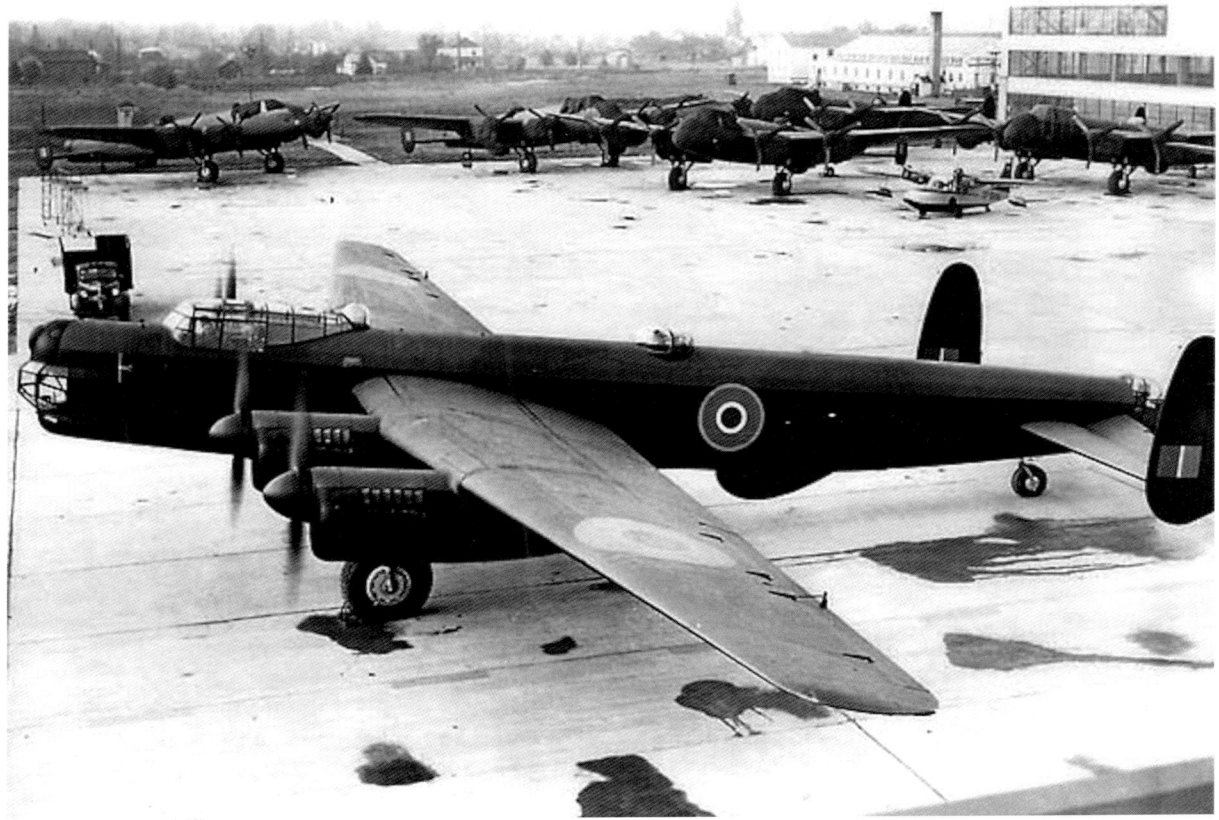

The long-span wings and much-revised nose of the Avro Lincoln are evident here, compared to the Lancasters arrayed behind. This is the one and only Canadian-built Lincoln XV, as plans to follow on Victory Aircraft's Lancaster X production were not followed through.

air attack. Nonetheless, that had needed more than 300 B-29s to be dispatched, each with a standard crew of 11 and a maintenance footprint to match. Now just one of these aircraft could unleash an equivalent level of immediate destruction, with, as became sadly evident within the subsequent weeks, profound and lasting consequences for those who had survived the initial explosion.

As far back as 1932, Stanley Baldwin had laid out the potential, if not the theory, for bomber deterrence.

"I think it is well also for the man in the street to realise that there is no power on earth that can protect him from being bombed, whatever people may tell him. The bomber will always get through, and it is very easy to understand that if you realise the area of space. Take any large town you like on this island or on the Continent within reach of an aerodrome. For the defence of that town and its suburbs you have to split up the air into sectors for defence. Calculate that the bombing aeroplanes will be at least 20,000ft high in the air, and perhaps higher, and it is a matter of mathematical calculation that you will have sectors of from 10 to hundreds of cubic miles.[5]

Imagine 100 cubic miles covered with cloud and fog, and you can calculate how many aeroplanes you would have to throw into that to have much chance of catching odd aeroplanes as they fly through it. It cannot be done, and there is no expert in Europe who will say that it can. The only defence is in offence, which means that you have got to kill more women and children more quickly than the enemy if you want to save yourselves. I mention that so that people may realise what is waiting for them when the next war comes."[6]

Of course, it had played out differently. For all the relative success of say, Operation Gomorrah at Hamburg in 1943, there was the counter example of the smashing of Bomber Command against the stubborn, sprawling rock of Berlin during the winter of 1943/44. The individual impacts might have been damaging, but the knockout blow remained elusive. The dramatic improvement in deterrent effect conferred on the bomber by the atomic weapon would firmly cement its place in the future order of battle, provided it remained the most suitable delivery system. For now, at the end of the war, it was the only option.

The UK had been a key constituent in the creation of the atomic weapon in the first place. In March 1940, Otto Frisch and Rudolf Peierls at the University of Birmingham had calculated that a uranium-based atomic weapon, of a size and mass that could be carried by an aircraft, was feasible. This relied on the relatively new theory advanced by Niels Bohr, that the rare isotope U-235 (just 0.7% by mass in a naturally occurring uranium sample) would be much more efficient than the common U-238 at capturing low energy neutrons.

Frisch and Peierls hypothesised the minimum size of a sphere of pure U-235, sufficient to maintain a chain reaction and hence a potential fission bomb. Using the best numbers available at the time, the startling result was a diameter of less than 5cm and a mass of just 600g, which would explode with the equivalent force of thousands of tons of conventional explosive. While some of the values used in the calculations were superseded as knowledge of uranium and nuclear physics improved, this almost inconsequential mass was in stark contrast to the many tons of natural uranium that had previously been thought necessary.

Fundamentally, this was the difference between a science project and an air-dropped super weapon. The implications were chillingly clear to the physicists involved; control of this technology would bring an end to a conflict, entirely in the favour of the side that managed to deploy it first. Subsequent American work in 1941 revised the mass required to around 8kg; significantly greater but still well within practicality. Reassuringly, this was based on measurements of actual (albeit microscopic amounts) of pure U-235.

The outcome of the study was typed by Peierls himself, recognising the sensitivity of his subject matter, in a two-part memorandum. The first part was a non-technical description of the likely nature of an atomic weapon, given that the majority of the decision makers would have been unfamiliar with the concept. It pulled no punches and the clarity remains admirable,

> *"The attached detailed report concerns the possibility of constructing a 'super-bomb' which utilises the energy stored in atomic nuclei… The energy liberated… is about the same as that produced by the explosion of 1,000 tons of dynamite… The blast from such an explosion would destroy life in a wide area [which would] probably cover the centre of a big city."* 7

The second part of the memorandum contained the scientific arguments and calculations, to make the case succinctly to sufficiently high-ranking government technical staff—specifically Sir Henry Tizard and colleagues—who would need in turn to sell the concept to many others. There were clear obstacles; for example, the means of separating the isotopes had only recently been developed and had never been used on uranium itself. But the first part concluded with a lucid insight into the future possibility of deterrence as a means of preventing war,

> *"If one works on the assumption that Germany is, or will be, in the possession of this weapon, it must be realised that no shelters are available that would be effective and that could be used on a large scale. The most effective reply would be a counter-threat with a similar bomb. Therefore it seems to us important to start production as soon and as rapidly as possible, even if it is not intended to use the bomb as a means of attack."*

The official response to this was the MAUD committee, surprisingly not an acronym, but instead named for a reference in a telegram from Niels Bohr to Frisch, in which he mentioned, 'Maud Ray Kent'. Assumed to be code, it was in fact the name of his Kent-born housekeeper. Ironically, their status as enemy aliens meant that both Frisch and Peierls were initially excluded from the committee, in spite of their significant contributions and undoubted talent. This situation was only rectified in March 1941, when a separate MAUD Technical Committee was spun off, to manage the work taking place at the Universities of Birmingham, Liverpool, Oxford and Cambridge.

Given the saga of Handley Page's Lachmann, whose similar status resulted in internment throughout the war lest he become directly involved in advanced aeronautics, it is ironic that this revolutionary and absolutely critical work was allowed to be conducted by German nationals. It cuts both ways of course: it may be indicative of the relative importance that the atomic bomb project began to assume, compared to the bombers that might be required to deliver it. It is also possible that, unlike Lachmann, the security services and MAP had no grounds—real or imagined—to consider Peierls, Frisch and their colleagues as anything other than loyal to the cause.

While not phrased directly as such in its terms of reference, the MAUD Technical Committee clearly had three problems to identify feasible potential solutions to: 1) How to isolate U-235 in sufficient quantities and in a practical timescale; 2) how to verify the underlying data for calculations, and 3) aside from the atomic element, the form that an actual weapon might take. The main committee's final report was unequivocal. Issued in the summer of 1941, it stated,

"We have now reached the conclusion that it will be possible to make an effective uranium bomb which, containing some 25lb of active material, would be equivalent as regards destructive effect to 1,800 tons of TNT and would also release large quantities of radioactive substances which would make places near to where the bomb exploded dangerous to human life for a long period." [8]

Here, less than a year and a half after Frisch and Peierls' memorandum, was a scientific report conducted ultimately under British government auspices that predicted clearly and accurately the type of events that would unfold in August 1945. Its findings could not be ignored. Having been brought into the loop by his scientific adviser Frederick Lindemann, recently ennobled as Baron Cherwell, Winston Churchill became the first national leader to authorise a programme specifically to develop an atomic weapon, on August 30, 1941.

The MAUD report had proposed continued cooperation with the (then neutral) United States, as a route to production of an atomic weapon in which the UK would be the key, if not lead partner. The British were sadly optimistic in two aspects of the process: the belief of how far they were ahead (and thus the time required to catch up) in the scientific understanding of the problem; and pragmatically, how they could hope to build the infrastructure necessary to produce the fissile material, in a bomb-ravaged country that was fighting for its existence. The reality was very different. The American effort gathered speed, with circumstances rapidly changing post-Pearl Harbor, while the innocuously named Tube Alloys project in the UK stalled. Peierls wrote,

"In January 1942… there was a full exchange of technical reports between the two countries. There was also an exchange of samples of material including samples of uranium highly enriched in 235 which had been produced in Berkeley and which were used in Liverpool and Cambridge… but the procedure for interchange of information changed at the end of 1942, and there was no exchange of reports or visits until August 1943." [9]

How could it have been that the flow of information between two allies, concerning a war-winning weapon, could have stopped for over eight critical months? One reason was the completely disproportionate effort that one side was putting in compared to the other. Reorganisation of command and a new emphasis on speed occurred in September 1942; while certainly Churchill and to an extent Roosevelt valued their transatlantic cooperation, the practical man at the helm, Vannevar Bush, had to find the quickest way through. With the US undertaking "ten times as much work as Britain",[10] it was a logical decision to (temporarily at least) abandon the exchange, on the assumption that any diversion of effort and requirement for secrecy would slow them down. By December, the US had decided to build its own full-scale plutonium and gaseous-diffusion plants, the development of the latter having been the key contribution that the British had been expected to make.

Late in December 1942, Roosevelt was made aware of an agreement signed between the British and Soviet governments in the previous September, which guaranteed information exchange between the two on both current and future weapons. Perhaps the secrecy of the atomic bomb project had become the enemy of UK progress; would such a wide-ranging agreement have been signed, had the implications on the relationship with the USA been clear? Roosevelt quite reasonably sided with Bush in making the call to restrict the flow of information to Europe; there was something to gain from a technical partnership with the UK, but far too much to lose if the information was freely supplied to the bastion of communism.

If the Americans were concerned about arming the Soviets, then they also felt the risk of what might happen after the war. Fundamentally, it appeared that they and they alone would be able to produce a bomb in time to use it in the current war. Had they required British help, then it would have been easily justifiable to seek it. If British assistance was not actually necessary and indeed might be motivated by the desire to have such a capability in a post war world, for military and civil purposes, then all kinds of moral, ethical and commercial factors came into play. At a simple level, why should the knowledge gained through millions of dollars worth of American taxpayers' money, be transferred to the chosen British instrument for atomic development, Imperial Chemical Industries (ICI)?

The Quebec Agreement of August 1943 set in stone both the principle of UK involvement in use of the atomic bomb, and the limits of UK influence. Remarkably, the understanding itself was between Churchill and Roosevelt only; on the British side, the Prime Minster did not involve the Cabinet or the Service Chiefs at all in this momentous call. The agreement mandated that

Two men that would play vital roles in the unfolding story enjoy some time out at the Potsdam conference, July 21 1945. Sir Charles Portal (holding fishing rod), Marshal of the Royal Air Force, and General Henry 'Hap' Arnold, Commanding General, US Army Air Forces had ultimately commanded the vast bomber offensives on Germany and Japan. NARA

the USA and UK would have to agree prior to the use of an atomic weapon during the war and that technology for that purpose would be shared. However, it also ceded control for post-war industrial purposes to the USA. Perhaps this was the best that the UK could achieve in that dark hour and with the hand available to play, but nonetheless it was a commitment that would return to haunt the UK within a few years.

Churchill's priorities were the availability of the bomb as a means to end the existential war he was leading the fight in, together with the ability to implement a deterrent based on it post-war. The Tube Alloys experience had shown that the former could not be achieved in the absence of the USA and would in any case require a reactor in Canada; any bomb from that source would be many years in the making. Potentially signing away the rights for commercial exploitation could be seen as a reasonable price to pay for security and peace, given the fresh memories of the desperate situation Britain had faced in 1940. Just a few days after the president and prime minister had signed the agreement, Bush sent a memorandum to the former to clarify the arrangements for renewed interchange on "Tubealloy".

"It would help if a top British scientist… could be sent here as chief liaison under Sir John Anderson… He should be of the calibre of Sir Henry Dale or Sir Henry Tizard.

In previous negotiations, difficulty was encountered because the British representative was an industrialist, Mr Akers of International [sic] Chemical Industries… Akers is a very able man, but not the one to handle this matter."[11]

The top-level US/UK discussions had seen emphasis from the former on the inability of the sitting president to commit a future successor to an international undertaking, lest it be done by a treaty. Here, in implementing the way forward, it was clear that Bush and his colleagues were still aiming to protect their investment from external post-war exploitation.

Was the future deterrent effect of the bomb perceived at that time, as we would understand it now? To an extent, the secrecy and 'off-grid' nature of the project make any assessment of the prevailing view a challenge. However, a fascinating insight appears in the published diaries of Sir Alan (Tommy) Lacelles, private secretary to King George

The Potsdam conference table, with its initial British contingent. The respective leaders are seated opposite their flags, and so the view is of the back of Churchill's head in the foreground, while Stalin and Truman are clearly visible. Foreign Secretary Anthony Eden is seated immediately to Churchill's right, while Clement Attlee is two places to the left. Both will become Prime Minister, but the latter will assume the role within the span of the conference.

VI. His entry for February 8, 1945, at which stage he was far better informed than the majority of the Cabinet, notes,

> "[Sir Edward] Appleton did not think that there would be time to use any sort of atom filled bomb against the Germans, but there might be an opportunity of doing so in Japan. I asked what would be the result. 'Oh,' he answered, 'a couple of them would end the war overnight—there is no doubt about that.' I said that it might be a good thing if humanity were given proof of the effects of these fearful engines, as it might convince it that any further indulgence in war would inevitably end in its own annihilation. He said that the deterrent aspect was an important one, and had not been overlooked; but, apart from its military side, the atom had immense commercial possibilities, and was destined to replace the world's already dwindling resources of coal and oil. I wish I understood these things better".[12]

Under the Quebec Agreement, Churchill's consent (as the UK prime minister) was required to permit the use of the bomb against Japan. This was given on July 2, 1945, recorded by his initials on a piece of foolscap paper. Events now moved rapidly. Two weeks later, on July 16, J. Robert Oppenheimer's leadership of Manhattan Project weapon development was validated via the successful Trinity test of a plutonium implosion bomb nicknamed Gadget in New Mexico. The new US president, Harry S. Truman, had been keen that the test took place prior to the upcoming Potsdam conference. Truman was therefore able to choose how to deploy the news to the best long-term effect for his country; he was fundamentally not bound by expedient agreements made under emergency wartime powers and possessed a much more sceptical view of Stalin than had Roosevelt.

Churchill, informed of the outcome of the Trinity test, immediately recognised that a US monopoly on A-bomb capability could temper Soviet ambitions in Eastern Europe. It was already distressingly clear that Stalin was acting well outside the agreements of Yalta earlier in the year, and that construction of a Soviet-controlled

All change—for the UK contingent at least. Clement Attlee's Labour Party landslide victory in the 1945 elections led to him becoming prime minister during the course of the Potsdam conference. Behind the leaders are the three senior foreign ministers, with Ernest Bevin second from the left, flanked by Admiral Leahy, Truman's Chief of Staff, and his US equivalent James Byrnes to his left.

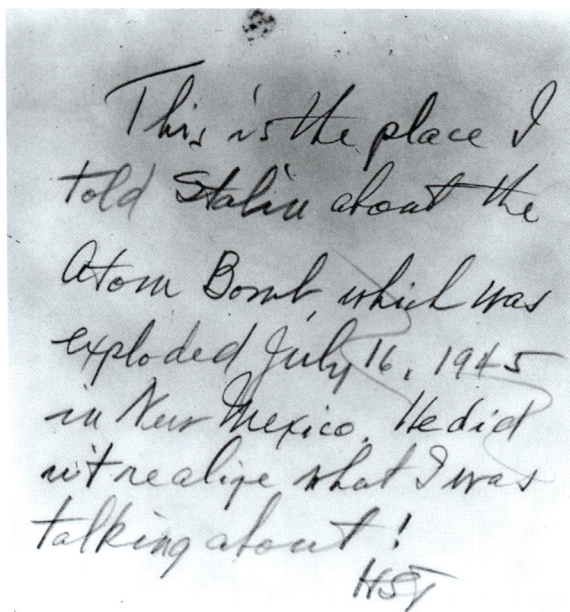

The back of a photograph of a scene from the Potsdam conference, annotated by President Truman. In reality, Stalin was well aware of the Manhattan Project.

The first detonation of a nuclear weapon occurred on July 16, 1945. Codenamed Trinity by J. Robert Oppenheimer, the plutonium implosion bomb was tested at Alamogordo, New Mexico. Apart from proving that such a bomb could work and demonstrating a yield of more than 20kT, the results were used to set the altitude of explosion for the air-dropped bombs.

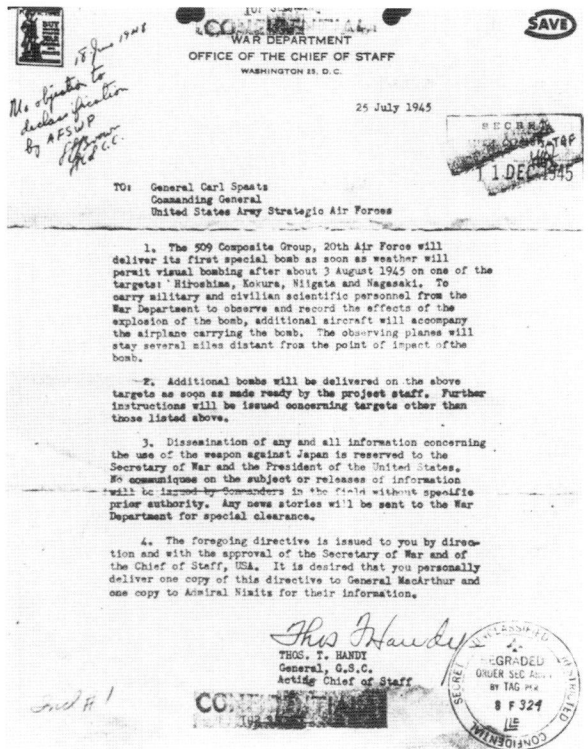

Nine days after the Trinity test, General Carl Spaatz received orders from Washington for the 509th Composite Group to conduct a nuclear attack after August 3, 1945. The initial target list was followed in order; it was only the presence of drifting smoke over Kokura that made visual bombing too difficult, that the B-29 flew on to the secondary target of Nagasaki on August 9.

communist buffer zone in the Red Army-occupied territories had begun. Churchill's planned post-war policy towards the USSR now had the teeth it needed; the Soviets be forced to consider the threat posed by this new weapon if they failed to comply with democratic norms in the 'liberated' lands.

Truman did indeed inform Stalin of the weapon's existence—but it was not news to him since the Manhattan project had been infiltrated by Soviet agents. Truman related that Stalin told him, "I hope you make good use of it against the Japanese".[13]

What may have seemed altogether more alien to the Soviet dictator, exercising near-total control over life and death in his country, would have been the replacement of Churchill and Foreign Secretary Anthony Eden mid-way through the conference. As the results of the UK general election were confirmed, the new Labour Party respective incumbents Clement Attlee and Ernest Bevin arrived on July 28 to continue negotiations. One of Churchill's final acts was to be a signatory of the joint Potsdam Declaration of July 26, which when issued required the complete and unconditional surrender of the Japanese, at the threat of utter destruction. This was not an empty threat, post-Trinity, but there was no official diplomatic response from the Japanese government and so the stage was set.

Thousands of miles away, at North Field on Tinian Island in the Marianas, the 509th Composite Group and its specially equipped 'Silverplate' B-29s awaited only the word.

Truman's post-Hiroshima statement on August 6 clearly sought to make maximum capital from the shock of the new paradigm, a city destroyed by a lone aircraft and a single bomb. He chose at this momentous hour, though, to acknowledge the key role played the USA's transatlantic ally.

"Beginning in 1940, before Pearl Harbor, scientific knowledge useful in war was pooled between the United States and Great Britain, and many priceless helps to our victories have come from that arrangement. Under that general policy the research on the atomic bomb was begun. With American and British scientists working together we entered the race of discovery against the Germans.

... In the United States the laboratory work and the production plants, on which a substantial start had already been made, would be out of reach of enemy bombing, while at that time Britain was exposed to constant air attack and was still threatened with the possibility of invasion. For these reasons Prime Minister Churchill and President Roosevelt agreed that it was wise to carry on the project here."[14]

Nonetheless, the implicit holding of the cards was included too. The atomic bomb was under American control.

"I shall recommend that the Congress of the United States consider promptly the establishment of an appropriate commission to control the production and use of atomic power within the United States. I shall give further consideration and make further recommendations to the Congress as to how atomic power can become a powerful and forceful influence towards the maintenance of world peace."

The USA alone could deploy the B-29 and the atomic bomb. Such a capability would surely intimidate any actor considering a role as the aggressor in another world war scenario. But would 'the bomb' maintain its status as the ultimate peacekeeper? And what would America's disenfranchised allies do next?

SECRET

SECRET

An Eye-Witness Account of the Trinity Shot on Monday Morning
at 5:30 AM - 16 July 1945

by

L. W. Alvarez

I was kneeling between the pilot and co-pilot in B-29 No. 384 and observed the explosion through the pilot's window on the left side of the plane. We were about 20 to 25 miles from the site and the cloud cover between us and the ground was approximately 7/10. About 30 seconds before the object was detonated the clouds obscured our vision of the point so that we did not see the initial stages of the ball of fire. I was looking through crossed polaroid glasses directly at the site. My first sensation was one of intense light covering my whole field of vision. This seemed to last for about 1/2 second after which I noted an intense orange red glow through the clouds. Several seconds later it appeared that a second spherical red ball appeared but it is probable that this apparent phenomenon was caused by the motion of the airplane bringing us to a position where we could see through the cloud directly at the ball of fire which had been developing for the past few seconds. This fire ball seemed to have a rough texture with irregular black lines dividing the surface of the sphere into a large number of small patches of reddish orange. This thing disappeared a few seconds later and what seemed to be a third ball of fire appeared again and I am now convinced that this was all the same fire ball which I saw on two separate occasions through a new hole in the undercast.

When this "third ball" disappeared the light intensity dropped considerably and within another 20 seconds or so the cloud started to push up through the undercast. It first appeared as a parachute which was being blown up by a large electric fan. After the hemispherical cap had emerged through the cloud layer one could see a cloud of smoke about 1/3 the diameter of the "parachute" connecting the bottom of the hemisphere with the undercast. This had very much the appearance of a large mushroom. The hemispherical structure was creased with "longitude lines" running from the pole to the equator. In another minute the equatorial region had partially caught up with the poles giving a flattened out appearance to the top of the structure. In the next few minutes the symmetry of the structure was broken up by the wind currents at various altitudes so the shape of the cloud cannot be described in any geometrical manner. In about 8 minutes the top of the cloud was at approximately 40,000 feet as close as I could estimate from our altitude at 24,000 feet and this seemed to be the maximum altitude attained by the cloud. I did not feel the shock wave hit the plane but the pilot felt the reaction on the rudder through the rudder pedals. Some of the other passengers in the plane noted a rather small shock at the time but it was not apparent to me.

I am attaching two sketches of the cloud which I made at the times noted. Mr. Glenn Fowler had made several sketches earlier in the development.

CLASSIFICATION CANCELLED
OR CHANGED TO Unclassified
BY AUTHORITY OF H. F. Carroll
BY B. Wire DATE 1-27-67

Luis W. Alvarez
Luis W. Alvarez

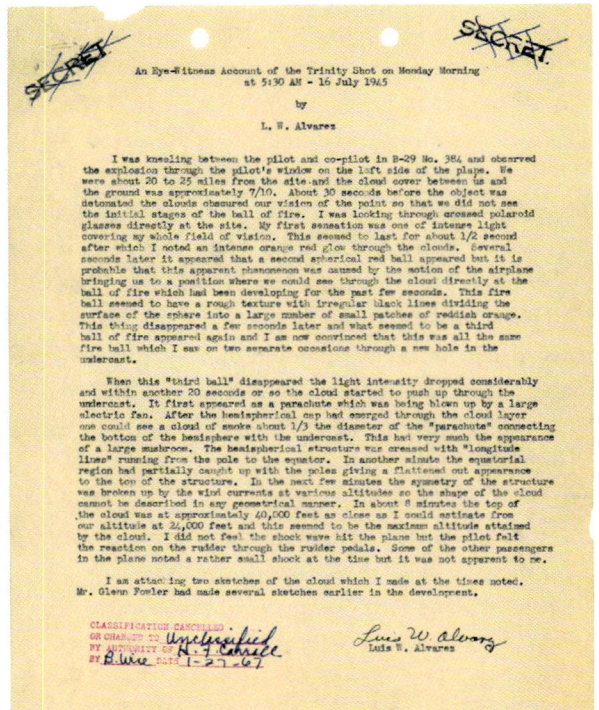

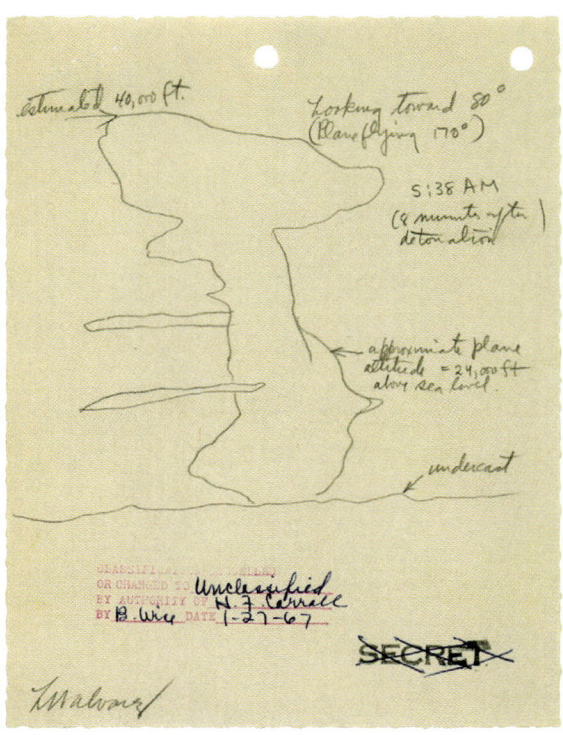

Luis Alvares was on board a B-29 tasked with observing the Trinity test — the first explosion of an atomic weapon. He described how the cloud reached about 40,000ft and completed sketches of what he had seen. While admirably descriptive, it does not capture the sense of relief that must have followed the successful test, on which so much rested.

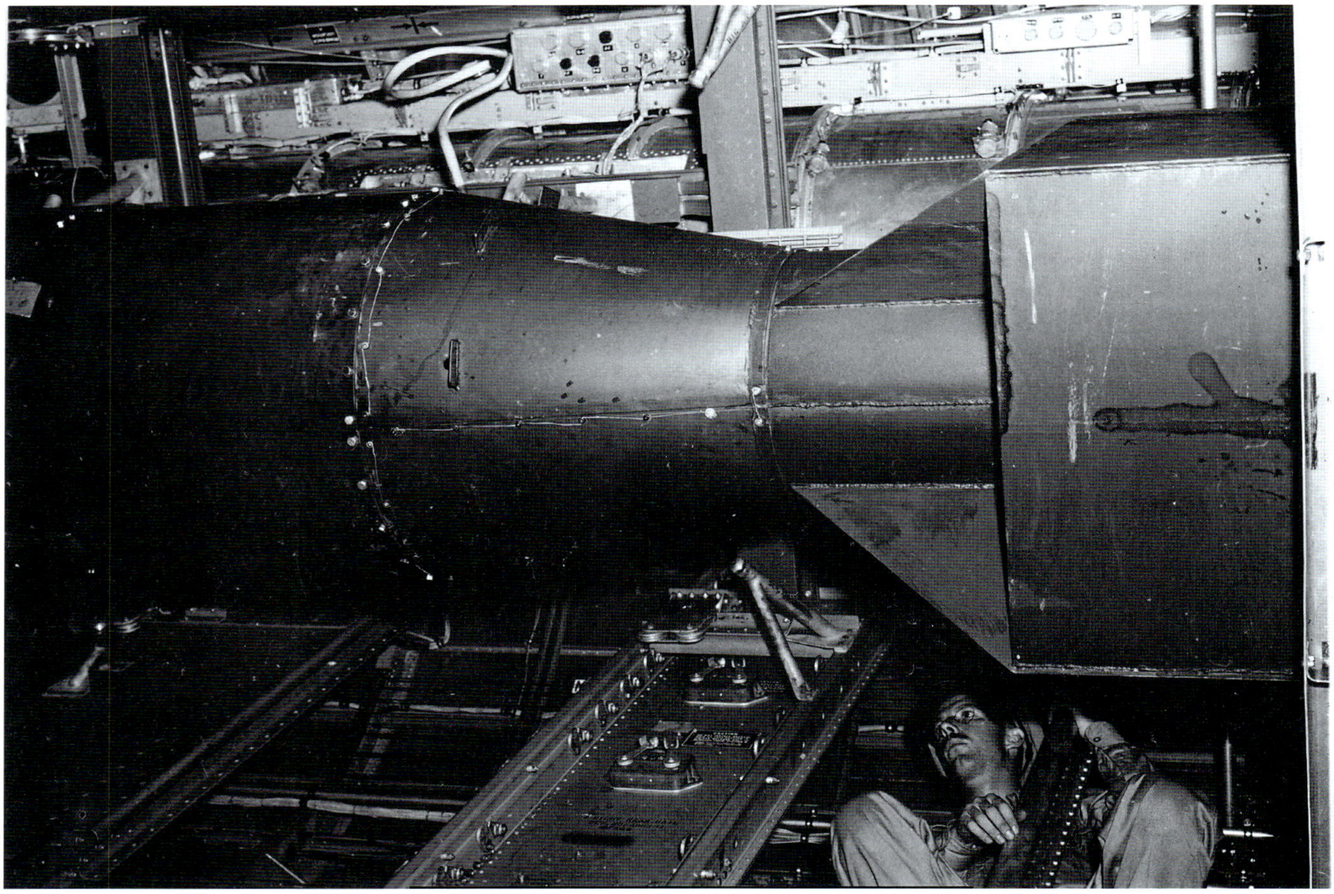

On board Enola Gay*, the Little Boy bomb is inspected prior to the Hiroshima mission.*

A flight that changed the world: the 509th Composite Group's commanding officer Col. Paul Tibbets returns his B-29 Enola Gay to stand on Tinian Island, after the atomic strike on Hiroshima, August 6, 1945. The Little Boy uranium bomb had never previously been tested as a complete weapon.

A 'Fat Man' type atomic bomb, as used for the second raid on Japan and dropped on Nagasaki. The unusual shape came from the need to enclose the implosion sphere, which was compressed by a surrounding layer of explosive. The UK bomb would be of a similar type, but in a more aerodynamic casing and hence much longer.

TOWARDS THE UK DETERRENT: WITH A UNION JACK ON TOP

For a prime minister often considered to have led a government with one of the most radical political programmes seen in the UK, it is sobering to consider that so many of Clement Attlee's immediate actions on winning power were associated not with establishing the NHS or democratising the means of production, but with war. The day after the Nagasaki attack, he sent a memorandum to the newly-constituted Cabinet Committee on Atomic Energy (GEN 75) in which he set forth his committed views.

> "…the modern conception of war to which in my lifetime we have become accustomed is now completely out of date. We recognised or some of us did before this war that bombing could only be answered by counter bombing. We were right. Berlin and Magdeburg were the only answer to London and Coventry. Both derive from Guernica. The answer to an atomic bomb on London is an atomic bomb on another great city".[15]

This was impossible to read outside the shadow of Baldwin's speech of 13 years earlier (see Chapter 1); the ideas he had expressed now tested in the most severe manner. For all of its resource and effort, the massive Bomber Command attack on Berlin had not delivered the hammer blow that Baldwin had predicted and Harris had anticipated. To paraphrase the commander-in-chief, it *had* been tried, and we had seen. The probability of success though, when only one bomber had to get through, to deliver a single bomb that could turn a city to glass and ash, was the clear factor that must have been at the core of Attlee's thoughts.

Churchill knew that the timescale for a purely British atomic bomb meant he would almost certainly not have to deliver it for this war himself. Now Attlee was committing himself and the country to a course of action that would deliver the independent deterrent—eventually. The environment in which he would do this was not entirely conducive to success either. Eight days following VJ-Day, Truman cancelled Lend-Lease and a country that had turned its entire economy over to war found that it could no longer make ends meet. Loans on an unimaginable scale from the USA and Canada would bridge the gap, but what status had been given up, and perhaps more importantly what was the anticipated quid pro quo, were questions that would take time to answer.

In the meantime, the unravelling of the loose consensus on the control of atomic energy would have enormous repercussions over the two years following VJ-Day. What did the Allies expect, and could it be compatible? Science historian Andrew Brown notes that,

Clement Attlee, British prime minister from 1945 to 1951.

> "The spectrum of opinion in the United States ranged from those who believed that the country alone possessed the secret of making atom bombs and should preserve its military advantage at all costs to those who realised there were no scientific secrets, just technical and industrial barriers, and who believed ultimate security would be best served by an effective system of international control."[16]

International control of all aspects (peaceful and military) of nuclear technology was an aim, viewed retrospectively, that could be held to be either noble or naïve. In fairness, the president had been advised by the Interim Committee, a group of civilian experts including Oppenheimer, tasked with making recommendations on how the Manhattan Project should practically contribute against Japan. On August 17, 1945, it provided a report to US Secretary of War Henry L. Stimson, on the more advanced question of nuclear weapons in a peaceful world. It noted in terms that there was no real countermeasure against an atomic weapon and that such devices could only be expected to become even more deadly as time passed. The stick might never be big enough, and Truman was advised to seek carrots.

> "We believe that the safety of this nation—as opposed to its ability to inflict damage on an enemy power—cannot lie wholly or even primarily in its scientific or technical prowess. It can be based only on making future wars impossible. It is our unanimous and urgent recommendation to you that, despite the present incomplete exploitation of technical possibilities in this field, all steps be taken, all necessary international arrangements be made, to this one end."[17]

Aside from the fact that he personally had accepted the responsibility for the terrible consequences of its use—reassuring a troubled Oppenheimer of this—it would have been hard for Truman not to find compelling the arguments of the very people who had created the weapon, that it should somehow be put beyond use. A system of international control, albeit after the Japanese demonstration, was seductive. A pre-requisite in the immediate post-war months was the transfer of authority domestically in the respective countries, away from the secretive and by necessity militaristic development organisations, to those within the fold of accountable government control. In the United States, Truman addressed Congress on October 3, 1945, emphasising the need to make decisions quickly.[18]

> "Almost two months have passed since the atomic bomb was used against Japan. That bomb did not win the war, but it certainly shortened the war. We know that it saved the lives of untold thousands of American and Allied soldiers who would otherwise have been killed in battle.
>
> ...The powers which the Congress wisely gave to the Government to wage war were adequate to permit the creation and development of this enterprise as a war project. Now that our enemies have surrendered, we should take immediate action to provide for the future use of this huge investment in brains and plant.
>
> ...I therefore urge, as a first measure in a programme of utilizing our knowledge for the benefit of society, that the Congress enact legislation to fix a policy with respect to our existing plants, and to control all sources of atomic energy and all activities connected with its development and use in the United States.
>
> ...The people of the United States know that the overwhelming power we have developed in this war is due in large measure to American science and American industry, consisting of management and labour. We believe that our science and industry owe their strength to the spirit of free inquiry and the spirit of

free enterprise that characterize our country…

…The hope of civilization lies in international arrangements looking, if possible, to the renunciation of the use and development of the atomic bomb, and directing and encouraging the use of atomic energy and all future scientific information toward peaceful and humanitarian ends. The difficulties in working out such arrangements are great.

The alternative to overcoming these difficulties, however, may be a desperate armament race which might well end in disaster. Discussion of the international problem cannot be safely delayed until the United Nations Organization is functioning and in a position adequately to deal with it.

I therefore propose that these discussions will not be concerned with disclosures relating to the manufacturing processes leading to the production of the atomic bomb itself. They will constitute an effort to work out arrangements covering the terms under which international collaboration and exchange of scientific information might safely proceed."

The president's aim at this point was to support passage of the May-Johnson Bill through Congress. This piece of legislation was the result of work by two lawyers, Kenneth Royale and William Marbury Jr., tasked with devising arrangements for post-war control of atomic energy. Their proposal was for a nine-strong commission, supported by boards of experts, operating outside of political influence and to an extent, with limited presidential power to remove its members. On at least two counts, it rapidly ran into vocal opposition. A powerful administrator and deputy would be created, who were widely expected to be drawn from the Army and Navy respectively, and to become dominant influences. In addition, draconian measures would be enacted to deal with the passing of information outside of the confines of what sounded increasingly like an extended Manhattan project. To the scientists who had laboured for years to bring the bomb to fruition alongside colleagues from Canada and the UK, the risk of a fine or even imprisonment for talking to them now was unacceptable.

Truman elaborated further on the issue of international peace for the future in a speech not in Congress, but in New York City's Central Park, on the occasion of Navy Day, October 27, 1945.[19] This time, he outlined a 12-point plan that defined the philosophy of the United States in the field of Foreign Affairs. The last of these points would become very important in future international discussions,

"We are convinced that the preservation of peace between the nations requires a United Nations Organisation composed of all the peace-loving nations of the world, who are willing jointly to use force if necessary to ensure peace."[20]

From a Western perspective, these statements were not controversial; indeed it seems if anything surprising that, within living memory, a sitting president of the United States of America should have felt it necessary to articulate,

"We seek no territorial expansion or selfish advantage. We have no plans for aggression against any other state, large or small. We have no objective which need clash with the peaceful aims of any other nation."

This address is also significant because of what he chose to imply: that information on technology pertaining to the atomic bomb would not be shared even with Britain and Canada in the short term.

"As I said in my message to the Congress, discussion of the atomic bomb with Great Britain and Canada and later with other nations cannot wait upon the formal organisation of the United Nations. These discussions, looking toward a free exchange of fundamental scientific information, will be begun in the near future. But I emphasize again, as I have before, that these discussions will not be concerned with the processes of manufacturing the atomic bomb or any other instruments of war.

In our possession of this weapon, as in our possession of other new weapons, there is no threat to any nation. The world, which has seen the United States in two great recent wars, knows that full well. The possession in our hands of this new power of destruction we regard as a sacred trust. Because of our love of peace, the thoughtful people of the world know that that trust will not be violated, that it will be faithfully executed."

He concluded with a hope that we know will be dashed, inevitably perhaps, but for all the cynicism that we would feel today, it is not difficult to understand how a war-weary nation might have clung to it.

"Indeed, the highest hope of the American people is that world cooperation for peace will soon reach such a state of perfection that atomic methods of

destruction can be definitely and effectively outlawed forever."

And therein lay the basis of a schism. It may have been achieved at the cost of more than two billion dollars and have been credited with ending the Pacific war, but the custodian of this unique capability felt it too dangerous to remain in existence. Surveying the ruins of Europe from closer quarters, Attlee was in no doubt that the genie was out of the bottle, and that an atomic bomb-based deterrent was the surest insurance policy against a repeat of the previous six years. In mid-November, he would journey to the United States by Douglas Skymaster—a 20-hour trip during which he was said to have slept well—in order to engage in serious talks with the president and the prime minister of the other party in the Manhattan Project—Mackenzie King of Canada.

Attlee grasped completely that the rules of the world had changed. In his briefing to the Cabinet prior to his departure, he had laid out his thesis and hence the arguments he would take to Truman. There was no defence against an atomic weapon and it held the capability of ending civilisation. "Power politics" and the "uneasy equilibriums" that they led to had to be consigned to the past.

"The only hope for the world is that we… strive without reservation to bring about an international relationship in which war is entirely ruled out."

If the idea of outlawing atomic weapons was a non-starter, Attlee highlighted the only alternative that had been proposed; deterrence through (atomic) strength—but through the international governing instrument of the United Nations Organisation.

"All Governments should be invited to become parties to a convention pledging them—
(a) not to use atomic bombs except in accordance with (b) below;
(b) to join in immediate and complete sanctions against any country making use of the atomic bomb in violation of (a) above. These sanctions to include the use of the atomic bomb by those countries which possess it. The agreement should remain in force indefinitely, and sanctions should be taken against any country attempting to denounce it;
(c) to enter into a full exchange of the basic scientific information relating to the use of atomic energy;
(d) to institute effective control of the use of atomic energy in their own territory."

Control through the UNO appeared superficially reasonable and defendable, but soon ran into problems. One such objection was the scenario where a country was invaded by an overwhelming conventional force, and chose to use an atomic bomb as the only way out. The proposal would then compel the rest of the United Nations to come together against the territory that had been wronged in the first place. Attlee recognised that,

"Britain is peculiarly vulnerable to attack by atomic bomb owing to her geographical position and her concentration of population. To accept this obligation would be to expose London to annihilation."

His solution was to endorse Truman's twelfth point and subsequently put his faith in the machinery of the UNO to keep the peace.

"I should point out [to the president] that it is essential that atomic weapons should be available to restrain aggression, and I should suggest that the best way of achieving this is not by any special convention, but rather by the determination of all those who develop atomic energy to live up to the principles and purposes of the Charter, and to back up its authority by using their atomic weapons against an aggressor if the occasion arises."

Attlee's planned commitment was that the world order be protected by the UN Charter, itself backed up by the atomic weapons of the individual countries that had them. He noted the fundamental problem with the idea of giving the Security Council direct control of the weapons: one of the Permanent Members (the only ones that in the medium term might conceivably develop nuclear bombs) would have the power not only to use them as the aggressor themselves, but to subsequently veto a response by the UNO as well. It simply wasn't credible, particularly as a means of restraining the Soviet Union, which he was sure would within the decade develop its own nuclear arsenal.

Immediately preceding the meeting of the three leaders, on November 7 the House of Commons debated the recent statements by Truman, and the implications for UK foreign policy. Unsurprisingly, it would involve a passionate speech from the hawkish Churchill.

"I think the speech of the president of the United States on October 27, is the dominant factor in the present world situation. This was the speech of the head of a state and nation, which has proved its

ability to maintain armies of millions, in constant victorious battle in both hemispheres at the same time. If I read him and understand him correctly, President Truman said, in effect, that the United States would maintain its vast military power and potentialities, and would join with any like-minded nations, not only to resist but to prevent aggression, no matter from what quarter it came, or in what form it presented itself. Further, he made it plain that in regions which have come under the control of the Allies, unfair tyrannical Governments not in accordance with the broad principles of democracy as we understand them would not receive recognition from the Government of the United States. Finally, he made it clear that the United States must prepare to abandon old-fashioned isolation and accept the duty of joining with other friendly and well-disposed nations, to prevent war, and to carry out those high purposes, if necessary, by the use of force carried to its extreme limits."[21]

The Foreign Secretary Ernest Bevin was critical of both the 'new' policy of denying the UK the atomic secrets to which it believed itself entitled as a partner, but also of the misunderstanding of the effect that this would have on the Soviet Union. He disagreed that fundamentally that the UK and the USSR would perceive the same risk from a uniquely US controlled atomic world.

"On October 27 President Truman not only issued his Twelve Points but introduced them in a speech which contained some matters of grave importance … For the moment I am only calling attention to the difficulties of reconciling the Twelve Points with the avowed policy which is at present being pursued by the United States. But most important of all was the statement that, not only would the secret of the atomic bomb be strictly confined to the United States of America, or, rather to its Government and its scientists; there was, further, the statement that it would not even be discussed with Britain, or even with Canada.

…I desire to draw attention to the psychological effect which that must have upon Russia. I agree that in excluding us all from the secrets of the atomic bomb, it can be said that no distinction is drawn between Great Britain or even Canada, or any other of our Dominions, and Russia. But all the world knows that the use of the atomic bomb against Britain or any one of our Dominions is not only unthinkable to us but unthinkable to anyone in the United States. There are several views about Russia, about her Government or her policy, but everyone seems to be agreed that Russians are realists."[22]

It should not be doubted then, that there was broad support from the leadership of both His Majesty's Government and Opposition, for the application of pressure to the United States to maintain an atomic capability as a bulwark against the Soviet Union. In turn, the expectation was for the UK to play an appropriate part in this, through both a UNO commitment and the use of its joint (albeit now inaccessible) technical knowledge. Nonetheless, there were fears over the destabilising effect of a world where a group of Western countries, well disposed to each other, held an atomic monopoly. What might this encourage the Soviets to do? Attlee clearly had a mandate to go to Washington and dissuade the president from surrendering the new super-weapon, whether to the UNO or entirely, while moving further to regain UK access to both technology and control lost through Quebec and Truman's subsequent measures.

In achieving his aims, Attlee's mission was a qualified success. Through the opportunity to address Congress, he would be able to set out the stall of his country in public, while in conferences over several afternoons, including cruising on the Potomac on the Secretary of the Navy's yacht *Sequoia*, direct discussions took place between the three heads of government. They issued an Agreed Declaration on November 15, 1945. This proposed that fundamental scientific principles and literature be shared with any nations committed to peace, while a specific UNO commission should be created with urgency, to develop and promote a multi-stage plan that would eliminate atomic weapons and advance instead the peaceful, commercial harnessing of this source or power.

"The Commission should be instructed to proceed with the utmost dispatch and should be authorised to submit recommendations from time to time dealing with separate phases of its work.

In particular the Commission should make specific proposals:
a) For extending between all nations the exchange of basic scientific Information for peaceful ends,
b) For control of atomic energy to the extent necessary to ensure its use only for peaceful purposes,
c) For the elimination from national armaments of atomic weapons and or all other major weapons adaptable to mass destruction,

d) For effective safeguards by way of Inspection and other means to protect complying states against the hazards of violations and evasions."[23]

Attlee reported back to the Cabinet a week later, from which it is recorded that he believed, "... it was satisfactory that the three Powers had expressed their intention to continue to co-operate in research and development on atomic energy…", but was forced to acknowledge that the terms of the Quebec agreement remained fundamentally in force. With his colleagues briefed, a debate in the Commons took place later the same day. Attlee reiterated his position: there would need to be an A-bomb.[24]

"In my view, it is impossible to isolate the problem of the atomic bomb from that of the use of other weapons of destruction. There was a time when wars were local, fought out with weapons which to us seem extraordinarily primitive. In those days the losses and destruction caused by war could often be made up in a few years, and great as was the misery caused, the thing was measurable. Sometimes even the losses were slight. Men in authority might count the cost of a war as worthwhile, for the advantages gained, though those advantages seem to us today very often trifling. Such an attitude towards war is impossible to our generation. We have seen two world conflicts, and we know the cost — or at least we know some of the cost — in human suffering and the destruction of the work of generations of men. The practical obliteration of great cities which took place in the last war as the result of shelling and bombing was bad enough. We know only too well what the effect of bombing was in London, and in Coventry, Plymouth and other cities; but anyone who has seen Aachen, or Stalingrad or Warsaw, knows how infinitely greater is the ruin on the Continent of Europe.[25]

The house was clearly still able to consider the problem in the round. Henry Wilson-Harris, incumbent of the soon-to-be-abolished seat for Cambridge University, was a long-term supporter of the League of Nations and now the new UNO proposal. His interjection would illustrate the profound difficulty of international relations in this confused period.

"I want particularly to address a few brief remarks to the House tonight upon a subject which has already figured in many speeches — the subject of Russia — and I do that because I believe in all seriousness that the question of our relations with Russia today is even more momentous than the question of the atomic bomb and its consequences. I say that for this reason: if our relations with Russia are right, then we need not be much concerned about the atomic bomb; if our relations with Russia are wrong, then the situation is most grave and menacing — atomic bomb or no atomic bomb."[26]

But what could be the 'right' relations with Stalin in the new world?

Truman's problematic attempts to find a formula for atomic control that would be acceptable, to both Congress and the Scientific community, would receive a significant boost within the month. On December 20, a new atomic energy bill proposal, sponsored by Senator Brien McMahon, himself only recently having been appointed as chairman of the Senate Special Committee on Atomic Energy, was introduced as a viable alternative to the stalled May-Johnson attempt. The Bill substantially removed the bias of control from the military, which enabled it to address the concerns of scientists. Control would pass from the Army to a new civilian organisation, with five full-time commissioners to oversee its responsibilities. The latter was a bone of contention since it was believed that the full-time basis would be less attractive to the highest calibre scientists.

It was debated extensively through into the spring of 1946 and, if passed in the form that consensus had built around at that stage, would fundamentally have allowed the collaboration between the Three Powers envisaged by the leaders and expressed from the polished deck of the *Sequoia*. It was not to be.

Back in September 1945, an unlikely event in Montreal would provide evidence of the profound effect a small number of people could have on history, effectively as an unintended consequence. A junior GRU (military intelligence) officer at the Soviet Embassy, Igor Gouzenko, fearful of the implications of an unexpected recall to Moscow, had decided to defect with his wife and young son. Equally concerned that the Canadians would not accept their need for asylum, Gouzenko had for months been removing documents of interest from the embassy, as an intelligence prize to give himself some bargaining power.

What these documents would actually reveal was the extent of Soviet infiltration within Canada, which had been targeted specifically due to the presence of work on advanced defence technology — including the atomic bomb — but relatively underdeveloped and overstretched security. Gouzenko's simple attempt to avoid being sent

President Harry S. Truman, British Prime Minister Clement Attlee and Canadian Prime Minister Mackenzie King at the Tomb of the Unknown Soldier in Arlington National Cemetery, November 11, 1945.

home, from an environment to which he had become accustomed to an uncertain and possibly even non-existent future in Moscow, would instead provide the perfect excuse for the restriction of US atomic knowledge to a domestic audience only. Among the many informants recruited by Gouzenko's boss, Colonel Nicholai Zabotin, was a British atomic scientist of considerable repute: Alan Nunn May.

May had become aware, via a report from the USA, that Nazi Germany was investigating the potential for an atomic bomb. As had been the case for the early UK assessment, while the challenge of producing such a weapon might prove in the fullness of time to be insurmountable, the risk associated with the success of such an endeavour could not be ignored. However, the report suggested that the passing of a much lower technical bar, the production of a 'dirty bomb', might well be achievable. This fitted chillingly with the slowing down of the German assault, together with the dehumanising of the Russian population by the invaders. Could they build and use one there? May thought so and towards the end of his life admitted,

"As I was a member of the Communist party in the UK at that time, I made use of my contacts with the Russians to issue a warning about the possible dangers of a dirty nuclear bomb, because at that time the Germans were vigorously attacking Russia. After I took this course of action, the Russians so to speak 'booked' me as an available source of information."[27]

May had been a member of the Cambridge heavy water team in the early war period, but in 1943 was sent to Canada as part of the effort to build a reactor there. He had of course been vetted and passed as secure by the British security services, so the revelation that these supposedly tight procedures, under the most serious circumstances, might have proven so inadequate had the potential to be a major embarrassment. Assessment of Gouzenko's contacts suggested that information being passed was generally of low technical value, but the real concern was the level of penetration; two of those implicated, for example, were Canadian MPs.

May was due to return to the UK in any case and the British were keen to arrest him as soon as he arrived, but there was concern about precisely what he might be charged with. MI5 thought it vital that he be prevented (via incarceration) from having the opportunity to defect, while the major concern in Canada was in keeping the Gouzenko affair in its entirety as quiet as possible.

Truman and Churchill in Fulton, Missouri, on the occasion of the former prime minister's celebrated speech in which he described the falling of an 'Iron Curtain' across Europe; March 5, 1946.

Successfully prosecuting May would require the interrogation of the informants named by Gouzenko, both in Canada and the USA. The strategy was discussed at the highest levels, between Truman and King, then later King and Attlee would agree to proceed with as little publicity as could be achieved. With delays growing, frustration developed among the respective security services, not least because of the ever-increasing risk of the Soviets working out what was happening. This concern was moot, if only they had known.

It was not Soviet agents in Canada that would pass on the information to the KGB, rather it was the very senior one already embedded in MI6. October 1945 came and went, while November saw the FBI occupied with a homegrown mole story. It was thus November 28 before the three powers agreed to act, to bring in those they thought could spill sufficient beans to implicate May. However, King had second thoughts and wished to take his concerns directly the Soviet ambassador, nominally again to maintain a cordial standard of relationship between the two countries.

Mackenzie King's hesitation sealed the fate of his plan, because by January 1946, it was clear that the United States press knew about Gouzenko and would publish the story. Realistically, this could only have been because of a planned leak, by either UK actors determined to force King's hand and move along with securing the position of May, or alternatively by J. Edgar Hoover's FBI, themselves with an agenda to push against communism and by implication, the USSR. The British may well have thought themselves home and dry, with spies caught and on trial, and the wrath of public opinion pointed ultimately at the Kremlin.

A month after the revelations of apparently widespread atomic espionage broke, the box office draw that was Churchill was introduced at Westminster College, Fulton, Missouri, by none other than Truman himself, on the occasion of the British statesman's award of an honorary degree. If May's arrest was the effective trigger of the Cold War, then Churchill's speech of that day would provide its first language and imagery.

"From Stettin in the Baltic to Trieste in the Adriatic, an iron curtain has descended across the continent... Warsaw, Berlin, Prague, Vienna, Budapest, Belgrade,

President Harry S. Truman signs the Atomic Energy Act on August 1, 1946 establishing the US Atomic Energy Commission. Critically, the contents of the Act precluded the exchange of confidential atomic information with other states, meaning that there was no prospect of a UK-US joint deterrent. The US decision meant that the UK would have to go it alone. Senator Brien McMahon himself is third from right, behind the president.

Bucharest and Sofia, all these famous cities and the populations round them lie in the Soviet sphere and all are subject in one form or another, not only to Soviet influence but to a very high and increasing measure of control from Moscow.

From what I have seen of our Russian friends and allies during the war, I am convinced that there is nothing they admire so much as strength, and there is nothing for which they have less respect than military weakness. For that reason the old doctrine of the balance of power is unsound. We cannot afford if we can help it, to work on narrow margin, offering temptations to a trial of strength."[28]

The deposed leader had brought the force of his oratory with him, to convey what was in the end a simple message. He had looked into the eyes of Stalin himself, and he knew that neither appeasement nor parity of military force would dissuade him from doing as he willed. But this was not 1938 and the cost of failing to confront a menace lay piled in rubble all over Europe. Only the overwhelming certainty of disaster for the USSR should it attack the Western powers could be an acceptable response.

"If the Western Democracies stand together in strict adherence to the United Nations Charter, their influence for furthering these principles will be immense, and no one is likely to molest them. If however they become divided or falter in their duty, and if these all-important years are allowed to slip away, then indeed catastrophe may overwhelm us all."

The United States government, in its various arms, had a duty to itself however, and now the factions that considered the UK and Canada to be less than reliable allies — that ran leaky ships — had the ammunition they had needed. McMahon's act was now subject to these forces. The result was a substantial strengthening of Article 10 of the bill, now applying a caveat that could never be met to the clause that covered dissemination.

"That until Congress declares by joint resolution

that effective and enforceable international safeguards against the use of atomic energy for destructive purposes have been established, there shall be no exchange of information with other nations with respect to the use of atomic energy for industrial purposes..."[29]

When eventually passed by both houses in August 1946, the door finally closed on a year of sometimes passionate negotiations—with Britain firmly excluded from direct information transfer regarding atomic weapons. The threat articulated by Churchill remained as close as ever though, and Attlee himself was determined to meet it. A bankrupt country would have to go it alone. In the same month, a commitment from a military perspective was made when the Chief of the Air Staff, Lord Tedder, issued the formal request for an atomic bomb. This was operational requirement OR.1001, and at a similar time the Air Staff's need for an aircraft to carry it was issued in OR.229. But still, there was no government decision to light the touchpaper.

The meeting of the GEN 75 committee, Attlee's instrument in this process, on October 25, 1946, has been a source of colour in this story down the years. Bevin, arriving late and in the words of Peter Hennesey, having been "lunching well", was astounded to find Hugh Dalton and Stafford Cripps firmly on the side of not building the necessary gaseous diffusion plant required for plutonium production. While the minutes of the meeting certainly do not record the exchange, subsequent recollections have been saved for posterity,

> *"That won't do at all... we've got to have this... I don't mind for myself, but I don't want any other Foreign Secretary of this country to be talked to or at by a Secretary of State in the United States as I have just had in my discussions with Mr Byrnes. We've got to have this thing over here whatever it costs... We've got to have the bloody Union Jack on top of it".*[30]

A year on, Bevin challenged the Cabinet to answer the effect of Alan Nunn May's treachery, and commit to the ultimate deterrence against a threat that might choose to become existential.

With a broad agreement to the bomb established within the GEN 75 group, consisting as it did of senior and influential Cabinet ministers, a new and specific committee designated GEN 163 would meet just once, in January 1947. This committee consisted of a very select group: Attlee, Herbert Morrison, Bevin and three others, eventually adding the Minister of Supply, who would be responsible for procuring the hardware. Pointedly, the ministers who would be responsible for financing it all were not present, although they had fundamentally acquiesced to the idea in the GEN 75 meetings.

Nonetheless, the constituting of a new group for the final call must indicate a need for secrecy, either externally in order to legitimately limit knowledge of the military technology, or internally, within government, to engineer the conditions in which the answer would be in the affirmative. In any case, the attention of the Chancellor and the President of the Board of Trade were elsewhere. It was deep into the worst winter in two centuries, and in an economically ravaged country, with the very means of production being squeezed by the simple demand for more coal.

The committee reviewed a proposal by Lord Portal, the Chief of the Air Staff who had guided the RAF through the Second World War, and had more recently answered his government's request for him to join the Ministry of Supply to lead the production of fissile material. Portal had been determined to retire at the end of the war, but felt he could not turn down Attlee's appeal to his patriotism.

Despite being sometimes criticised by the technical team actually working on the nuclear projects for his insubstantial contribution, in the end all endeavours of this nature need a trusted political champion plugged in at the right level. It was Portal who judged the lay of the land to be appropriate and pressed for a green light specifically on a weapons programme. Whether the conclusion was foregone or not, the outcome was indeed the unequivocal decision to proceed on the research and development of an atomic bomb.

It was January 8, 1947.

3

DEFINING THE BOMBER

THE EFFICACY of Britain's nuclear deterrent, when it had one, would depend entirely on the ability of the system to convince any potential enemy that it could place sufficient atomic weapons on target. The resolve of the country to actually use such weapons would be a political matter; in military and engineering terms, the answer to the calculation had to be that the bomber would indeed get through. What would it have to be able to do? This depended to a large extent on what defences the enemy might be able to mount. And by the time of the decision to proceed with the independent UK weapon, it was absolutely clear that the enemy in question, at least in the near term, would be the Soviet Union.

An attacking force in 1947 would be confronted by anti-aircraft artillery and fighter aircraft. As had been discovered in Germany, the cost of deploying extensive static AA systems was eye-watering and directly subtracted from the production of offensive weapons. The feared 88mm flak guns were accurate up to about 26,000ft, which when coupled with radar guidance had placed the vast majority of Allied heavy bombers within range. The USAAF B-17 and B-24 formations flew higher than the RAF's night bombers, but even then 25,000ft was typical. However, to successfully engage at higher altitudes, even larger calibre and expensive guns were required. In the longer term, German research had again shown the path to the future, in terms of surface-to-air guided weapons. Just how far ahead the effective deployment of these systems might be was another matter for the crystal balls of the intelligence community. The timespan estimated for the development of nuclear-capable British bombers, a decade or so, made extrapolation of Soviet capability into the 1960s a requirement.

The standard Soviet AA gun was of 85mm calibre, although from 1951 an altogether more fearsome 100mm weapon was known to have entered service. By 1955, the numbers of these guns available was roughly equal, with about 3,000 of each in use. The May Day parade that year gave unwelcome confirmation of a further development however, with a 122mm gun being shown off.

Although only about 50 of these were thought to have been deployed, it was estimated to be able to effectively engage targets at up to 45,000ft. All of these guns worked with formidable gun-laying radars as part of integrated fire control systems. The WHIFF radar, in widespread use from 1951, was based on the American SCR-584—derived from plans obtained by espionage. Its parent had proven extraordinarily successful in countering V-1 flying bomb attacks when rushed to the UK in June 1944. It can only have been considered then that, despite the expense, the AA defence of major Soviet targets such as Moscow and Leningrad would present a lethal threat at altitudes up to at least 40,000ft.

But how high and how fast would be enough to avoid such defences? This question was tackled scientifically by Don Hallows at the Royal Aircraft Establishment in

A Soviet KS-19 100mm anti-aircraft gun. Beginning to be deployed from 1948, this weapon could engage targets above 40,000ft.

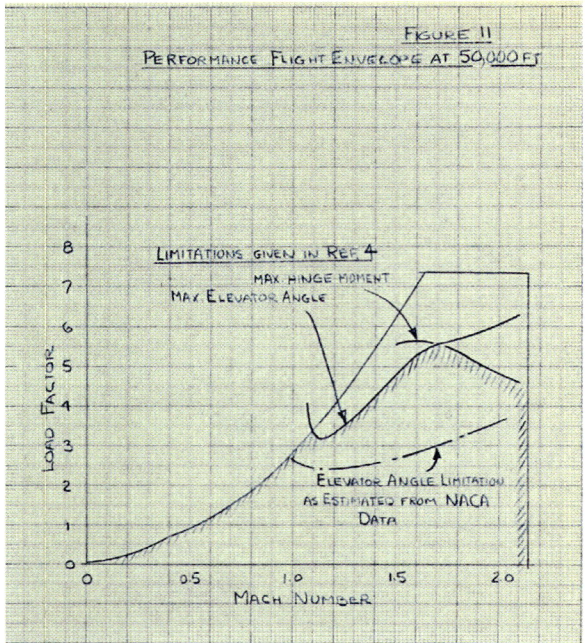

A hand-plotted chart from the development of the Avro Canada CF-105 Arrow, a fighter that might have been fielded in the 1960s. Despite its supersonic speed, various limits came into play to restrict the achievable load factor and so manoeuvrability. Engaging a bomber would be far from simple; at M = 1.5, little better than the Vulcan's 2.5G might be available at 50,000ft.

1945.[31] He had actually been looking into the opposite problem, that of understanding what capability an interceptor fighter would require in order to defeat a modern bomber. Taking the technology available in the immediate future, a Rolls-Royce Nene-powered fighter with a top speed of the order of 600mph, his conclusion was that a bomber flying at 500mph and 50,000ft (M = 0.76) would evade interception on 70% to 90% of occasions. While the performance of the fighter in terms of speed might have been better than that of the bomber, this and other work would demonstrate that in order to perform a level turn of a reasonable radius and get behind the target, its speed and ceiling had to be substantially superior.

Because there were many attack profiles that might be considered, an analysis needed to make simplifications. Assuming that the fighter would need to engage in turning flight to bring its weapons to bear on a cruising bomber, then a portion of its aerodynamic force would need to be devoted to lateral acceleration. To maintain level flight, with lift equal to weight, then the total force—equivalent to the lift in level '1g' flight, would need to increase by the factor of 'g' used in the turn. Either the fighter would need to be flying below its service ceiling, or it would lose height in the turn. A paper published in 1955 by P. L. Sutcliffe of Avro would describe the impact of this:

> "The implications of these unique values in terms of the ability of the jet fighter to attack the jet bomber are also interesting. It is seen that if the fighter is to attack the bomber at the 1g ceiling altitude of the latter and have a manoeuvrability of, say, 2g in steady level flight at that altitude, then the 1g ceiling of the fighter at its attack Mach number and engine setting must be some 14,500ft higher than that of the bomber at the bomber's own cruise Mach number.
>
> The overall problem of interception and attack is, of course, more complex than this, but it indicates that the bomber's best defence lies in its absolute altitude or 1g ceiling… The difficulty of maintaining the superiority of the fighter increases with increase in the cruising altitude of the bomber…"[32]

While Avro may have had good commercial reasons to be championing the prospects of a high-altitude bomber, with supersonic fighters just around the corner, the analysis itself was factual. It also helps to illustrate the basic difference between the conclusion drawn previously by the RAE in the P13/36 debate of 1938 with Handley Page and company chief designer George Volkert who had, as mentioned previously (see Chapter 1), proposed an unarmed high-speed bomber as an alternative to turret armed but correspondingly slower Halifax and Manchester that were being specified by the RAF to meet their medium bomber requirement.

This was one of several proposals that ultimately culminated in the procurement of the de Havilland Mosquito. The RAE had argued, quite reasonably, that any advanced performance technology applied to a bomber must equally be applied to a defending fighter; if this notion had been valid back then, why not still in 1947?

Before the war, the increment of speed that development might confer on a new aircraft still left it in the same aerodynamic regime. That is to say, adding 50mph to the top speed of an aircraft previously capable of 300mph largely left everything else still working in the same way; travelling at 350mph just wasn't that different to travelling at 300mph. This was certainly not the case if the bomber was already firmly transonic however. That regime of itself was difficult enough to master, and there really could be substantial advantage in the bomber using its physically larger size to enclose and contain systems, engines and payload, thereby sidestepping the aerodynamic interference issues as far as possible.

A transonic fighter, smaller in size and more difficult to

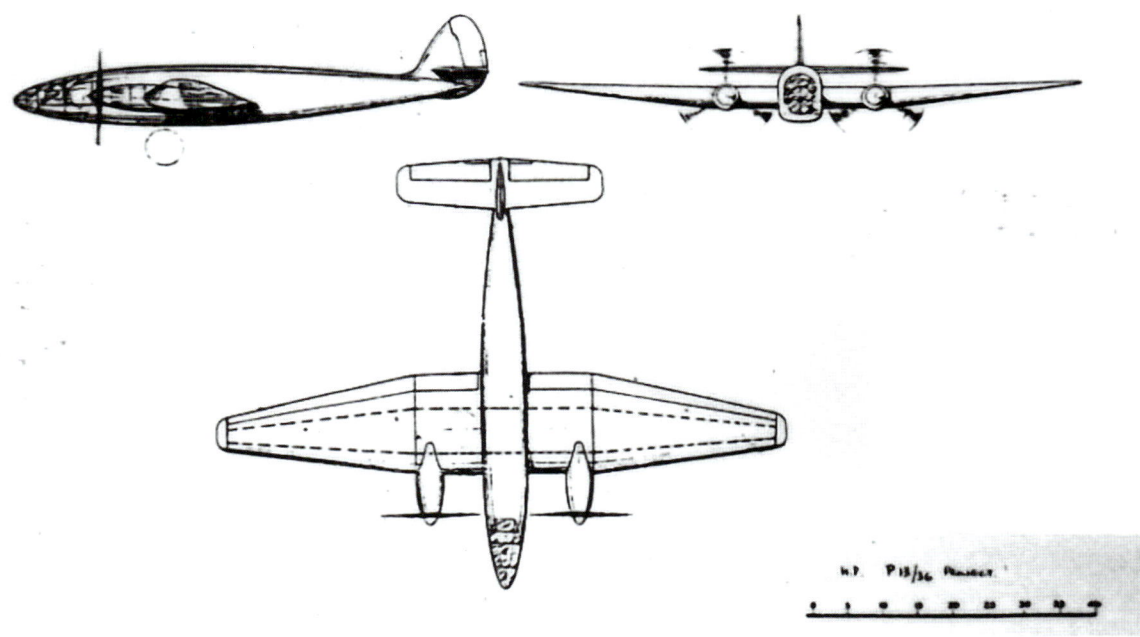

Volkert's alternative P13/36 proposal from 1937, an illustration of what the Handley Page chief designer believed achievable with contemporary technology, if a high-speed, unarmed philosophy was chosen. Confidence in the efficacy of the available defensive armament won out, with the Halifax and Manchester the end result. VIA RALPH PEGRAM

In retrospect, a very optimistic view of the potential success of the original Halifax, in this case wearing daytime camouflage and safely returning from its mission over occupied Europe. Like all of the RAF's heavy bombers, the defensive armament was no match for the Luftwaffe in its prime, and the cover of night was necessary.

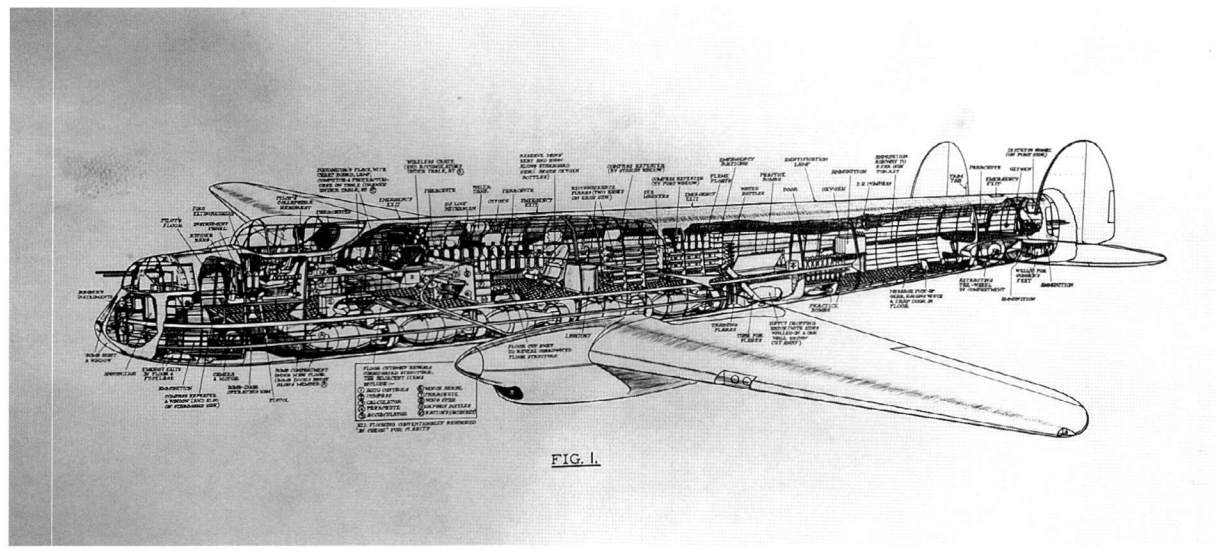

An illustration from Avro's tender document to the P13/36 specification, showing the aircraft that would become the Manchester, as originally proposed. It is interesting that some features are reminiscent of the Armstrong-Whitworth Whitley, another product of the Hawker-Siddeley group. The original shape of the nose and the fin/tailplane configuration fall into this category. On the other hand, the large bomb bay would remain a vital feature. FRANK PLESZAK

package, was proving hard enough to make work. A supersonic fighter—almost certainly essential to provide the worthwhile performance step needed to take on the transonic bomber—was a technology that did not exist. Indeed, it was not until ten months after the V-bomber specification had been issued that Chuck Yeager would achieve supersonic speed in level flight in the Bell X-1. Such technology was hardly practical for an air defence system; there was clearly a margin of time where the transonic bomber might reasonably be expected to succeed, prior to the defences catching up.

B35/46

Roughly two weeks after the cabinet decision to proceed with the independent nuclear deterrent, the Ministry of Supply issued specification B35/46 to the selected aircraft manufacturers, on January 24, 1947. Classified 'secret' it concerned, "the design and construction of a medium-range bomber for worldwide use by the Royal Air Force…" which would be "suitable for economic production of at least 500 aircraft at a maximum rate of not less than 10 per month". The main body of the document was, in retrospect, curiously contractual and incorporated items that seem almost trivial, such as the colour of the cockpit interior (matt black). It was only in the appendices that the meat was found. The fundamentals of the requirement were captured in Appendix B, clause 1.

"The Air Staff require a medium range bomber land plane capable of carrying one 10,000lb bomb to a target 1,500 nautical miles from a base which may be anywhere in the world. The aircraft will be required to attack targets at great distance inside enemy territory and it must be assumed that it will be plotted by radar and other methods for a large part of its flight. It must, therefore be capable of avoiding destruction by making the inevitable attack from ground and air launched weapons difficult. To this end, it must have:

A high cruising speed—the cruising speed shall be such that attacking fighters will have to fly at a speed at which they will tend to become uncomfortable.

Manoeuvrability at high speed and high altitude—the design must be such that the aircraft can turn rapidly without loss of height or much loss of speed when at maximum cruising height.

A high cruising height—the cruising height must be such that ground launched weapons can only be guided at long range and such that the design of the intercepting enemy aircraft will be difficult.

Capacity for carrying adequate warning devices—these will be needed to detect the approach of ground launched weapons and the proximity of opposing aircraft. To be effective the warning device must have long range and may therefore be large. Adequate provision must be made for mounting such a warning system so that it can scan the required field.

Capacity for carrying defensive apparatus—such as proximity fuse exploders and homing or guided missile jamming devices."

DEFINING THE BOMBER

A famous image of USAAF 6th Bomb Group B-29A 44-61784 'Incendiary Journey', over Osaka on June 1, 1945. This dramatic view highlights many of the features of the Superfortress and the way in which it was used. The cleanliness of the sleek fuselage and high aspect ratio wings were responsible for, respectively, low zero-lift and lift dependent drag. The two upper fuselage turrets are visible, mounting a total of six 0.5in machine guns and apparently trained on the same target, which the defensive weapon system could accurately engage at more than half a mile. The very high altitude is clear, as is the stain of an oil leak from the number three engine, indicative of the peril of flying with the underpowered and initially dangerously unreliable Wright R-3350 engines.

To these philosophical and conceptual requirements, specific numbers were added—and these would, in time, drive the unusual solutions that the Handley Page and Avro designs represented. These clauses included:

"The aircraft must be capable of cruising at maximum continuous cruising power at heights from 35,000ft to 50,000ft at a speed of 500kts.

The maximum speed in level flight should be as high as possible, but it is not essential that it exceed the cruising speed.

The flying characteristics must not become dangerous when the speed temporarily rises above the top-level speed in the course of combat or other manoeuvres and Mach number increases to a maximum of 0.90, but a speed restriction is acceptable below 25,000ft.

The final approach speed should not exceed 120kts and good manoeuvrability must be maintained at this speed.

The Maximum operational radius of action with a 10,000lb bomb load must be 1,500 miles. To attain this a still air range of 3,350 nautical miles at a height of 50,000ft with a 10,000lb bomb carried on the outward and return flight is required."

An extremely important requirement concerned field performance and a method of indirect control that the Air Staff sought to deploy.

"It must be possible to operate the aircraft from existing H. B. [Heavy Bomber] type airfields and the maximum weight when fully loaded ought, therefore not to exceed 100,000lb. The Air Staff is to be informed if the weight will be exceeded".

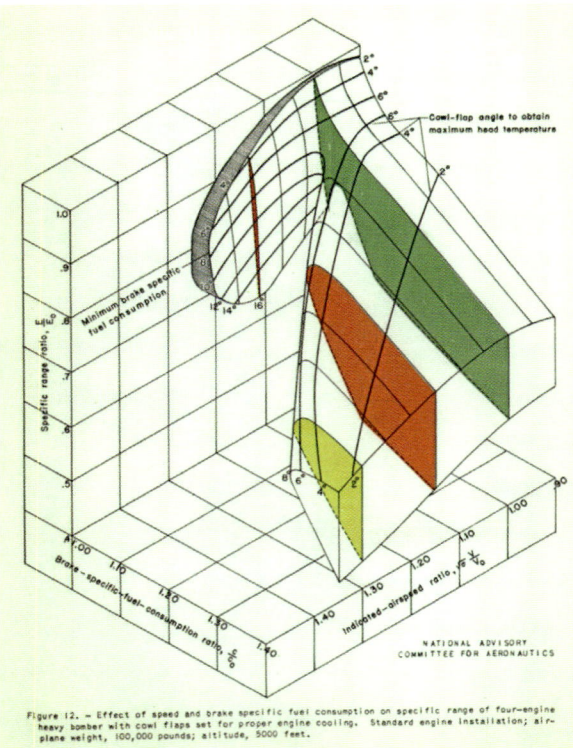

A design surface diagram, showing contours of the three variables of specific range, fuel consumption and speed for a theoretical 100,000lb bomber with four 2,000hp engines. Such tools allowed the capabilities of modelled future aircraft to be assessed. In this case, it is clearly a B-29 class machine.

While this would make it very challenging to provide the required performance—particularly range—there was clearly logic in attempting to use the vast number of concrete runways already to be found throughout Lincolnshire and East Anglia, such that an economically bereft country might have the deterrent it was perceived to need. The high ceiling specified also implied from the outset that a low wing-loading would be necessary, contributing at sea level to the relatively low approach speed and field length. Nonetheless, with a Lincoln (previously Lancaster IV) tipping the scales at about 80,000lb and a fully-loaded B-29 in the region of 130,000lb, the weight limit was far from generous. The trend towards compactness extended throughout the specification; looking back, the requirement that "the pressure cabin must be as small as possible" might have elicited a wry smile from a Vulcan crewman. The Avro team was apparently determined to give the Air Staff that wish, at least.

Once in the air, how was the aircraft to find its target? Just half a decade earlier, the RAF had struggled to place bombs within five miles of a target at night. From 50,000ft the target might well be visually obscured even in daylight, by cloud if not simply distance, as the RAF had found during abortive high-altitude bombing trials in 1941 using the American B-17C. At 500kts, how would forward throw and the ballistics of the precious weapon be handled? The answer could not be, except in extremis, the entirely manual methods of earlier days. It was not written into B35/46 itself, but the capability required by the Air Staff was a blind bombing accuracy of 200 yards from these cruise conditions. As ever, a web had been woven over the wartime years that would lead to the answer.

One of the reasons that the émigrés Frisch and Peierls had been available to conduct their atomic research at the University of Birmingham, was that it kept them away from another major defence project being conducted elsewhere at the same institution. The physics faculty, led by Australian physicist Mark Oliphant, also included a research fellow and former GEC researcher named John Randall. Along with his colleague Harry Boot, Oliphant allocated the two men a thorny problem posed by the Admiralty: the generation of microwaves that might form part of an airborne radar system.

Short wavelengths were essential to resolve targets of the size of a submarine's periscope, but the generation of sufficient power in a suitably compact installation was beyond the technology of the time. Randall and Boot investigated a more obvious solution, the klystron—an electron tube able to amplify or oscillate at microwave frequencies—but by the outbreak of the Second World War had understood that it would not provide the power that would be required.

By November 1939, they had switched to working on the split-anode magnetron, which was based around widely used cathode technology. The breakthrough that the two men made was to use the output of the magnetron within a number of resonators, an arrangement which allowed the required amplification. By February 1940, the prototype cavity magnetron had generated a kilowatt at 9.8cm wavelength, rapidly demonstrating the promise of sufficient power output for an airborne radar.

With the Blitz underway, the most immediate and pressing application of the cavity magnetron was to improve the Airborne Interception (A.I.) radar systems used in Bristol Beaufighter night fighters. The millimetric wave system, A.I.Mk.VIII, was a major advance on the preceding A.I.Mk.IV by virtue of the narrow, focused beam that it created—vastly more controllable and at low altitude in particular, finally viable despite ground clutter due to its high resolution.

Observations that radar reflections from the Earth's surface produced varied patterns according to what was on that surface, would in turn open up the possibility of

With the gap between the capability of the Avro Lincoln and threat it would face widening post war, the RAF operated 88 ex-USAF Boeing B-29As for a period. This is WF502 (formerly 44-62231) of 90 Squadron at RAF Hooton Park, Cheshire in September 1952. RUTHAS

ground mapping radar, with the cavity magnetron again making the size and power requirements acceptable.

From the mid-1943, the Avro Lancasters and Handley Page Halifaxes of RAF Bomber Command could be navigated with star shots taken by sextant from the astrodome, much as James Cook had done sailing the Pacific, or by the white-hot technology of their H2S radar sets, tracing eerie, flickering green representations of the ground features passing below. Truly, a remarkable advance in the most trying of circumstances, which had included the deaths of several senior engineers when an H2S-equipped Halifax caught fire, lost a wing and crashed during a demonstration flight on June 7, 1942.

H2S had reached Mk.IV by the end of the war, and although the latest variant was not ready to deploy, it would equip the Avro Lincoln in the immediate post war period. By the time of B35/46, it was recognised that the provision of a radar on its own was no longer sufficient. The specification called instead for a long range automatic D.R. [Dead Reckoning] system and a long-range fixing system. The D.R. system would,

> *"comprise a type of navigational and bombing computer to which navigational data is supplied by an H2S set of long range and moderate definition, the bombing information being supplied by separate high definition equipment, which may operate on a different frequency. In addition at a later date a system of measuring track and ground speed utilising the Doppler principle may be coupled to the N.B.C.*
>
> *For fixing positions there are two requirements:*
> *a) H.2.S. To pinpoint and identify over or near land*
> *b) Astro A completely automatic astro system will be developed during the life of the aircraft and an optically correct glass panel in the roof of the aircraft above the navigator's station will be required."*

The integrated NBS system was nothing if not ambitious, viewed from the perspective of 1947 bomber technology and the capability of existing aircraft. Remarkably however, the doppler system described was indeed developed in time to equip the V-force. The Radar Research Establishment (RRE) produced a device named Green Satin under the 'rainbow code', a system which married a randomly selected colour with a noun taken from a pre-drafted list. Looking directly downwards from a position aft of the main undercarriage bay in the Vulcan for example, Green Satin aligned wave guides to the aircraft's track and measured the doppler shift between the front and rear guides to compute ground speed.

Finally, what was the aircraft required to carry to its

target? B35/46 detailed a total load of 20,000lb and identified four types of bomb. Three of them were conventional high explosive weapons, in 10,000lb, 6,000lb and 1,000lb sizes. The fourth was a Special Bomb, of 60in diameter and 290in length, although the latter dimension was offered as a first estimate only. The aircraft would not, however, be required to carry that great burden of the Lancaster: defensive armament. The B35/46 bomber would rely entirely on radar countermeasures and its "speed, height and evasive manoeuvre" for the protection it needed. It was reasoned that this would be enough.

SLIDE RULES, GIRL COMPUTERS AND THE MARK 1*

In October 1949, a lecture was given to the Royal Aeronautical Society by Ivan Driggs, the director of research at the United States Navy's Bureau of Aeronautics (BuA).[33] Titled 'Aircraft Design Analysis', it provides a fascinating insight into how the conceptual design of aircraft was conducted at the time of the V-bombers. Driggs described the BuA method for assessment, which,

> "arose directly from the need for quick, reasonably accurate and simple means for studying a broad series of aircraft designs leading toward fulfilling an overall military requirement."

He noted that the parameters to be considered were all linked and had to change together. Fuel consumption was linked to engine thrust, which might give an estimate of fuel weight for a particular range. But the size of the aircraft needed to carry that mass of fuel would be linked back to the thrust required, invalidating the first assumption.

Once a technology has reached a reasonable level of maturity there is usually a bank of data from previous solutions to fall back on. During the Second World War, the Royal Aircraft Establishment began to collect and tabulate data on the aerodynamic performance and structural parameters of new aircraft as they were tested. By 1946, this covered a correspondingly wide range of types, which by the nature of development in the period would often provide multiple data points against similar requirements. In terms of heavy bombers for example, the RAE had access to information on the Halifax and Lancaster that had outgrown their original specifications and intruded on the territory of the Stirling. This could be coupled with that for the Fortress (B-17), Liberator (B-24) and Superfortress (B-29); all of them using basically similar technology, built to meet similar structural requirements and operating in similar environments and speed regimes.

Extrapolating from these data points would allow the designer of a future type to meet similar goals with a degree of confidence in estimating component masses and realistic aerodynamic drag, for example. There was usually a viable place to start, rooted in previous experience, important for both the manufacturer and customer. However, as with all empirical analysis, the accuracy of estimations for solutions significantly far removed from the dataset are suspect. In 1946, there was no existing information pertaining to an actual transonic jet bomber. The solution hung off the end of the graph.

Driggs described a method that used a combination of similar empirical data along with values derived from first principles. Specific slide rules were used with this data, in order to speed the process of calculating the weights and performance of many different configurations quickly. Attending his lecture were representatives not just from the airframe manufacturers, but the end users too. The airlines, for example, needed to independently verify that the data given to them by the manufacturers made sense. They also had to understand what might be possible beyond that and what these possibilities might reasonably cost. C. Dykes of BOAC was certainly interested in the level of resource that might be required, or indeed saved by using this new method.

> "Mr Driggs was notable in that he had done so much work with a staff of only eight to simplification of methods. It provided a valuable lesson for everyone, and it was one which he had frequently impressed upon his own staff. It was fatally easy to adopt some refinement of method which did not sufficiently improve accuracy, but which greatly increased the work involved."

Dykes presumed that the eight engineers of Driggs' staff were supported by "a small number of computers; his own experience showed that one girl computer to each engineer was necessary".[34] Driggs' reply is clearly of its time and, unsurprisingly, responded in kind to the presumably predominantly if not exclusively male audience:

> "Mr Dykes' experience that one girl computer to each engineer was necessary was interesting. Was this necessity one of morale or of proper accomplishment of the job? If the former, he agreed most heartily, but if the latter were the case, they had found in the Navy that the logarithmic performance methods were considerably more effective, in that they kept the engineers more closely in touch with the progress of a design."

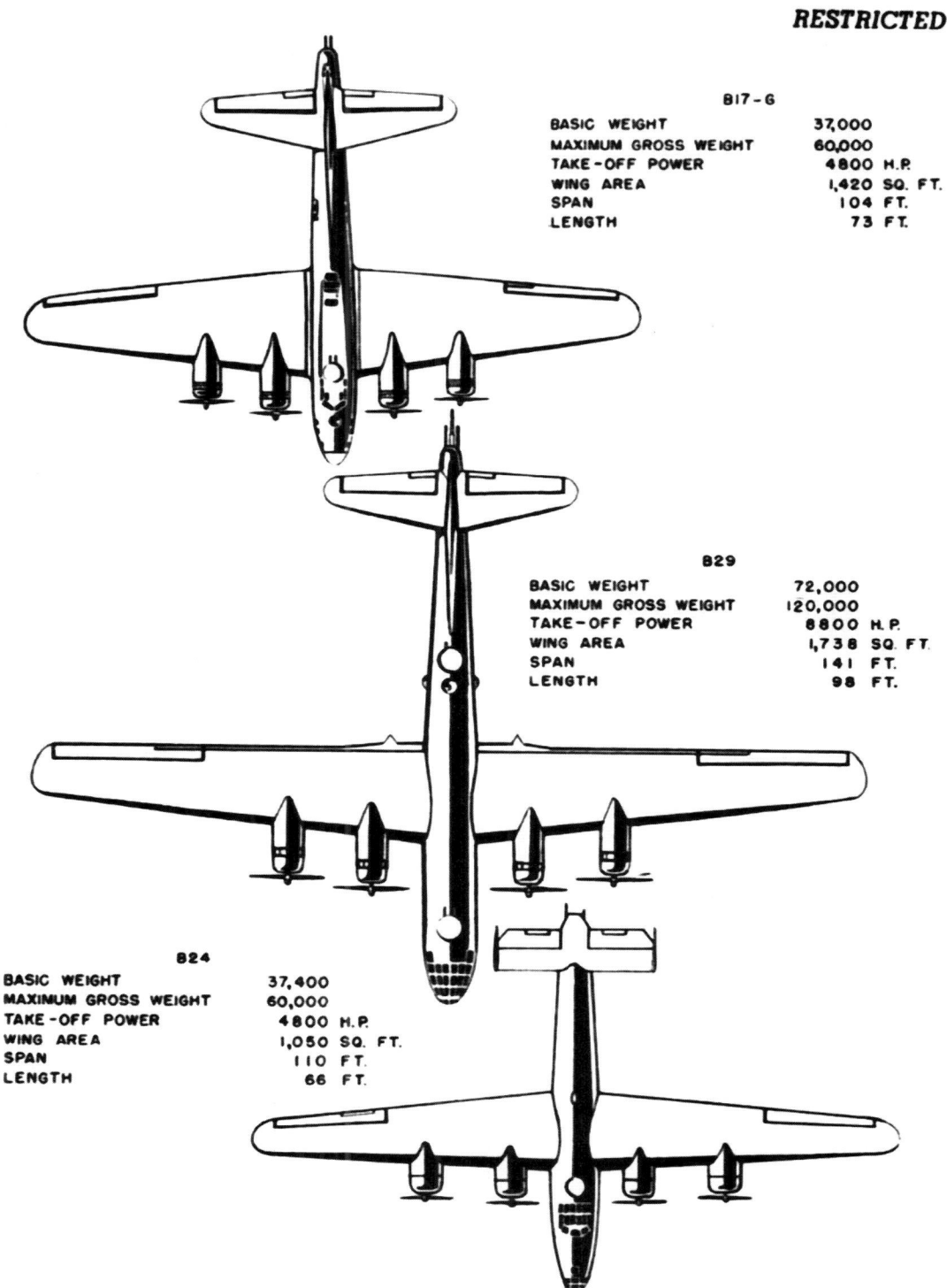

COMPARISON B-17, B-29 & B-24

Boeing comparison of the B-17G, B-24 and B-29. The Superfortress had approximately double the maximum take-off weight, but a wing area that grew by less than a quarter compared to the earlier Boeing design. The trade-off between the efficiency of a high aspect ratio with a smaller wing area as used by the B-24, and the operational considerations of manoeuvring at high weights and altitudes, is also brought into focus here. These would be of great importance in designing the V-bombers.

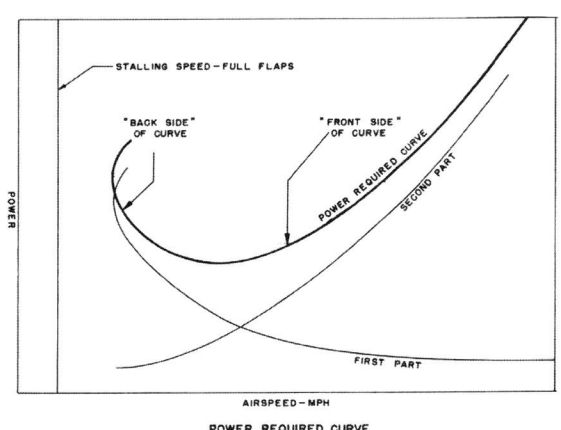 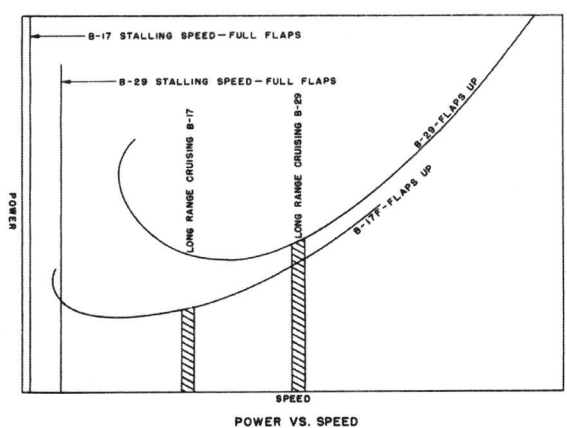

The complexity of the B-29 and the need to extract its performance fully, would mean a new level of information being provided to its crews. These illustrations from the Pilot's Notes show the derivation of the aircraft's optimal cruising speed. Power was consumed by lift-dependent drag, which reduced as lift coefficient was traded for speed (labelled 'first part'), while skin friction drag continued to increase with speed. Adding these together showed an airspeed in power required was minimised. The 'back side' of the curve was an unstable region; if the aircraft slowed down, it would need more power and hence continue to diverge from the intended flightpath. Plotting these curves from the B-29 and B-17F showed the impact of the low ski-friction drag, high aspect ratio design and high wing loading of the new bomber. It would have to fly more quickly for lift to equal weight, but it could do this much more efficiently.

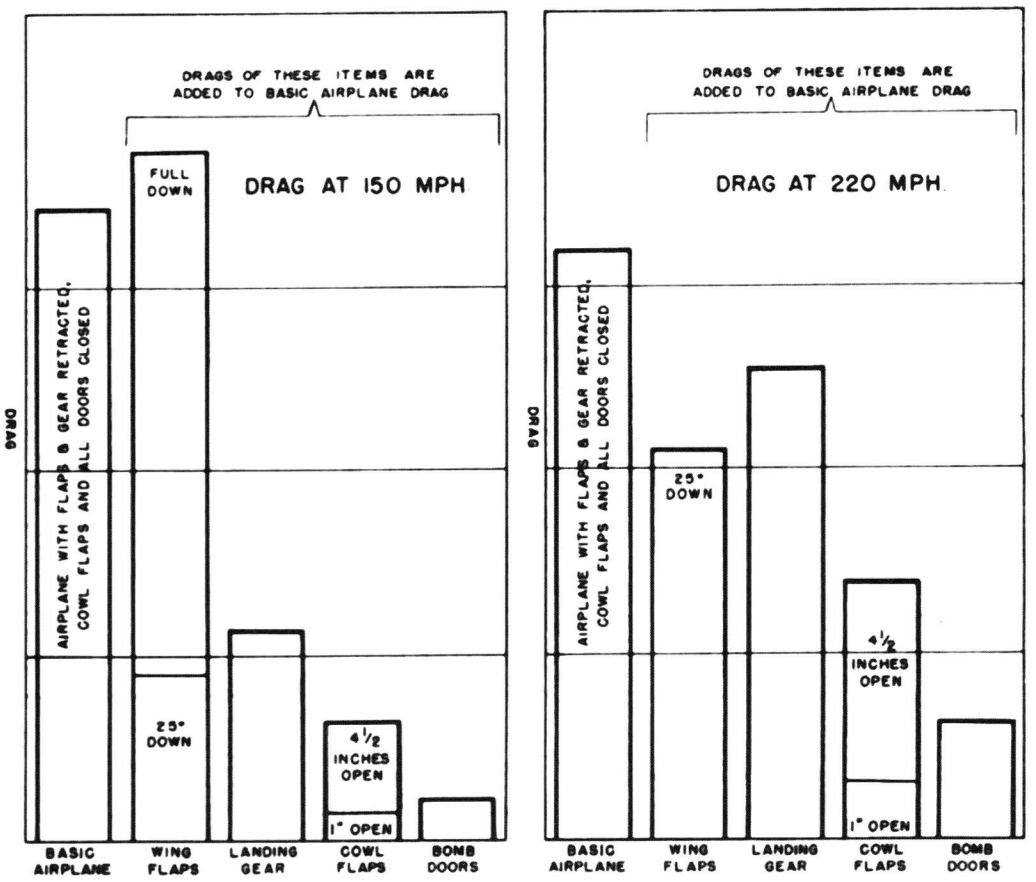

The drag build-up for the B-29, aiming to inform the crews as to the impact of their choices. Some of these were forced upon them; in the right-hand chart for 220mph airspeed, the effect of fully open cowl flaps adds about one third to the clean aircraft drag. However, the cooling of the engines was initially so poor that, in the long climb to altitude, otherwise excessive settings were needed. NACA's aerodynamic work in 1944 made a huge difference to this.

A 'computer' at NACA Langley, in the period in which the B35/46 bombers were being developed in the UK. She is viewing a transcript of raw data, which may come from wind tunnel or flight test, which she will subsequently process using the Friden calculator.

At this time engineering teams involved in aircraft design in general, and aerodynamic research in particular, frequently consisted both of engineers whose roles included the conceptual thinking and definition of the mathematical problems in terms of algebra, and of other workers literally termed 'calculators'. These calculators, often women, would generate the actual numbers needed.

This engineer/calculator partnership applied to both design in the first instance and data reduction from testing afterwards, including the large tables of numbers that would emerge from the wind tunnel. The casualness of 'girl computer' as a term and the familiarity of gender roles at the time, are an inescapable part of the social context of the V-bomber story. Peter Lissaman joined Handley Page as an aerodynamicist after completing his degree studies at both Cambridge and Caltech, together with time at the Bristol Aeroplane Company. He would recall,

"The plant was in a Dickensian factory building in the London suburb of Cricklewood. The aeroplanes were built in great echoing hangars at ground level while upstairs was a rabbit warren of tiny random offices. There were few corridors, so one had to go through different rooms to get anywhere. One of these, the Calculator Office, was commanded by Mrs Whitley. These were pre-computer times, and calculations were done by a team of twittering girls, operating massive green, dented Friden mechanical calculators.

They would set rows of ten-digit numbers by hand, and pull a heavy lever, like that on a Las Vegas one-armed bandit. It went Kerchunk, Kerchunk and the little numbers on dials rotated like those on an odometer. The girls were neat and proper, wore no make-up, and were under the severe matriarchal control of Mrs Whitley. They had graduated from the local secondary schools, and could have been waitresses or shop clerks. Pairs of girls made identical calculations for each step, and each checked the other's results. They NEVER, EVER made a mistake. It was legendary."[35]

Despite this assertion, it is clear that the laborious work

Measurements of pressures wind tunnel models in this period were indicated on banks of liquid tube manometers, each connected to an individual surface pressure tap. These women, working as calculators at NACA Ames in 1943, are collecting the data from a particular run of a model, for which conditions of M = 0.8 and α = 6° are indicated.

was neither quick nor easy. One of Lissaman's recollections concerned the development of the revised wing profiles for the extended span Victor B.Mk.2, working for Lachmann:

> "I, a boy fresh out of graduate school at Caltech in the USA, was ordered to do this man's job. It involved designing a new airfoil. The airfoil, called an aerofoil in England, is the beautiful streamlined cross-sectional shape of a wing. Using what were then very advanced mathematical methods, the theory of Küchemann and Weber, I drew up a shape that I thought would be OK. But the pressure had to be checked. Massive calculations are needed to find the pressure distribution on an airfoil, a job for Mrs Whitley. I carried the long list of numbers defining the shape, called the ordinates, into her office and asked her to complete these computations. She allowed as she might have it done in two weeks if one of her girls recovered from the flu.
>
> Two weeks later Mrs Whitley presented me with a large table of hand written numbers, within which the pressure values were embedded. There was nothing in the way of a print-out in those days, and graphics was just an impossible dream."

Also present at Driggs' lecture was Handley Page's Godfrey Lee, deeply involved in the aerodynamics of the HP.80/Victor. His recorded comments therefore provide a direct understanding of the situation at Handley Page and their ability to investigate conceptual aircraft design at the relevant time.

> "He thought it would be of interest to report on the experience that Handley Page Ltd had had when actually using the apparatus that Mr Driggs had kindly placed at the disposal of the Royal Aeronautical Society. They had used the 'slide rule' both in its 'power' form as applied to a piston-engined aircraft and also in its 'thrust' form as applied to jet-propelled machines. They could not give any detailed comments on the method as they had only worked out one example in each case, but from what they had done, it appeared that the method did everything that the lecturer said it would and they thought that it was a technique with considerable possibilities.
>
> It seemed that the lecturer's method for the calculation of jet-engined aeroplane performance would be useful in doing climb calculations when the distance covered and the fuel used in climb to a specified

The computers of the NACA Ames 16ft wind tunnel, January 1943.

altitude had to be worked out. This had to be done in assessing the range performance of a jet-propelled aeroplane and was quite important as an appreciable quantity of fuel was consumed on the climb. It seemed to them that this method might be considerably quicker than the, more or less, straightforward methods that they had had to use up to the present. Some people had seemed to give the impression that firms were not concerned with studying performance of a series of different aircraft.

While it was true that most of the work in a firm was concerned with the performance of one specified machine, it was also true that in the early design stages, when an attempt was being made to meet a given specification, a whole series of different aircraft had to be studied in order to arrive at the optimum design. For this purpose, the method proposed by Mr Driggs would undoubtedly be very valuable, even if it were not used greatly for the more detailed subsequent calculations."

In retrospect, we know now just how much the designs of both the Avro 698 and HP.80 evolved between specification and first flight, and in fact continued to do so into their respective mark 2 variants and beyond. What Lee was describing was the difficulty in rapidly adapting some the certainties of the piston-engined world—the data that had been established over so many years—to the new characteristics of the jet aeroplane. Clearly, the solution space that required exploration was large, and any tools to speed the process were to be welcomed. Driggs' reply was supportive.

"He was very glad that Handley Page Ltd. had applied the logarithmic performance methods and had had favourable results. He had no reason to believe that other trials, whether by Mr Lee or by other engineers, would not give like comparisons. Mr Lee evidently approached the problem of designing an aeroplane to meet a given specification, much in the same manner as was done in the research division—namely, by designing a series of aircraft more or less on a trial and error basis until a satisfactory design could be chosen. As pointed out previously, the logarithmic performance methods lent themselves ideally to that type of problem."

The latter stages of V-Bomber development had the chance to benefit from another technology for which Britain was in the vanguard. The Ministry of Supply had purchased a Ferranti Mark 1* computer and installed it at Fort Halstead, Kent. Beginning in 1953, programmers from Avro were given access to this machine, which anticipated the delivery in late 1955 of their own at Woodford.

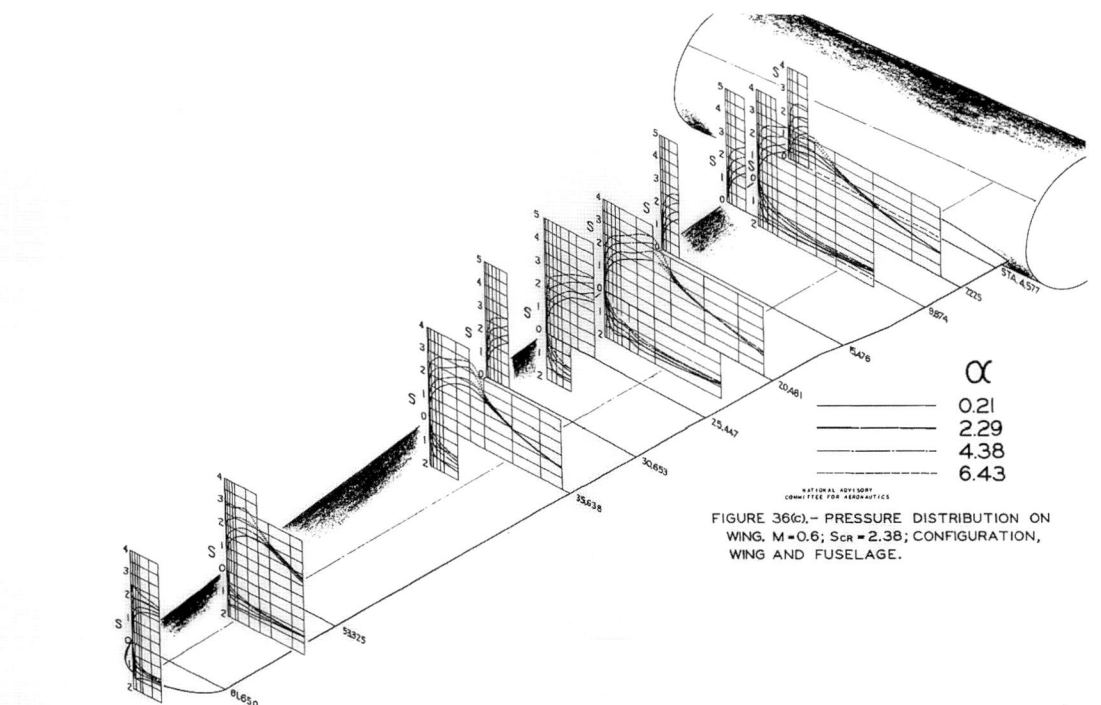

In 1943, NACA Ames conducted wind tunnel tests on a model of the B-29, in order to identify the effect of Mach number. While this might seem remarkable for a large bomber aircraft at the time, the combination of high cruise speed and high altitude (where speed of sound is low) meant that a dive could generate local supersonic flow. This hand-plotted chart shows the stepped pressure distributions implying the presence of a shock wave, in the mid span region of the wing, for dive speed of M = 0.6.

In 2016, Avro mathematician Joan Smith documented her experience of using this new capability in practical aircraft design.

"Many thousands of calculations are necessary to arrive at the optimum design for an aircraft; they were previously carried out on analogue devices when the design engineers would laboriously crank a handle to achieve the results. Two teams of mathematicians were engaged to write the necessary programs (software) for the computer: one group for stress, the other for aerodynamics. Ferranti's provided some basic software, including subroutines which could form part of a program written by a mathematician. Everything was in machine code; there was no such thing as autocode at the time. Moreover, the machine operated in fixed point, integers only. To achieve floating point a short subroutine had to be written and included in the program. These early programmers were the pioneers."[36]

The impact of the new electronic computers on the speed with which aerofoil and wing design calculations could be conducted, might be indicated by the disappointment barely concealed in a report from 1958. Calculating the velocity distributions over examples of swept and delta wings, E. York of NPL concluded,

"The time taken was approximately 150 hours. The efficiency was not high partly owing to the difficulties imposed by the large demand on storage space. It is roughly estimated that the computation was done at 60 to 70 times as fast as it would have been on desk machines, whereas this factor is more commonly of the order of 200 to 300."[37]

Despite the progress in that case not being as rapid as hoped, the productivity increase that the new methods allowed was impossible to ignore. Optimising many more points together on an aerofoil profile, or many more stations on a three-dimensional wing, enabled large gains in aerodynamic efficiency.

THE REYNOLDS NUMBER PROBLEM

Despite the critical importance of identifying the drag divergence Mach number and hence being able to compare configurations during development, it is apparent from the contemporary reports that scale effects caused substantial difficulty. In a world where high subsonic Mach number wind tunnels were a relatively recent innovation, the ability to generate transonic flow and hence an understanding of shock wave formation was a clear priority.

However, in the presence of extensive laminar boundary layers the effect of shock induced separation would be overpredicted, as relatively weaker shocks could provide the trigger. The provision of Reynolds number similarity concurrently was not feasible and, realistically, required cryogenic wind tunnels that were decades in the future.

The generation of aerodynamic data therefore relied on the application of experience and judgement to the raw results, the forcing of transition to simulate flight Reynolds number, or most expensively, full scale flight testing. The existence of this problem helps to explain the extensive provision of dedicated test aircraft for the V-Bombers: the Avro 707 series for the Vulcan and HP.88 for the Victor.

In late 1946, a matter of months prior to the issue of B35/46 and contemporary with HP's tailless bomber proposals, the considered view from NPL was one of profound uncertainty.

"At high Mach numbers when shock-waves are present in the flow between the leading and trailing edges of the body, there is considerable evidence that the nature of the interaction of the shock-wave and the boundary layer depends on the nature of the boundary layer upstream of the region of interaction. Thus, not only the nature of the shock-wave itself but also the flow downstream of the wave are dependent on the Reynolds number. It, therefore, appears that the Reynolds number may have an important influence on the distribution of normal surface pressure near and downstream of the interaction, on the normal force and moment coefficients, and also on the drag coefficient. The experimental evidence which is at present available is sufficient to support these general statements, but is not yet sufficiently conclusive to enable a precise description of the influence of Reynolds number to be made."[38]

Similarly, the Royal Aircraft Establishment found that although it might have been using technology that was relatively modern, it still could not deliver all that was required for the generation of transonic aircraft being developed in the early 1950s. William Perring had led the design of the RAE's major high-speed research tunnel during the war, and would be promoted to RAE director in 1946, from which position he would have the opportunity to make a critical impact on the B35/46 programme.

"Before the 1939-45 war the research tools needed by the aeronautical scientist were beginning to increase rapidly in cost and size. I remember those of us brought up on small wooden atmospheric wing tunnels feeling that we were entering a strange new world when the 'high speed tunnel'—the first big pressurised tunnel for high subsonic work at Farnborough—came into being.

For some years models could be tested in wind tunnels at low speeds or fully supersonic speeds, but tests at or near the speed of sound—the so-called transonic region—were impossible because of the difficulties in setting up steady flow conditions in the wind tunnel near sonic speed… it was finally decided to convert the old 'high speed tunnel' at Farnborough to transonic working by greatly increasing the horsepower in conjunction with a completely redesigned working section…

As originally conceived it was a variable density tunnel in which tests could be done at Mach numbers up to 0.8, at Reynolds numbers of about 1.4×10^6 per ft; at lower speeds, Reynolds numbers of as much as 5×10^6 per ft could be obtained. The working section was 10ft x 7ft. During the life of the tunnel speeds were increased by sundry internal modifications, by modifying the model support from struts to sting, and by reducing model size. Mach numbers exceeding 0.9 were thereby achieved, and nearly all our best-known British designs went through this tunnel."[39]

The high-speed tunnel had been commissioned in 1942, enabling testing of the final generation of high-performance piston engine fighters and the first jets. The recognition that testing at Mach numbers straddling the speed of sound was essential led to a major upgrade programme, commencing in October 1954 and leaving the tunnel out of action until its recommissioning (as the transonic wind tunnel) in April 1956. This 'blackout' period marked the final stage of development of the Vulcan and Victor, during which the first production aircraft were being constructed and issues discovered during flight testing of the prototypes were still being ironed out.

RAE reports from 1955 describing high speed tunnel tests on the Vulcan demonstrate this clearly. It was noted that an appreciable drag rise commenced close to M = 0.87, which was accompanied by a rolling moment. This was attributed to tip stalling, but not considered a serious problem as the mechanism was the separation of a laminar boundary layer. At full scale Reynolds number, transition at 60% chord as was being observed was not plausible, but equally this behaviour was crucial to the performance of the aircraft. It therefore required the application of judgement, rather than simply a reading of the data. Another contemporary illustration of the problem was given by the Supermarine Swift. High speed tunnel tests in 1948 explored the design's performance up to M = 0.92.[40] Some of the findings were disappointing, particularly relating to maximum lift and tip stall, which would restrict effectiveness as a fighter at altitude. The study concluded,

> "The following effects observed at the test Reynolds number (0.45×10^6 based on wing mean chord) should not be so marked at a higher Reynolds number;
> a) the rise in drag with increasing Mach number below say M= 0.8 (e.g. in the tests, for CL = 0, CD is 0.005 higher at M= 0.825 than at M= 0.5.)
> b) the backward movement of the aerodynamic centre with increasing CL, particularly near CL = 0.2
> c) the low CL (0.45) for tip stalling."

Flight testing by both the manufacturer and the A&AEE found that the elevator control became ineffective above M = 0.9, along with a nose down trim change, while pitch up when pulling relatively low 'g' severely limited manoeuvrability. The Swift was incapable of generating sufficient lift to fly and fight at 40,000ft and above, while longitudinal control was difficult until the introduction of a variable incidence tailplane on the F.4. All of this was evidenced in the RAE's wind tunnel findings quoted above, years earlier, but they were not able to make the categorical statement that the full-size aircraft would be similarly afflicted.

The trends of more positive characteristics with increased Reynolds number were called on to do much heavy lifting in the limited assessment possible. The Swift was a disaster as an interceptor, but very effective as a low-level reconnaissance machine, where maximum speed was limited by dynamic pressure before the Mach effects became dominant.

Testing of the Victor prototype configuration in the RAE high speed tunnel in the period April 1951-July 1952—effectively the sign-off of the design as aerodynamically suitable to meet the B35/46 requirements—revealed significant sensitivity to surface roughness. On the positive side, it is clear that these effects were recognised and exploited, with the tests being more representative of full-scale conditions than would otherwise have been the case. Put simply, a rougher surface could be used to trip the otherwise laminar boundary layer into transition at an earlier stage than would naturally occur, but unless at full scale it was expected to happen at the leading edge for all incidences, it would only be correct for one condition.

At this time, the RAE tended to use natural transition as the default for tests, while in the late 1950s better methods of artificial transition became the norm. Deep into the 1960s, on aircraft that are well-known such as the Lockheed C-141 Starlifter, the effect of transition and what is today termed Shock-Wave Boundary Layer Interaction (SWBLI) would cause severe problems (and embarrassment) in terms of performance prediction.

The engineers of Avro, Handley Page, the RAE and other stakeholders were capable and committed. They had the best technology and tools at their disposal, and a national imperative behind them. It is clear though that at the start of 1947, the maturity of the scientific and engineering understanding of the solution to the problem was suspect. They would be doing much more than merely shouting into the dark, but it wasn't absolutely certain where the light was, or if they could get there at all.

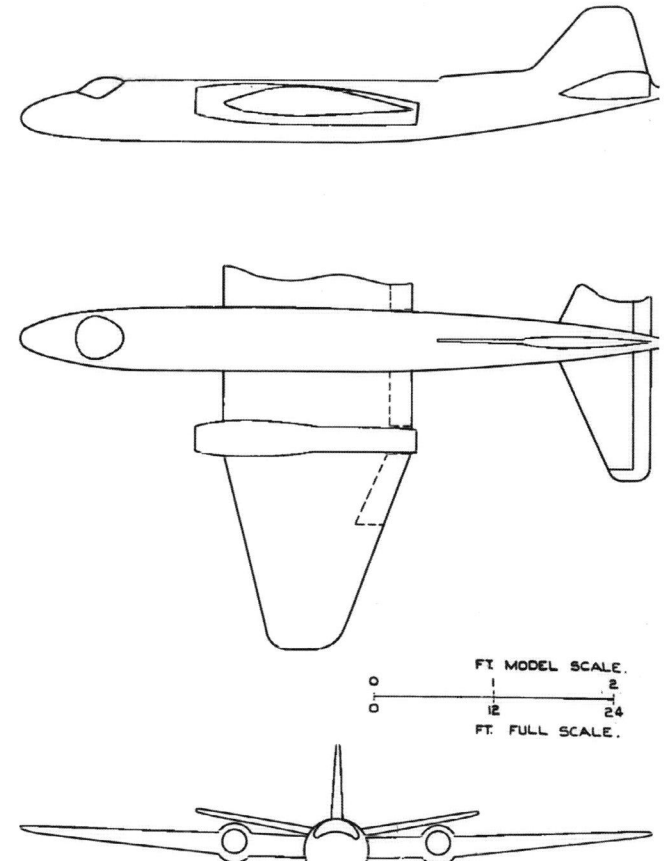

FIG. 3. ENGLISH ELECTRIC B3/45.

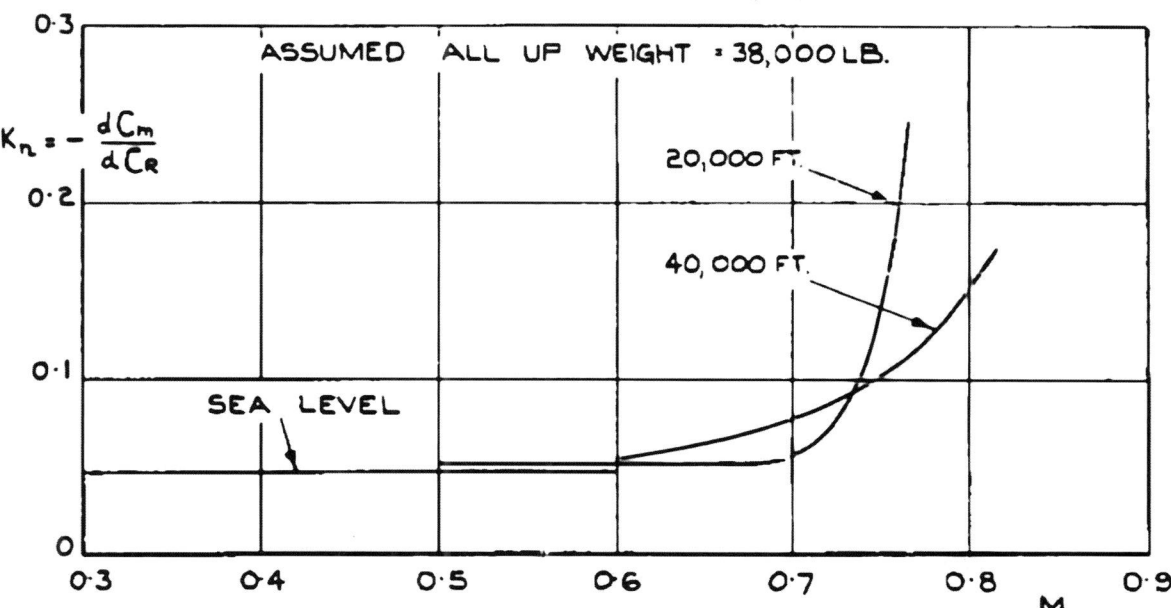

FIG. 19. B3/45 EFFECT OF MACH NUMBER ON STATIC MARGIN.

The English Electric B3/45 bomber, as wind tunnel tested by the RAE at 1/12 scale and reported on in September 1947. The measured static margin and its change with Mach Number was plotted; a parameter that was critical to stability and control. However the report noted, with worrying implications for the B35/46 programmes, that it could not be measured accurately.

THE BUILDING BLOCKS

To have any hope of meeting the B35/46 specification, exploitation of aerodynamic technology specifically designed to operate in the transonic, high Reynolds number, regime was required. This was at or beyond the state of the art in 1947. In the later words of Handley Page's Godfrey Lee,

> "In some designs the aerodynamics plays a dominant, even obsessive, role throughout. It was so in the case of the first high subsonic swept-wing designs undertaken in the period 1945-50: aircraft like the Boeing B.47 (sic) and the Handley Page Victor bombers. Here we were working on the frontier of aerodynamic knowledge and the whole design was dominated by the need not to fail in this field; problems of drag, stability and control, lift, flutter etc. being involved."[41]

The understanding and demonstration of swept wings was in its infancy, while suitable wind tunnel facilities were also in short supply. The low-speed requirements were implied by the relatively stringent field performance, allowing operation from standard RAF Bomber Command airfields. These, coupled with a desired maximum gross weight of 100,000lb, would become the primary drivers towards the unusual configurations of the final designs. What was the basis of the knowledge and tools that the engineers to Avro and Handley Page, together with their colleagues at the Royal Aircraft Establishment and other research agencies, had to work with? In this chapter, the importance of several specific technologies will be discussed, along with the existing experience that these eventually successful firms would be able to bring to bear.

AEROFOILS

In 1947, there was no such thing as a two-dimensional aerofoil shape specifically designed with transonic Mach numbers in mind. Instead, aerodynamicists and designers sought to adapt the knowledge that already existed to meet—in the short term—the demands of the new regime, in which regions of local supersonic flow would form cells around the profile. Where did they start?

The process of wing design in the period immediately prior to the issue of B35/46 rested on the ability to translate two-dimensional aerofoil section information into a three-dimensional world that accounted for planform effects. For classical straight-wing aircraft, the calculation of the aerodynamic loads to compare wing planforms during design involved a perhaps surprisingly new but well-established technique. For the wings of moderate to high aspect ratio in use for typical piston-engined bomber aircraft, using available catalogued aerofoil sections provided adequate, documented aerodynamic performance together with structural pragmatism—and at an acceptable level of research and development cost too.

When Spitfires clashed with Bf 109s over the Kent countryside during the long summer of 1940, it might have

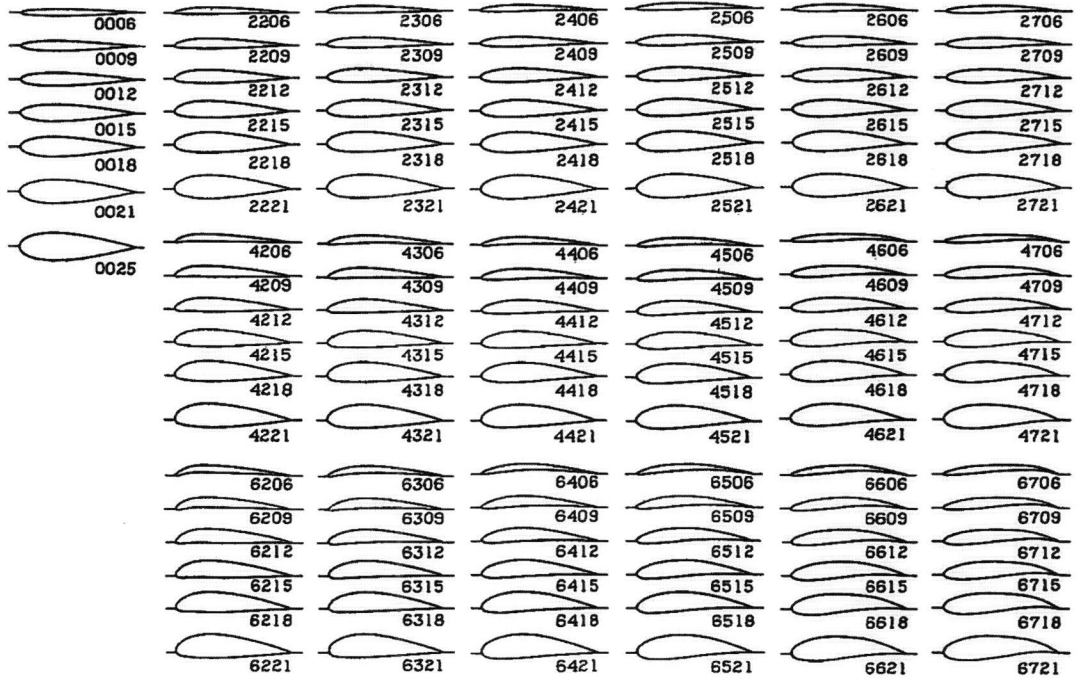

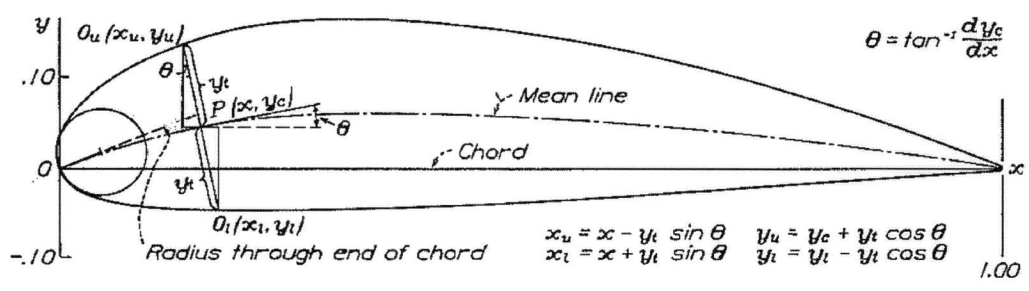

Sample calculations for derivation of N.A.C.A. 6321

The 78 systematically-designed aerofoil shapes tested in the first such NACA programme, including the symmetrical, 10% thick 0010 that would be the basis of the Vulcan's wing. The derivation of the 6321 profile was used as an example in the report. NACA-TR-460

come as a surprise to all the pilots concerned that a key piece of the aerodynamic technology used in both aircraft had come directly from the then-firmly neutral United States of America. The aerofoil sections of their wings were the result of a systematic investigation conducted at the expense of the US federal government, by the National Advisory Committee for Aeronautics (NACA) at its Langley Memorial laboratory at Hampton, Virginia.

The flow of German aeronautical engineering talent and information to both East and West in 1945 is well known, but the achievements of German aerodynamicists during early part of the 20th century, particularly by Ludwig Prandtl and his team, also made them prime targets. In 1920, German aerodynamicist Max Munk, a protégé of Prandtl, began work as an advisor to the NACA. This had itself required a special order to be issued by President Wilson, allowing Munk as a very recent enemy, to come to the United States and work. Munk developed aerofoil theory, but crucially conceived and recommended the construction of a wind tunnel facility that could replicate the conditions found in flight at speed and altitude. NACA's Variable Density Tunnel (VDT) was commissioned in 1922 and, run by Eastman Jacobs from 1928, began to contribute to Munk's project: a parametric understanding of aerofoils, providing experimental data that would allow engineers to select appropriate designs from a database.

The physical shape of an aerofoil could be described in terms that included how thickness related to length (chord), where the maximum thickness was positioned

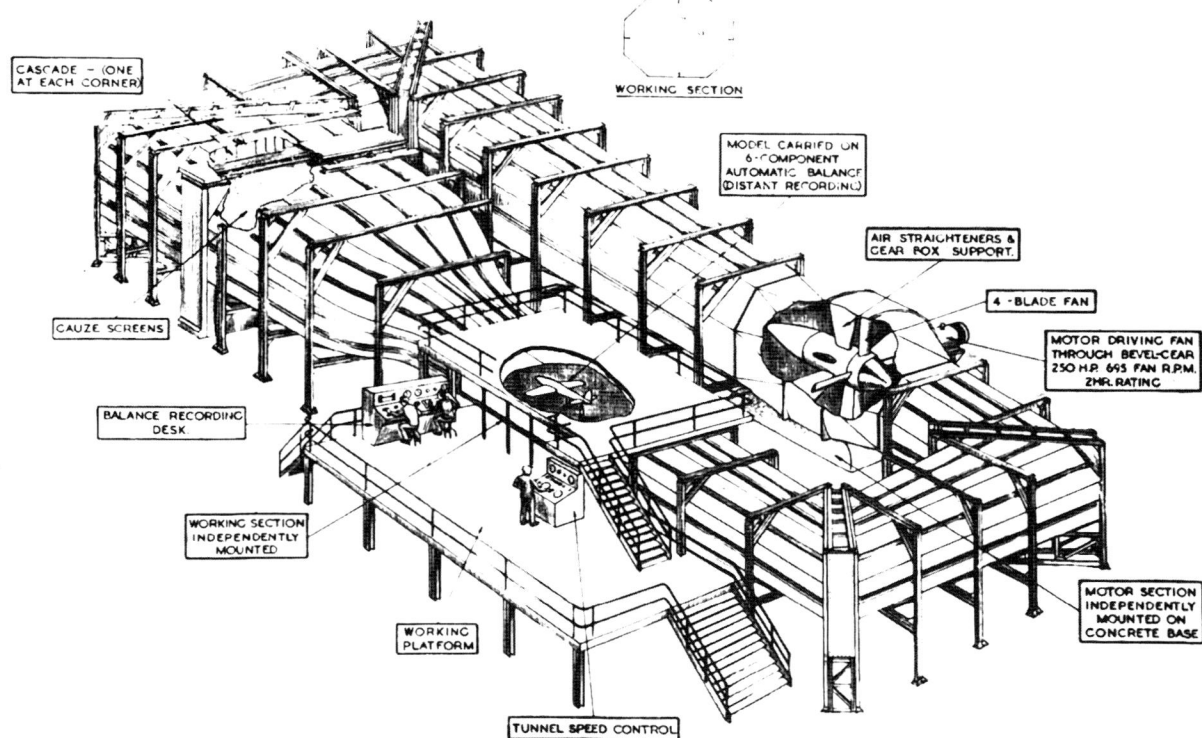

In the late 1940s, the Society of British Aircraft Constructors recommended a wind tunnel design of 9ft x 7ft cross-section for low speed industrial work, up to speeds of about 200ft/s. English Electric at Warton and Avro at Woodford built such facilities, using steel rather than concrete or wood, as supplies were easier. Avro's tunnel was used heavily used in the Vulcan and Blue Steel programmes, and after eventually moving to the University of Manchester, conducted work on Woodford's final new aircraft design, the RJX regional airliner in 2000.

along the chord and the curvature of the mean line (camber). NACA's programme resulted in a major report in 1933, which described the results of 78 variations on a basic theme.[42] Importantly, the geometry changes were easily decodable from a name made up of four digits, which gave camber in % of chord, location in tenths of chord and finally thickness in percentage chord.

Because the tests illustrated the aerodynamic effect of varying these parameters independently, for example the effect of reducing thickness at constant camber between a 2412 and 2410 shape, engineers could understand how these trends impacted their designs. It was obviously still possible to individually create an aerofoil for a new application if that were desirable, but the granularity of the published data and the ease with which it could be interpreted made the report revolutionary. Jacobs and NACA did not rest on their laurels, but swiftly moved on to target particular design lift coefficients through alignment (the five-digit aerofoils), and then on to what they perceived to be the big prize, the low profile drag potential of laminar flow, in the 6-series.

The five-digit shapes were available by the time that Avro came to design their twin-engined medium bomber that became the Manchester and ultimately the Lancaster, which used the 230xx series ranging from an 18% thick root to 12% at the tip. At some point, the shape would have passed through 15% and hence been a 23015, which was the root section adopted by Focke-Wulf for its Fw 190 fighter, used in the second half of the war to defend against attacks by RAF Bomber Command's Lancasters.

The NACA four- and five-digit series were fundamentally exercises in geometric design, in which variations of shape from a logical, working starting point were assessed for their performance. This involved the wind tunnel tests in facilities such as the VDT, its expensive replacement the NACA Low Turbulence Tunnel (LTT) and the UK's equivalent Compressed Air Tunnel (CAT) at the NPL, but also theoretical efforts to understand the flow mathematically.

By the early 1940s, engineers on both sides of the Atlantic were using these methods to design bespoke aerofoils that were predicted to deliver specific pressure distributions. In particular, there was a deep interest in the potential of low profile drag from the exploitation of laminar flow. Jacobs reported in 1939 on preliminary tests in a new wind tunnel that was itself a model to validate the extensive facility he wished to build. During this period of

Woodford's extensive wind tunnel model shop, c.1960. Avro claimed, "...Extensive experience in the construction of all types of models in wood, metal and plastics, has been accumulated over a number of years. The design of models suitable for testing in any of the tunnels is undertaken in the wind tunnel drawing office. Again, extensive experience in this specialised field has been obtained by a team of model design draughtsmen".

change, NACA's theoretical giant Theodore Theodorsen and his team were working through the methods of inverse design, that is geometry resulting from a specified pressure field, while Jacobs and his group continued to propose geometry that met their hypothesised criteria for extensive laminar flow, subsequently checking the results against the work-in-progress theory. Starting from a 0009 symmetrical aerofoil, leading edge radius, position of maximum thickness and other parameters were modified, all with the aim of moving the position of minimum pressure as far rearwards as possible. In doing so, a moderately favourable pressure gradient which might stabilise a laminar boundary layer was promoted. The performance and geometry targets were similar to those that would have been appropriate for a high-altitude bomber, with the benefit of high efficiency.

These effectively interim low drag aerofoils were superseded by the definitive implementation of the combined Theodorsen/Jacobs work, the 6-series. Although they retained a nomenclature that described some geometric aspects, the code now described in the main the aerodynamic performance characteristics rather than the shape. For example, rather than the position of maximum thickness, the calculated position of minimum pressure was stated. Other parts of the code specified the design lift coefficient and also the range above and below this design value that the low drag region extended. However, the implementation of these aerofoils during the second half of the war did not necessarily give the improvements hoped for.

In his 1939 paper, Jacobs had emphasised that new aircraft configurations might be necessary to exploit, or even realise, low drag benefits. He considered that low profile drag would need to be matched with high aspect ratio to give a corresponding reduction in induced drag, because the quadratic model of aircraft total drag showed the maximum lift to drag ratio would be achieved when they were equal. He also suggested that the presence of nacelles and extensive regions affected by the propeller slipstream would inevitably destabilise a laminar boundary layer. A pusher prop layout would therefore be necessary, which helps to explain the unusual design of the Northrop B-35 and Convair B-36 bombers. What was inescapable was that NACA's entire focus in terms of aerofoil development throughout this period was on the exploitation of laminar flow for low drag.

The state of the art as the Second World War drew to a close and the atomic age began. The enormous Convair B-36 Peacemaker was designed to fulfil a mission that would not have been remotely possible with pre-war technology: intercontinental bombardment.

The emergence of an additional dimension to the problem, that of compressibility, was a challenge to this order. The linearity of the aerodynamic loading with dynamic pressure disappeared as sonic flow was established, continuing to alter in a manner that was certainly much more difficult to predict, as the discontinuities in the pressure field allowed by the presence of shock waves themselves moved with Mach number. This was not to say that information on flows affected by shock wave formation was not available; on the contrary, high-speed piston-engined fighter aircraft deployed towards the end of the Second World War encountered the effects of compressibility and some results from pressure plotting in flight tests had been generated. While attention at NACA was focused on laminar flow aerofoils for reducing drag at high speed, the importance of Mach number effects (coupled with some scepticism about the practicality of achieving laminar flow in service) was seen as most significant in the UK. To some extent, this was driven by the increasing possibility of achieving transonic cruise speeds, through the advent of gas turbine propulsion, the progress on which was only revealed to NACA's aerodynamicists in 1943. Combining these two physical regimes would prove impossible, but NACA's work would certainly not be wasted.

The new problem therefore was to design aerofoils that would behave in a manner that was sufficiently linear to be useful, while minimising the wave drag caused by the entropy rise associated with the presence of shocks. Ultimately, this was another problem of calculating a pressure distribution, and in the UK it would be Brian Squire and Alec David Young at Farnborough (the former with experience at the Aerodynamische Versuchsanstalt—the AVA—at Göttingen, Germany) and Brian Thwaites at NPL, who would be recognised for making contributions to the art equivalent to those of their NACA contemporaries. A very similar progression was evident: Squire in particular had conducted pioneering research (for the UK community) on 'low-drag' section, from about the time of the outbreak of the war. Young wrote of him,

"He was almost certainly the first man in this country to think to good effect about the relation between section shape and pressure distribution in this context. He produced a simple first-order theory relating pressure distribution to thickness distribution and this enabled the so-called inverse problem of designing sections with different types of pressure distribution to be tackled.

Since the end of the war aircraft speeds have been steadily increasing... The designs of wing sections suitable for flight at close to the speed of sound had therefore become important, and the techniques that had previously been used for the design of low-drag wing sections were readily adapted to this later problem... Brian was one of the first to tackle [it] and he was responsible for the design and much of the testing of a family of high-speed wing sections that were later widely used..."[43]

In the words of NPL's Professor D. W. Holder, looking back on this period in 1964;

"Largely because the sections for which compressibility effects were first studied were designed on the basis of low-speed performance, they were thick and unsuitably designed for high-speed operation. The notorious difficulties of a direct theoretical attack, even for inviscid flow, were thus reinforced by viscous and scale effects that are now known to be large ..."[44]

His colleague Herbert Pearcey (the same Herbert Pearcey who would write to Dr Robert Pleming in 2015) commented on the value of aerofoil research and how it should be targeted;

"... it should be aimed at the derivation of chordwise distributions of thickness and loading which in combination will produce a favourable chordwise distribution of velocity and at the same time meet the structural and lifting requirements."[45]

The use of symmetrical aerofoil sections rather than those 'properly' cambered and aligned for the design lift coefficient was in keeping with the advice of the RAE. The approach of designing an aircraft with full plan view symmetry (for which symmetrical aerofoils were an enabling technology) was promoted as a mitigation for the trim changes found, as sonic conditions were encountered.[46] Symmetrical aerofoils were 'known' to give the lower profile drag at low lift coefficients, smaller change in pitching moment with Mach number and a numerically small value of C_{mo}.

This conclusion had also been drawn in the USA, with a NACA summary paper on high-speed aerofoils from 1948 concluding that camber was not beneficial for L/D at flight above M = 0.73. In turn, this analysis was partially based on German data, which itself showed that L/D continued to increase in this high subsonic range, as camber became negative. Remarkably and as freely admitted by its author

John Becker, the progression beyond symmetrical to negative was assumed to be impractical and was ignored. It was thought that the relatively forward location of the upper surface shock wave on a symmetric aerofoil occurring in a thinner boundary, inherently more resistant to separation, was the mechanism at play.[47]

Back to NACA's preoccupation: the firmly subsonic, low drag aerofoil. The 6-series had achieved its aims convincingly in the wind tunnel, while proving reasonably effective in real aeroplanes. NACA's theoretically designed shapes had evolved towards relatively sharp, low radius leading edges and peak thickness well aft, aiming to generate a shallow, favourable pressure gradient for as long as possible. When the low thicknesses were employed, these shapes proved effective too in the transonic regime, as the efficient generation of supersonic leading-edge flow was assisted by this relaxed upper surface in producing acceptable wave drag.

Jacobs' wartime work can be judged in many ways; industry may have been scathing towards the end of the war when considering the unrealised promise of the 6-series (as it was seen then), but he had rapidly and with minimal tools created the shapes that would contribute hugely to the North American P-51 Mustang. The ultimate realisation, however, would need to wait for an entirely different problem that its characteristics would accidentally solve.

Consequently, Avro and Handley Page would have used the best information that they had available, while accepting that it was incomplete, coupled with advice from the government agencies in their aerofoil selection. It is known that HP's aerodynamicists were familiar with the virtues of the NACA 6-series, having intended to apply this to the HP.65 'Super Halifax'. It would be no accident that the starting point for the HP.80's aerofoils would be of very similar form. What could certainly be said about all of the sections considered was that, in spite of the relatively good high Mach performance of the NACA 6-series and the analogous British 'Squire' series (later the RAE 100-104), they were designed firmly with incompressible operation in mind. They worked to mitigate the effects of compressibility, rather than exploiting the potential of the flow physics that was found in the transonic regime. In the words of Dr M. A. Ramaswamy, head of the aerodynamics division at National Aerospace Laboratories, Bangalore, from 1960-1985:

"The main disadvantage of a conventional aerofoil is that the development of supersonic region on the aerofoil is associated with strong shocks and consequently there is a rapid drag rise beyond [the critical Mach number] and also buffet due to separation caused by shock boundary layer interaction"[48]

The unpleasant discovery later of the inadequate buffet boundary on the final Vulcan design showed the importance of this distinction and, as shall be described, highlights the advanced thinking of Pearcey's underlying flow physics in Ken Newby's solution for the Vulcan.

PLANFORM DESIGN AND THE GERMAN INFLUENCE

Transonic cruise was not a prerequisite for Mach number-associated effects to become problematic, as the Westland design team discovered when working on the Welkin high-altitude fighter in the middle of the Second World War. This large twin Merlin-powered machine, designed to combat high-flying Luftwaffe bombers or reconnaissance aircraft, resembled a high-aspect ratio version of the earlier Whirlwind. To provide enough structural depth to make the slender wing sufficiently stiff, the root depth used was 21% — comparable to that of the much bigger and heavier B-29.

The sections used were of the NACA 230xx series, similar to those on the Lancaster. Diving tests from high altitude using early production example DX297, conducted by the Aeroplane and Armament Experimental Establishment (A&AEE) in 1944, found pitching oscillations at M = 0.7. Even in level flight it was possible to trigger similar behaviour by small increases in incidence, which pointed directly towards shock-induced separation of the flow over the wing.

Once this aerodynamic situation had developed, the recovery of the flow to a more benign regime was not directly possible, in other words it displayed hysteresis. In attempting to pull out of the dive, the pilot would find that the available maximum lift coefficient was very low, of the order of 0.5. To put that into context, a similar lift coefficient would be a high end, but not unreasonable, value for a cruising airliner today. Tests in the RAE High Speed Tunnel reported in November 1943 showed an even larger loss of maximum lift than the flight test result, possible due to the lower Reynolds number. The A&AEE concluded that the reason for the "really violent" pitching oscillations encountered was,

"...the exceptionally thick wing, and no large improvement in behaviour at high Mach numbers can be expected unless the wing thickness ratio is reduced."[49]

The lead engineer of the Welkin team, W. E. W. 'Teddy' Petter, left Westland eventually to set up the in-house aircraft design capability at English Electric. The new A.1 bomber that his small team designed was intended to cruise at altitudes similar to where the Welkin's interceptions would have taken place. Mindful of having been bitten once, Petter gave the new bomber an unusually low aspect ratio, with a very long chord root that provided both structural (dimensional thickness) and aerodynamic (non-dimensional thickness-to-chord ratio) parameters that were appropriate respectively. The Canberra, as it would become, could sensibly cruise at about M = 0.76, but was not fundamentally designed for transonic speeds. It took the conventional straight wing layout about as fast as was possible—unfortunately well below the requirements of B35/46.

Merely achieving very high ceilings implied correspondingly high lift-to-drag ratios, and Westland was not alone in attempting to use aspect ratio to drive this. It became particularly important as typical wing loading increased, in turn demanding increased lift coefficients and consequently more lift-dependent drag. The B-29 would be the ultimate exponent of this design philosophy at the end of the Second World War, able to cruise above 30,000ft and configured around a wing with aspect ratio of 11.5. If power were available, then what did limit the performance of the B-29 at high altitude? The answer was revealed in high-speed wind tunnel tests, which demonstrated that the flow over the wing near the outboard nacelle would become supersonic, that is to say achieve its critical Mach number.

Accounting for the weight of the aircraft, it was shown that this condition was encountered in level flight at 32,000ft, for a wing loading that was typical of combat weight. To fly higher while avoiding the power-sapping and stability-complicating shock wave effects, the B-29 needed either more wing area, or new (and inevitably thinner) aerofoils that delayed shock wave formation. Either option would increase structural weight.

It was not incidental that the presence of the engine nacelles, and their influence on the local pressure distribution, correspondingly reduced the local critical Mach number and triggered the problem. A cleaner wing, unspoiled by external protrusions, would provide a way of postponing this to high Mach numbers. In turn, the benefit of complete aircraft configurations that could accommodate engines and payload within the mould lines of an optimised external shape were clearly attractive. If the high aspect ratio straight wings of the B-29 were a dead end for high speed at extreme altitude, what might be the answer?

The Allied nations had the opportunity to observe two high-performance aircraft featuring swept-back wings during 1944. Neither employed this feature as an engineered means of transonic drag reduction however. In the case of the Messerschmitt Me 163, the tailless design compelled a longitudinal extent of wing for stability and control; measured along the length of the aircraft, the wing tips had to be some distance from the centre of gravity to be able to exert a moment about it.

In the case of the turbojet-powered Me 262, the swept outer wings allegedly corrected a centre of gravity problem resulting from a requirement to install heavier than expected engines. Whether or not the reason for the latter was speculated on and identified, the rocket-powered aircraft's planform was entirely consistent with its tailless nature; it would have made sense to an informed British engineer on that basis, with no obvious direct connection to its high speed. Remarkably, the concept had been articulated in a public forum—the Volta Congress on High Speed Flight, in October 1935.[50]

It was here that Adolf Busemann had presented his ultimately famous paper concerned with the calculation of lift and drag theoretically, for wings travelling in purely supersonic flowfields. He showed that the lift to drag ratios achieved would be poor, compared to the performance of the same wing in a subsonic flow.

He then proposed a solution: sweeping the wing back, so that the oncoming flow might be considered to split into two components. One would be at a right angle (normal) to the wing's leading edge, and therefore encounter the effect of the section's shape, creating a pressure distribution and hence lift (and drag). The other, parallel to the leading edge, would not see any curvature for a wing of constant section and extruded shape, therefore having little effect. Busemann referred to it as, "…that useless velocity component along the span". The key was that the normal component only had velocity in the proportion that the wing sweepback angle and trigonometry would give it. It would appear to have slowed down.

The lecture was focused entirely on the supersonic problem and because Busemann emphasised this, it is entirely possible that the assembled mass of aerodynamic prowess in the room simply did not appreciate the inherent potential of applying the same principle at subsonic speeds. Guided by the futuristic and theoretical take on the concept, in a world where supersonic flight was an engaging fantasy and not even a fully accepted possibility, it is easy to see how that might have been.[51] Years later, the independent American developer of the swept wing concept, Robert T. Jones, would be moved to write,

"How could this have happened? Clearly, Busemann's thinking was ahead of his time. Perhaps also, as a true scientist, he had emphasised too much the limitations of his theory."[52]

There is evidence of another missed opportunity in the UK—and one with proximity to bomber design. In 1937, Handley Page's Lachmann presented an in-depth lecture to the Royal Aeronautical Society on the subject of high aspect ratio, tapered wing design. The discussion at the end moved to experiments that another researcher had conducted on a wing with swept back trailing edge. Lachmann opined that,

"...because if one had a parallel wing and the resultant flow was not normal to the leading edge, it could be resolved into a component normal to the span, and a component which was span-wise..."[53]

There is nothing to show that this otherwise far-sighted engineer made the jump to understanding the significance of this at high speed, in that the wing would see a flow that appeared to be at a lower Mach number. By this time, Busemann's swept wing work had been classified by the German government, and so the limited interest that might have been kindled in the minds of foreign researchers by his ideas was not provoked further. However, to make progress domestically, his new role at the Luftfahrtforshungsanstalt (LFA) in Völkenrode, Braunschweig, left him extremely well positioned. The facilities were world-class and growing in capability, while his contacts with the AVA at Göttingen would lead to a productive relationship with Albert Betz there and a patent specifically for subsonic applications.

Busemann's ideas would remain theoretical until the commissioning of a planned subsonic-supersonic wind tunnel at the AVA in 1939. At this time, Betz was unaware of Busemann's Volta lecture and supersonic application concepts, but had independently decided to assess a swept wing as one of the first cases in the new tunnel. Betz wrote to Busemann to discuss his ideas on November 10, 1939. He had been motivated to submit a patent when Messerschmitt AG declared an interest in them, but clearly recognised that he would not be awarded priority. Busemann was able to respond within a few days, letting Betz know that he too had discussed the matter with Messerschmitt (and Heinkel), and was preparing his own set of wind tunnel tests at the LFA in conjunction with them.[54]

Having agreed that a joint patent was desirable and an appropriate form of words formulated, it fell to Betz's nominee Hubert Ludwieg, of the High Speed Aerodynamics branch of AVA, to make what are now recognised as the first ever high speed swept wing wind tunnel tests. The plan was simple, elegant and complete. A known Göttingen wing section of 12% thickness, at around 30% chord was used, with a mildly tapered, trapezoidal planform as the baseline wing. The swept version was formed by (conceptually) splitting this on the centreline and rotating the two halves of the trapezoid. While the section was thicker than would be chosen typically for high subsonic work and with the position of maximum thickness also further forward, that meant an amplification of the favourable effects of sweep, in the event that they emerged as predicted.

Ludwieg's results were considered sensational by the few that were privileged to be brought into the fold immediately; both Busemann and Prof. Messerschmitt receiving them in December 1939. By then there were more immediately pressing matters to attend to, however.

THE SLENDER DELTA

Just over five years later at NACA Langley, Jones had made a mental leap that would be part of a revolution. He had been considering low aspect ratio configurations as part of his work on guided bombs but concluded that the standard methods of analysis for high aspect ratio wings were unsuitable. Instead, he resorted to methods described two decades earlier by Munk, and more recently by Hsue-shen Tsien, a researcher at CalTech now attached to the headquarters of the USAAF as a scientific adviser. Specifically, he would apply these methods to a shape for a 2,000lb glide bomb, which had been proposed by Roger Griswald and we would recognise today as being of slender delta form. Work done, the analysis literally sat in his desk drawer for several months, believed to be of little consequence.

Jones would come to understand the true importance of this, in his own word, "crude" analysis when he connected it to another part of his work: compressible, supersonic flow. If the shape of the wing were sufficiently narrow that it could fit inside the shock cone formed at the nose, then the flow over it would be subsonic. It could then be analysed by the method he had devised for the glide bomb. Jones was certainly curious enough to investigate further, moving in the opposite direction to the earlier (but unknown to him) German work, which had gone from delta planforms to sharply swept wings.

By now recognising that he had identified a key aerodynamic mechanism (and method of analysing it for aircraft design), Jones wrote up his notes and discussed with both

The LFA facility at Völkenrode, May 9, 1945. From the left, Hugh L. Dryden (NACA), Ben Lockspeiser (MAP), Theodore von Kármán (CalTech/USAAF) and A P Rowe (TRE).

NACA Langley colleagues and others there on secondment from the USAAF. Encouraged by their reaction, he took the findings to NACA's director of research, George W. Lewis, on March 5, 1945; at a similar time, Tsien and Theodore von Kármán himself became aware through the USAAF channel, well in advance of their travel to the German aeronautical research centres in the aftermath of their capture. NACA Langley wind tunnel test results were reported on May 11, 1945, demonstrating by virtue of the lack of movement of the aerodynamic centre, that the slender wing within the shock cone at M = 1.75 was indeed operating as if in an incompressible flow.

Jones had independently conceived and demonstrated the effect of delta and swept wings on the forces and moments that would be generated in supersonic flow — all by May 1945. The previous day though, Boeing's chief aerodynamicist George S. Schairer had written home with news of what he had discovered in Germany. Schairer's letter was addressed to Boeing colleague Ben Cohn and was written on May 10 at Völkenrode; clearly, the contents could not wait. He noted,

> "The Germans have done extensive work on high-speed aerodynamics. This has led to one very important discovery. Sweepback or sweep forward has a very large effect on critical Mach number... thus the critical Mach number is determined by the airfoil section normal to the wing and by the sweepback".

The movement of the Allied forces through Germany in the spring of 1945 would lead to shocking revelations about advances in high-speed aerodynamics, and the weapons that might have resulted from that research. From the UK side, William Farren, the first director of the Royal Aircraft Establishment, was at the forefront of assessing German research and experimental material. It is said that his investigation was on the direct orders of Churchill, but it is certain that it included George Edwards of Vickers-Armstrong and Reginald Stafford of Handley Page. Both would be very significant figures in the development of their company's respective V-bombers.

To the amazement of those dispatched on research gathering expeditions into occupied Germany, the British had control of the Völkenrode facility, in theory at least. However, the first Allied troops to occupy it were American, and prior to the handover it is clear that at least some knowledge and information was transferred across the Atlantic. An early notable to visit was Ben Lockspeiser, the Air Ministry's Director of Scientific Research, after,

> "...he had received a telephone message to the effect that the army had discovered some wind tunnels in a large forest and he was asked to go to Germany to find out what it was all about."[55]

By his own admission, he was amazed by what he found there: "The finest aeronautical establishment he had ever

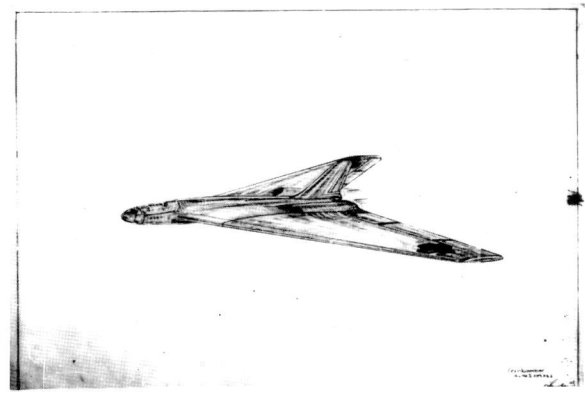

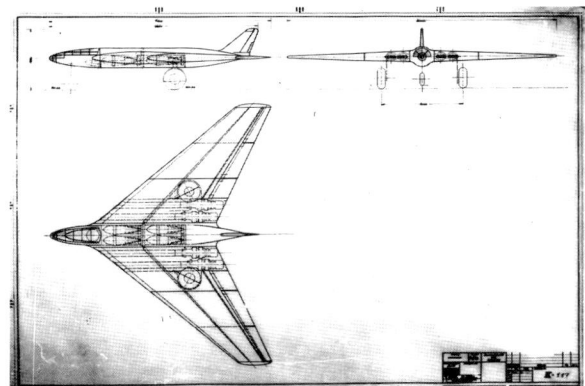

Messerschmitt's P 1108 tailless bomber design was reviewed by Farren. DS COLLECTION

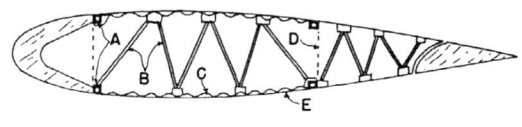

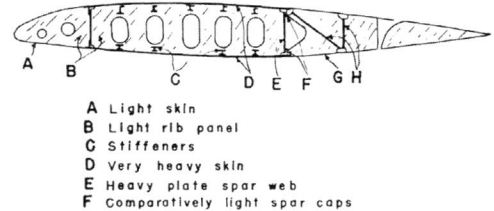

The progress in structure for wings, which was essential to the use of flexible, swept geometry. The B-17 style structure used a heavy internal truss, while in the B-47, very thick skins carried the torsional loads within the central box. A pre-requisite was the capability to machine these skins, which did not really arrive in the UK until the era of the Buccaneer in the mid to late 1950s.

seen." The attention that had been given to hiding the facility was substantial; the buildings were largely indiscernible or otherwise innocuous when viewed from air, while even the provision of transport links to the site was planned with deception in mind.

Busemann at the LFA was, as described earlier, had been the joint originator of the concept of wing sweep as a mitigation for the impact of compressibility on efficiency in subsonic flows. The LFA in turn developed, over its brief existence, some major facilities for the assessment of designs in these regimes. One such was the 2.8m high speed wind tunnel, capable of at least M = 0.85 when containing a model, and hence equivalent to the RAE's High Speed Tunnel at Farnborough. How had this translated to the front line, during the total war in progress? The situation was aptly described by Smelt in October 1946:

> "It is surprising that very little of the experimental work described above had any great influence on aircraft in operational use at the end of the war. The later jet-propelled fighters had symmetrical thin aerofoil sections, adopting the results of the aerofoil work… but large angles of sweepback had not appeared on any operational aircraft when the war ended. There was ample evidence of its universal adoption in high-speed aircraft designs at the prototype or project stage, however."[56]

One such project examined by Farren and his team was a tailless bomber at Messerschmitt, the P 1108. It was reminiscent on some ways of much-enlarged Me 163 B, but as a long-range aircraft, the latter's rocket propulsion was replaced by four HeS 109-011 turbojets mounted internally at the trailing edge of the wing root, and fed from slots in the leading edge. As with the earlier interceptor, the wing sweep served both a longitudinal stability and control purpose, together with alleviating transonic effects, as Messerschmitt had been aware of since Ludwieg's experiments of 1940. The aircraft was never built and—as would become known later in the UK programme—would likely have run out of control authority at the Mach numbers required by B35/46. Nonetheless, knowledge of such configurations was obviously now available to the right people in the United Kingdom.

A B-47 making a jet-assisted take-off at Portsmouth AFB, New Hampshire, in February 1956.

In the USA, the result of Schairer's letter was an extensively redesigned B-47 with a 35° swept wing, submitted as a proposal to the USAAF in September 1945. At first retaining the four forward fuselage-mounted engines but adding an additional pair in the rear fuselage, the safety concerns were finally put to rest with the concept that would become familiar, all six engines being wing mounted. Mounted high on the fuselage, the almost constant 12% thick mainplane contrasted with the 22% thick root of the earlier B-29 design; an enormous structural challenge given the additional torsional loads it would need to carry. Coupled with this limited depth in which to package structural spars, was the high aspect ratio of 9.4.

The engineering of a long, flexible wing with distributed masses hung from it was a non-trivial challenge. On the one hand and in a static sense, the outboard mass helped to reduce the bending moment at the root and more easily balance the aircraft, as the engines were close to the CG longitudinally. On the other, without close attention, the flutter speed of the wing would have been reduced, which in turn would have begun to negate the positive effect of sweep and taken the design back towards square one. The necessary answer was careful tailoring of the position of the engine CG with respect to the wing's elastic axis, and an understanding of the dynamic coupling of the two as loads were applied. This meant mathematical modelling and dynamic wind tunnel testing, both expensive and time-consuming.

The study of aeroelastic phenomena probably assumed a greater importance during B-47 development than with any previous aircraft design, requiring new methods of testing to be devised. For example, a flexible, dynamically scaled model (that is, one that reacted to applied loads and hence deflected in the same way that the full-size aircraft was expected to) was mounted to a vertical pole via a sliding gimbal. It had freedom to move vertically and rotate in pitch, with the model trimmed for the appropriate wind tunnel speed such that lift was equal to weight. By steadily increasing wind tunnel speed, any flutter could be identified and action taken to provide damping.

Superficially, the high aspect-ratio swept wing with separate fuselage seemed a rather basic solution; the reality

SHAPING THE VULCAN

Wind tunnel testing of the XB-52 configuration at NACA Langley, 1950. The view of the tufted model was taken at M = 0.875 and CL = 0.3, similar to the Victor's planned cruise condition. Extensive shock induced separation can be seen over the rear half of the wing chord. NACA-RM-SA51C16

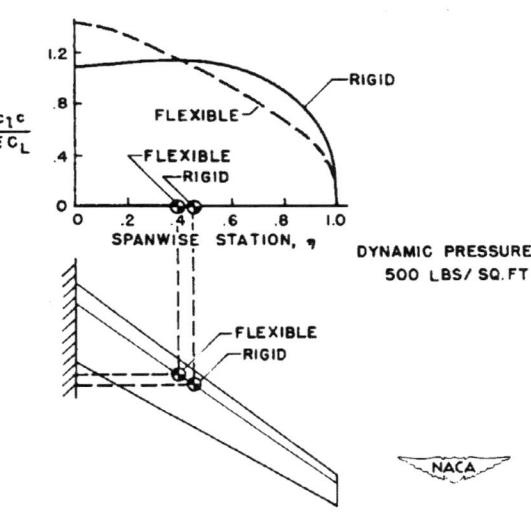

NACA diagram illustrating the favourable effect of static aeroelastic deflection. As the aircraft went faster and so more force increased wing bending, the flexible wing retained a more elliptical loading and the aerodynamic centre moved forward, in the opposite direction to the transonic effect.

Handley Page were obviously content to connect their high-technology future in the 1950s, with the swept back wings of Blue Bird, HP's first aeroplane, in this contemporary advertisement.

was a vast amount of specialised engineering. Even then, neither the B-47, nor ultimately the related technology used in the B-52, could offer the high Mach number performance required by the UK specification. Both offered peak efficiency well below M = 0.8 and fell off rapidly, such that cruise at M = 0.87 and 50,000ft was not viable.

In 1953, as the prototypes of the B35/46 contenders made their early forays skywards, Jones himself would comment,

> "These new developments illustrate... the fact that the disturbance fields at transonic... speeds are essentially three-dimensional phenomena. It was not long ago that our ideas concerning the wing section—which had their origin in the older incompressible flow theory—had to be relinquished because of the predominating effects of the wing planform. Now we must learn how to design the wing and fuselage together."[57]

In fairness, this was in the context of area-ruling, a step beyond what was being considered during the development of these immediate post-war concepts. However, it was absolutely the case that in order to meet the specification, the B35/46 aircraft would need to be designed in three-dimensions at once, and that was a very new thought. The more challenging aerodynamic demands of the higher speed and altitude compared to the United States's requirements, together with the difficulty in designing and physically manufacturing the swept wing shapes, helps to explain why the B-47's deceptively simple 'tube-and-wing' configuration was not considered appropriate. Instead, the hints seen in the ruins of Germany would support the ideas of minds in the UK, without actually telling them the complete answer.

TAILLESS AIRCRAFT

Tailless or flying wing aircraft were of interest to a variety of constructers between the First and Second World Wars. Captain G. T. Hill's Pterodactyl series (associated with Westlands) springs to mind, but these machines had built upon the experience of John William Dunne during the earliest days of British flight, and his laudable aim of building an aeroplane that would protect itself from stalling. For the increasingly mature aeroplane, the attractiveness of a tailless arrangement lay in the possibility of eliminating the parasitic drag, mass and engineering effort that came with the development of a tailplane, together with the use of the wing's internal volume to carry the payload. In the words of HP's Godfrey Lee,

> *"In discussions on the relative merits of the orthodox and tailless types the specious argument that the tailless layout must 'obviously be better because it is not encumbered with a tail' is sometimes put forward. A little consideration will show that it is not possible just to remove the tail of an orthodox aeroplane, for to do so would make it uncontrollable in flight. It is thus clear that, if the ordinary tail be removed, something else must take its place and it is not at all obvious, a priori, that the alternative will be any better than the tailplane originally provided."*[58]

However, the mounting of a tailplane on a long moment arm behind the wing was a simple way to provide longitudinal stability and pitch control; some means of generating aerodynamic force and damping at distance from the aircraft's centre of gravity was still necessary. Sweeping back the wings enabled this, as lift was generated along a diagonal axis relative to the direction of flight, with force at the tip significantly rearwards of that inboard. As well as a rolling moment, a control surface at the wing tip would now create a useful pitching moment too, allowing it to provide the lost functionality of the tailplane's elevator. A group of swept wing aircraft therefore developed, for entirely different reasons than those of dealing with high Mach numbers.

A major problem was the generation of high lift, because the strong nose-down pitching moment caused by the deployment of trailing edge flaps had to be trimmed by a nose up moment, and this had to come from the outboard, rearwards part of the wing. In other words, lift was increased significantly inboard and reduced outboard, a story of diminishing returns. On a conventional aircraft, not only was more of the wing span available for flaps, but the tailplane generated the required pitching moment to trim, by multiplying a small lift increment with a long moment arm. If the maximum lift coefficient was much lower than for the conventional equivalent aircraft, then the recourse would have to be an increase in wing area, potentially negating the weight and parasitic drag improvements that the tailless configuration was intended to provide.

Nonetheless, for a bomber in the cruise and assuming that sufficient runway had been available, for a given weight the tailless design ought to have a better spanwise lift distribution (when the wing and tail were taken together, for the conventional case) and less induced drag, so of the two should have achieved higher altitude and longer range. There was clearly much to play for, and potential performance to be unlocked.

One way out of the dilemma, not necessarily a satisfying avenue for the purists, was the 'tail first' layout. A stabilising surface forward of the mainplane had the advantage of countering the nose-down moment by providing one that was nose-up, that is by lifting. Unlike the tailplane then, the foreplane could contribute positively to the total lifting force and so for a given lift coefficient, the combination of the pair might be smaller than the conventional layout. Handley Page's research concluded that this solution was less than ideal, as the effect of the new lifting surface was to bring the aircraft's neutral point (the aerodynamic centre of the whole aircraft, as opposed to the wing alone) forward, which in turn meant that the centre of gravity would move correspondingly forward and require lifting support from the new foreplane.

The company's designers perceived that another spiral of diminishing gains might occur in this case, as either the lift coefficient required of the foreplane would be excessively high, causing an induced drag problem, or

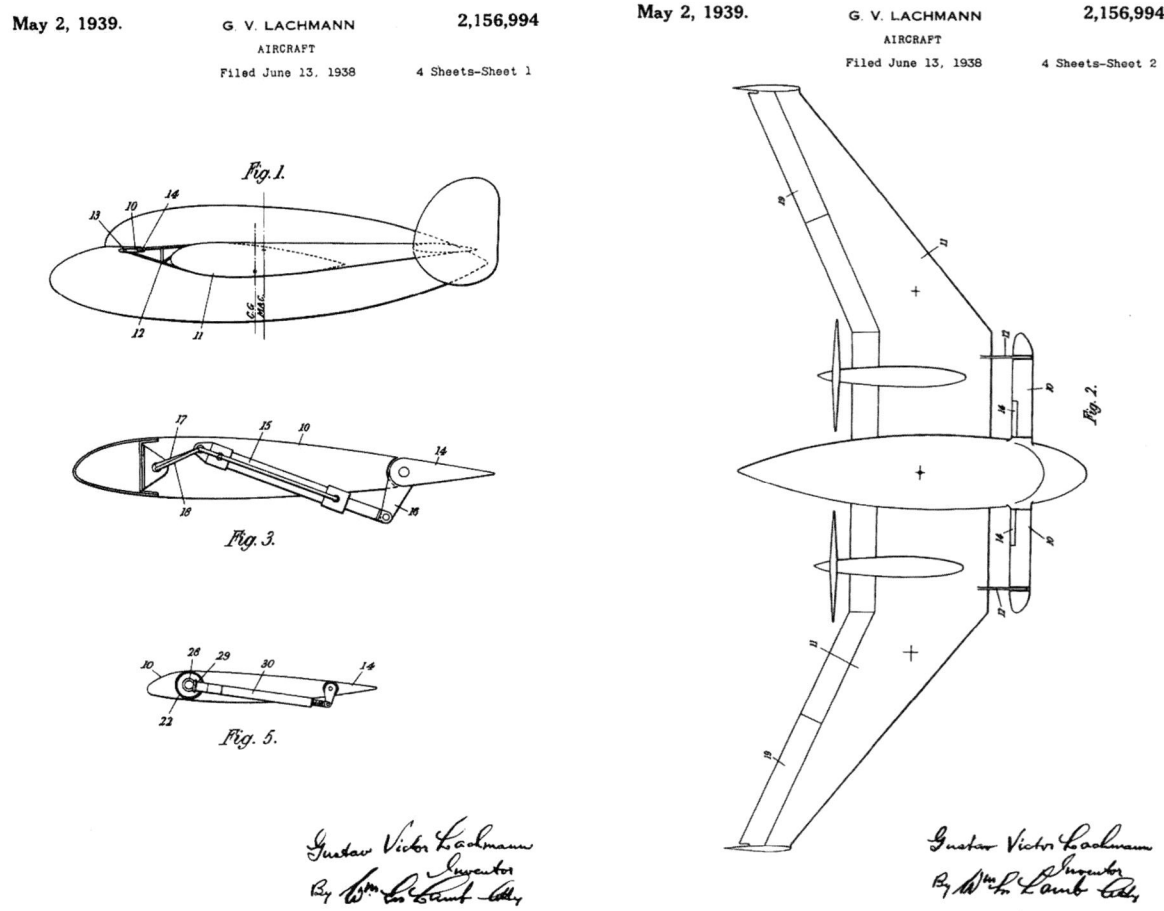

Lachmann's Rider Plane aircraft proposal, patented in 1939.

else its area or span would become so large that the design was back where it had started. "Carried to the limit, this argument results in having so enormous a front plane that the original main wing has shrunk by comparison to an orthodox tailplane."

The proposed solution to this was another exercise in Lachmann's conceptual brilliance. Imagine that instead of being rigidly attached to the nose of the aeroplane, the foreplane was pivoted and could independently change its incidence. By attaching auxiliary trimming surfaces to it which would be under the control of the pilot, the angle of incidence could be set relative to the flightpath, completely independently of any incidence (or lift) changes of the main wing. Because the pivoted foreplane's lift would remain unchanged, all of the overall lift change could be considered to act on the main wing; the neutral point would be unaffected.

HP described the pivoting foreplane as, "essentially a small aeroplane pivoted at the nose of the big one" and called it the Handley Page 'Rider-Plane'. Unsurprisingly, the suggestion was that its size be limited, as far as possible, by the use of leading- and trailing-edge slotted devices.

By design, it would operate in a neutral, minimum drag state in the cruise, being sized predominantly by the trim requirements of high lift. As such, its operation could be automatic and scheduled (presumably mechanically) with flap deployment. Based on the maximum lift requirement, published results demonstrated that a Rider-Plane configuration offered the smallest total area (wing + stabiliser) against any of the competing conventional, tailless or canard configurations studied.

Lachmann, on behalf of Handley Page Ltd, applied for a patent for the Rider-Plane in July 1937; close to the time when Volkert was attempting to persuade the Air Ministry of what they really needed for P13/36. The relevant document showed an illustration of an aeroplane that was unusual by definition, with a short, teardrop shaped fuselage, a long-span wing with a straight central planform with a pusher propeller and engine each side, and swept outboard panels with elliptical tip fins. The very high aspect ratio Rider-Plane fitted immediately in front of the mainplane, and in the drawing was mechanically connected.

A similar application was filed in June 1938 in the

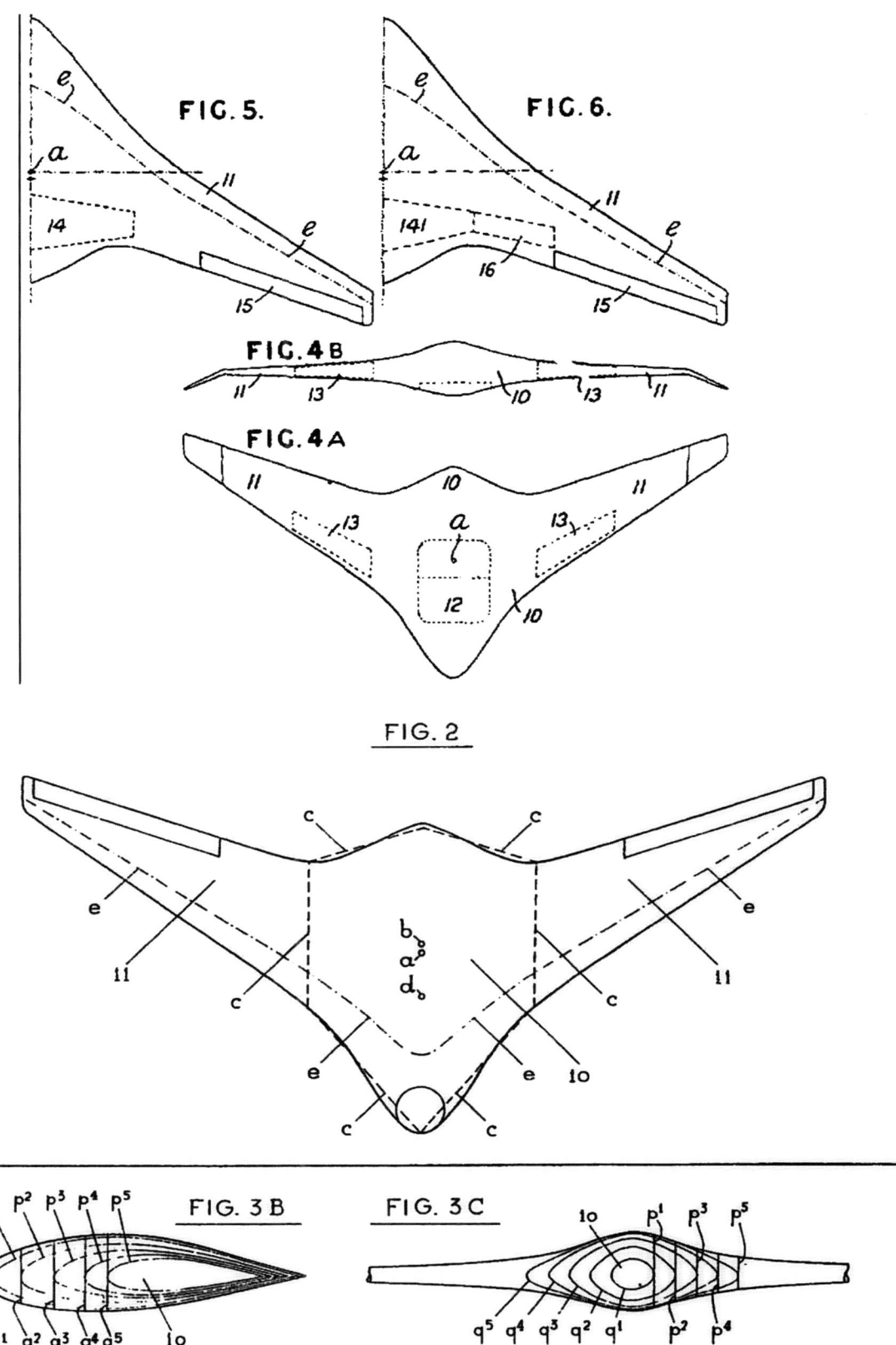

The genius of Lachmann is evident in the detailed thinking behind this blended wing-body aircraft concept, part of a patent document in which the design of a viable (non-circular) pressure cabin in such a machine was described. The application was made in 1944, while he was interred on the Isle of Man but in contact with both Handley Page and their patent agent, Messers Griffith Brewer.

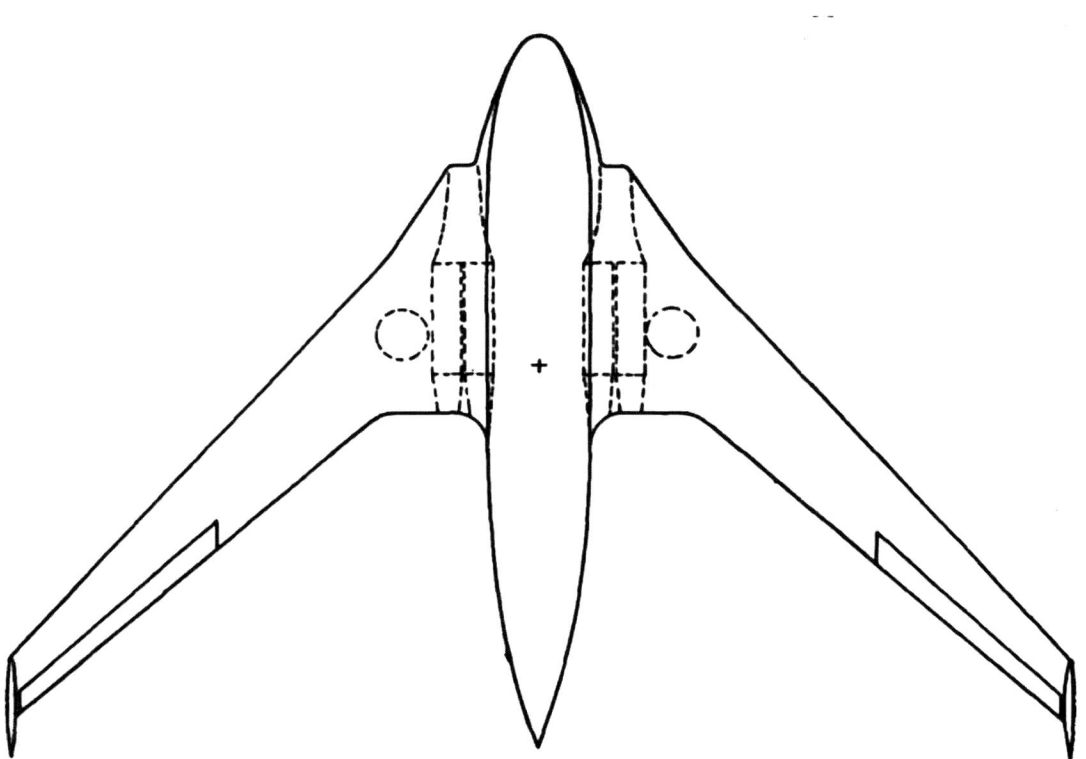

Lee's tailless bomber proposal, May 1946.

United States, showing that HP continued to take the concept seriously as a commercial prospect. Progress towards a flying demonstration was painfully slow, for obvious reasons given the events of 1940. The first proper test flight of the resulting H.P.75 Manx did not happen until June 1943; it would carry out about 30 hours of flying before it was put into storage in April 1946, the Rider-Plane never having been fitted. Without this crucial addition, the aircraft unfortunately proved little that was new.

In the event, Handley Page would conceptually design a number of tailless bombers in the mid-war period within two distinct contexts. The first was the requirement for a Halifax and Lancaster replacement, which would actually lead to the Hastings and Lincoln. The second was a MAP 'suggestion' that industry consider bombers of 70 tons and 100 tons gross weight, which would have been not just physically enormous, but a similarly impressive revision to the weight-limited policy that had hitherto caused so many problems.

In both cases, HP deployed the Rider-Plane to make their tailless designs work, while also comparing them directly to conventional layouts to meet the same specification. At 157,000lb, more than twice the weight of any bomber that RAF would actually field in the Second World War, it was the conventional tailed design that crept outside of the orthodox and suffered accordingly. To make this machine balance, the proposed Napier Sabre powerplants had to be mounted well back within the wings, connected to the tractor propellers along the leading edge by long shafts. The penalty of powerplant installation and weight was relatively small for the tailless alternative, along with large savings on fuselage weight helping to reduce zero-lift drag down to about three-quarters of that for the tailed aircraft.

Given that the overall weight of the two types was the same, there was an induced drag cost from the tailless aircraft's smaller wing, but when speed was high and lift coefficient low, performance was projected to be extraordinary: 429mph TAS at 30,000ft with a 20,000lb bombload aboard. The large reduction in structural mass meant more fuel and much greater range. Handley Page quite clearly therefore, in the second half of the war at least, had conducted relatively detailed design (albeit still at the conceptual level) of what a state-of-the-art piston-engined heavy bomber would be like, and the Lachmann Rider-Plane gave the edge.

HP would provide a third design to MAP, which pointed to where the world was headed. Using the same wing loading target (72lb per square foot) and sweepback (22.5°) as the Sabre-powered tailless bomber, the substitution of eight Metrovick F.2 axial-flow jet engines and increased fuel was expected to yield a bomber capable of cruising at 500mph at 40,000ft. Compared to the earlier proposals, wing loading was increased as aspect ratio reduced to 7.6, an indication of the potential reduction in the impact of induced drag both for a thrust-dominated design and at higher speed but constant weight in particular. It would likely have needed an assisted take off of some kind, but like the piston-engine version, was stressed to carry a staggering 48,000lb bombload over short ranges.

It is not known to what, if any, extent the problems of compressibility had been addressed in these machines. From the company's concurrent work on the HP.65 'Super Halifax', it is certain that HP's designers were familiar with the NACA 6-series aerofoils, and Lee documented that the assumption of laminar flow to 20% chord was made. They would not have known at the time of the mechanisms of cross-flow transition and attachment-line contamination, both of which would restrict extensive natural laminar flow to almost unswept wings. However, relatively good characteristics up to the moderate M = 0.76 target seem a reasonable aspiration, achieved within the next few years by the Canberra using similar sections and without sweep, albeit at a much lower wing loading. By 1944 then, Handley Page's engineering team, in the visible form of Volkert and Lee, together with less overt inputs from the interned Lachmann, had conceived of a 500mph+ turbojet powered, swept wing tailless bomber, albeit utilising the unvalidated Rider-Plane concept. The team's lack of experience with mitigating the effects of high Mach numbers suggests that it would not have worked as schemed, but it would be a superb jumping off point for the future.

June 1945 saw Sir Frederick compose a confidential memo to Lee and chief designer Reginald Stafford, requesting that they investigate two new bomber projects: One of 100,000lb all-up weight and powered by four of Rolls-Royce's projected AJ.65 turbojets, while a smaller 60,000lb design would be powered by only two. The brochure for the former would be complete in the spring of 1946, forming the ultimate basis for the large jet bomber that they would actually build.

STALL PROTECTION AND HIGH LIFT

As the wing loading of aircraft increased with higher cruise speeds, so the importance of being able to adapt wing geometry to provide adequate lift at low speeds (for field performance) grew. Both HP and Avro had needed to consider this in their P13/36 contenders (eventually the Halifax and Manchester respectively). Avro's design used deceptively primitive split flaps, but as the brochure submission indicated, this was in order to provide a capability that would never be used in service: dive-bombing.

"In order to obtain the high speed which this design provides we have employed a high wing loading. This wing loading raised the problem of the type of trailing-edge flap which would be used… in the end [we] decided to use the simple split trailing-edge flap, in spite of the fact that flaps of, say, the Fowler type, would enable a still higher maximum KL to be obtained.

…these flaps have to be used in the dive, in order to reduce diving speed, and after investigating the matter thoroughly, we have concluded that they are the most suitable, from a constructional point of view, to withstand the heavy loads.

It will also be noted that a rather high structure weight has been estimated for the wings… from a consideration of the forces… under the diving conditions with the flaps down".[59]

Handley Page were in a very different place, as the world experts on slotted wing technology. Concerned by both safety and the need to get the most out of a flying machine, Frederick Handley Page had sought to replicate a low aspect ratio flow on a relatively high aspect ratio (AR = 6) wing, by introducing a series of chordwise slots, as if they were local tips. Experiments in the company wind tunnel showed that the angle of incidence for stall was increased, however there was no corresponding increase in maximum lift; the overall effect had been to shift the no-lift angle. Possessed by the desire to introduce "live air" to the suction surface however, he next tested spanwise slots. It was here, in time, that he would strike gold. Outlined in an initial patent application in 1919, HP's slotted wing and with it the basis of modern high lift systems had been legally established. Which is to say, it would have been, provided nobody else had come to the same conclusion at the same time.

Still in Germany at this time, Lachmann had been inspired to pursue similar methods of ensuring safety as a result of crashing his own aircraft after a stall.

"The idea occurred to me to sub-divide the wing into a number of part aerofoils, separated from each other by curved nozzle-shaped slots like the wings of a closely-staggered multiplane, but all contained within the contour of a thick aerofoil… The leading edges of these part-aerofoils would be situated on the contour of the bottom surface of the enveloping master aerofoil, their trailing edges on the contour of the upper surface. I reasoned that at small angles of incidence of the master aerofoil not much air would flow through

Detail of the Handley Page Gugnunc, G-AACN, at the Science Museum, South Kensington. The deployed leading-edge slats are clearly visible. This machine was designed and built specifically to take part in a competition organised by the Daniel Guggenheim Fund for the promotion of aeronautics, and a share of the $150,000 prize pot. Although beaten into second place, the winning entrant from Curtiss used a slotted wing, in contravention of the Handley Page patents. The matter was eventually settled out of court, with Curtiss subsequently paying royalties.

the slots separating the part-aerofoils but, as the angle of the master aerofoil increased, the part aerofoils would contribute lift to an increasing degree."[60]

In February 1918, he submitted a patent application, based on data from slotted wing model made by a carpenter, which he then used with a hairdryer as a source of air and tobacco smoke for flow visualisation. Sadly for him, these ad hoc experiments would not pass muster; it would be necessary for more substantial proof of concept to be provided before his application would be accepted. He paid for a similar model to be tested by the master himself, Prandtl at Göttingen. The single page report that he received back some weeks later was emphatic: maximum lift had been increased by 60%. The idea could no longer be dismissed by the German patent office, and was granted in 1920.

It is clear that the concept of the lateral slotted wing had been developed concurrently by both Lachmann and HP, which resulted in a thorny problem for the latter in terms of the priority of the German patent. HP's practical solution, that would in time reveal itself as a feat of commercial genius, was to take Lachmann on board and then support him in developing the concept further within Prandtl's laboratory. An agreement was drafted and hands shaken at a meeting in Berlin, in August 1921. Lachmann would later say that it was, "not grand from a financial point of view. But nobody could foresee at this early stage the ultimate commercial success and… it would take almost a decade… before the fruit could be harvested." HP, with his willingness to strike out in inspiring directions and take risks, may have been more open to imagining a future of higher wing loadings and speeds than Lachmann gave him credit for.

It did indeed, in the 1920s, take something of a visionary step to understand quite why the slotted wing might become important. With low wing loadings, aircraft landing and take-off speeds (related to stall speed) were low enough that high-lift systems per se were not relevant to the biplanes of the day. Controllability close to the stall in turning flight, whereby separation at the tip might render ailerons ineffective, was an easier sell in concept, but would need the slats to be rapidly deployed in response to an out-of-control event. The next step would be less visionary than fortuitous.

Handley Page's automatic slat offered the first realisation of the commercial potential of the joint invention, exploiting the increased suction available close to the stall (at maximum lift), to deploy a spring-loaded device. Frederick Handley Page himself was the commercial driver, with

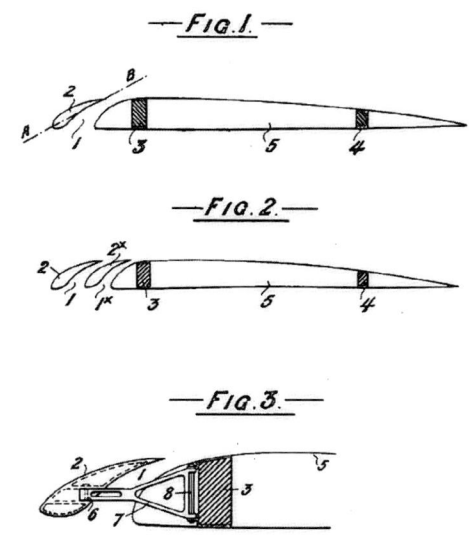

It would take time to mature, but the eventual backbone of Handley Page's intellectual property was established early in the company's history. These diagrams are taken from the US patent application for the slotted wing.

Lachmann the continual innovator, particularly following his joining proper of the firm in 1928. With the coming of the streamlined monoplane, the leading-edge slat would eventually find its place, but the slotted trailing edge flap would become extremely important in the interim. When compared to the simple split flap (patented by Orville Wright himself), the slotted flap offered improved lift to drag ratios for take-off, and hence increased weight to be flown for a given field length.

HP's strategy was to licence the rights in respect of their patents to foreign governments where possible, as opposed to individual aircraft firms. This would ironically result in royalty payments being made to the British company by the Nazi regime until just six weeks prior to the outbreak of the Second World War. It is an interesting question as to whether this advanced the capability of the enemy machines that the Royal Air Force would have to face, but the fact was Junkers was a major user of the slotted trailing edge in its own designs, and it was only a court case in Germany that ruled the firm to be in breach of Lachmann's 1918 patent.

Both the Hampden high-speed medium bomber to B9/32 and the later Halifax used the slotted trailing edge flap technology, while the former also used leading edge slats to protect the outboard wing. This home-grown, advanced technology allowed a smaller, more efficient wing for the cruise (in theory) while maintaining field performance. It would only become more important as cruise speeds increased, and the optimum wing area continued to fall.

By the time he came to design the Me 163 B, Alexander Lippisch had spent almost two decades thinking hard about tailless aircraft, and the solving of inherent problems with such configurations was evident. A fascinating — and essential — feature was the manner in which the propensity to tip stall was solved. While the use of sweepback was a logical method of addressing the absence of the tailplane for stability and control, it brought its own problems in terms of aerodynamic robustness.

With the isobars over the wing, and in particular the peak suction line, nominally following the geometric sweep, the boundary layer flow would always see a lower pressure region outboard. This favourable spanwise pressure gradient in the direction of the tip had the effect of causing an outflow of the boundary layer, consequently thinning inboard and thickening, reducing its health, outboard. Stalling first at the tip would reduce lift behind the centre of gravity, causing pitch up and worsening of the stall. As it would be unlikely to take place symmetrically, the chances were that a rolling moment and potentially a spin would be the consequences. For a fighter using high lift to manoeuvre, this was a highly undesirable characteristic that would be detrimental to combat effectiveness.

Lippisch's solution to this problem was to protect the wing tip, such that it would stall at a greater angle of attack than the root. For an aircraft that was trimmed approaching the stall, the initial loss of lift forward would have the result of pushing the nose downwards and giving longitudinal stability. He achieved this by using fixed slots in the outboard leading edge, on the basis that at low angles of attack typical of normal climb and cruise, the pressures at the upper and lower openings of the slot would be close to equal. Conversely, at high lift, the movement of the stagnation point around the leading edge to the lower surface, together with the high level of peak suction on top, would drive flow through.[61] Lachmann, with whom Lippisch had discussed the potential and pitfalls of tailless aircraft at the Royal Aeronautical Society in 1931, had provided the patented means to safely realise Lippisch's Komet, and in almost exactly the way he had considered in his first experiments.

For once though, Handley Page could not count on the receipt of royalty payments from the supervising government in respect of Lippisch's use. Nonetheless, the confidence with which the HP company was able to use the demonstrated slotted wing technology, for both high lift and stall protection purposes, was a driver of their approach to a B35/46 solution. It facilitated their high aspect ratio, tailless approach. Avro could not be said to have been invested in this direction in early 1947, having not made use of either modern high-lift or tailless technology. It was a point the company would pass through while considering the options, but would be quickly discarded.

AXIAL-FLOW TURBOJET PROPULSION

Frank Whittle's successful demonstration of a working gas turbine in 1937 had subsequently spurred the Royal Aircraft Establishment into action, waking up a field of their research that had lain dormant for six years. In March of that year, the engine sub-committee of the Aeronautical Research Council (ARC) recommended that,

> "… the Air Ministry should take up the question of the development of the internal combustion turbine as a matter of urgency and make all possible arrangements for its production at the earliest possible moment."[62]

This was a remarkable turn of events for those involved in the defence R&D community, and perhaps for none more so than a scientist historically perceived as having stood in Whittle's path. Alan Arnold Griffith had been an early proponent of the axial flow compressor, which in his mind would provide the efficiency required for a turboprop powerplant that might usurp the piston engine. While such a machine could and had been made to work, of a fashion, it was the 'efficiency' part that was the stumbling block.

Early in the century, steam turbine manufacturer Parsons had experimented with compressors, drawing on experience of what was effectively the same concept in reverse. However, a turbine accelerated the flow, encouraging the fluid in a direction that was energetically favourable. The compressor has all of the challenges of a wing slowing the air over its aft part and returning it to ambient conditions. The blades are subject to stalling, losing both efficiency and stability.

The axial compressor was analogous to a sailing yacht

tacking into the wind, reliant on careful alignment through a limited working range, whereas the turbine was in the relatively carefree regime of having the wind behind it. The work by Parsons and others could relatively simply show an efficiency of 60%, whereas a contemporary centrifugal fan would be well above 80%. In a weight critical application such as an aircraft, this could not be ignored. The key would be the application of the developing field of aerodynamics—which barely existed as an engineering discipline at the turn of the century—to the design of compressor blades.

In 1926, Griffith published a paper which described this fundamental problem and speculated that by solving it, the gas turbine might become a viable (and in time superior) aircraft powerplant, in conjunction of course with a propeller.[63] The enticing possibility was investigated through further theoretical studies and rig tests, but in 1930 the ARC decided that there were still sufficient doubts over the potential viability of the concept that large scale funding leading to a prototype engine could not be justified. They did recommend that the rig testing and other work be continued by the RAE, in order to ensure the UK had a foot in door for when or if success seemed more likely in the future. Amazingly in retrospect, this simply didn't happen; both Griffith and Whittle independently concluded that focusing their efforts elsewhere was their only option, until the time was ripe again.

While RAE might have been confident in their aerodynamic capability, subsequent events showed that they understood progress might be more rapidly achieved in conjunction with an industrial partner that was experienced with high temperature metallurgy and possessed the general production capacity to iterate that would be needed. The ARC's recommendations were accepted and in June 1937 an RAE delegation along with W. S. Farren, who was with the Air Ministry at this time, would visit the Trafford Park works of Metropolitan-Vickers, known to all as Metrovick.

Griffith left the RAE in 1939, with an invitation to pursue axial compressor development at Rolls-Royce. In his place, his colleague Hayne Constant would move to lead the turbine engine effort within the RAE Engine Department, which in turn he would be promoted to lead in its entirety in 1941. Posted to the team towards the end of this period, Peter Lloyd recalled,

> "Some eight of us shared a room of modest size and off this Constant had a small glass cubicle of an office with bare space for a table and chair. Looking back, it is hard to conceive how we managed in these cramped quarters, especially as one of Constant's habits was to stalk slowly up and down wrapped in thought... He was then 36, tall, rather gaunt, with light hair and a dark complexion, a hard mouth but a rather gentle withdrawn look (he did not seem to change much with the years); he wore the mantle of a natural authority. For one quickly recognised his intellectual mastery, his formidable dialectical powers, his laconic and often mordant wit."[64]

Constant too had suffered the gas turbine hiatus of the early 1930s, and had left the RAE for Imperial College. On his return in 1936, he had resumed his interest in axial compressor technology and, in his capacity as head of supercharger development for the engine division, supervised the design of an axial test rig named 'Anne'. The 8in diameter device was the first multi-stage axial compressor to be built at Farnborough, but due to bearing failure came close to destroying itself on its first run in February 1938.

It would be November before the modified rig produced useful data, but this was enough to reward the faith placed in it. Griffith, meanwhile, had headed off down a complicated development path to design a contraflow axial engine, in which each stage of the compressor was individually driven by its own outer ring of turbine blades. It seems likely that his theoretical analysis led him to think a conventional axial design might never work at off-design conditions, hence his determination to strike out in a profoundly innovative but undoubtedly mechanically complex direction. Initially supported by the RAE and then at his new employer in Derby, it was never a successful concept.

Data from the Anne compressor showing efficiencies of over 75% at pressure ratios of around two. Performance was still not as good as seen with contemporary centrifugal fans, but certainly an improvement over the prevailing state of the art with clearly more to come. By 1940, a proposal for an RAE-designed compressor named Freda, or Scheme 'F', was with Metrovick, as the basis of an engine to give just over 2,000lb static thrust. All effort was now focused on the pure jet rather than turboprop, in a further evolution of RAE thinking towards the direction of Whittle's design. Looking back, Constant would recall,

> "We climbed down, or rather climbed over, and backed the jet because we thought it an easier job than the propeller turbine and one which might be carried through in time to affect the war..."[65]

Having joined the team from Cambridge in 1938, Alun

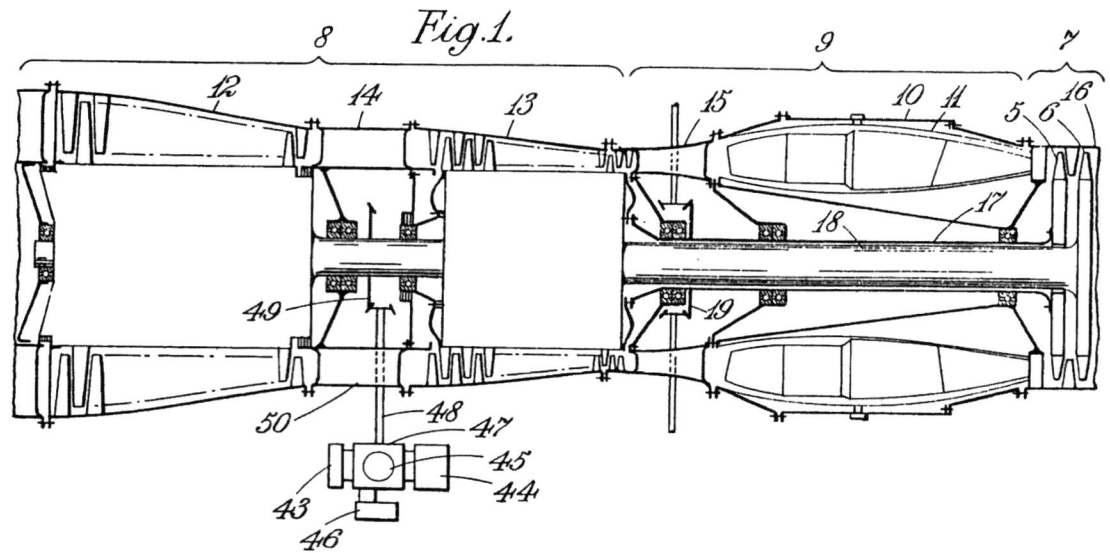

Bristol engines patent diagram from 1947, in which the claim is for a twin-concentric shaft design with starter engaging only on the high-pressure system.

Raymond Howell collated the known data and thinking for the design of axial compressors that was available and it was published in a report in June 1942. The fundamental method of working was one in which the results of various blade geometries tested in the RAE rig were correlated to performance and efficiency. The result was a set of design guidelines and a framework by which performance, including at off design conditions, could be predicted in advance of testing.

The experiments determined that a small number of parameters could characterise the aerodynamics: nominal outlet flow angle, space to chord ratio and Reynolds number. Conspicuously absent was the camber of the blades, which for a reasonable Reynolds number and within the admittedly narrow range of blade setting angles tested, proved only weakly sensitive. Howell's analysis was an extension of correlative rules proposed by Constant, and ultimately a strategy of calculating flow deflections, velocities and pressure rises through a cascade of adjacent blades, as advocated by Griffith. The important step was to document this in a way that could be readily used, together with the support of large quantities of relevant data.[66]

Metrovick, with an engineering team led by D. M. Smith, worked closely with the RAE on the 'flight-ready' version of the latter's design, referring to it as the F2. Even then, the first unit assembled in November 1941 was excessively heavy, a quality inherent in the basic design, perhaps related to the steam turbine background of the mechanical designers and a desire to robustly gain running experience.

The Ministry of Aircraft Production's solution to this unwelcome development direction was to encourage an established aero-engine manufacturer to come on board with the project. The best placed of the four 'ring' companies at the time was Hawker Siddeley's Armstrong Siddeley, brought on board from January 1942.

The F2 and a more powerful development, the F2/4, would between them power one of the prototypes of Britain's new jet fighter, the Gloster Meteor, and do much to demonstrate the capability of RAE's compressor design techniques. However, the relationship with AS did not survive for long; by the late summer of that year it had broken down and each organisation went its own way. AS would continue with its own axial flow programmes which, in the fullness of time, would lead to production turboprops, while Metrovick failed to produce a genuinely production-ready engine.

All service Meteors were powered by Whittle-derived engines. However, with MV's efforts painfully close to yielding gold, MAP chose to sidestep any attempt at production of the F2/4 and channel the firm's hard-won experience into a new design for a 7,000lb thrust turbojet that would challenge the Rolls-Royce AJ.65 project. Incorporating some US-derived design features in its compressor, the resulting F.9 was eventually to emerge as a winning combination of advanced and robust technology.

Metrovick, however, had decided that it was unlikely ever to receive a production order and planned to divest itself of aero-engines. After considering de Havilland, the F.9 (named Sapphire) was ultimately passed on to MV's difficult partner of half a decade earlier, Armstrong Siddeley. In time, AS would make the F.9 the success it

Armstrong-Siddeley used the dramatic shape of the Victor in this advertisement in Aeroplane *magazine of October 23, 1953. Perhaps carefully worded to avoid any suggestion of superiority over fellow HS group product, the Avro Vulcan!*

Bristol advertisement emphasising the thrust and altitude capabilities of the Olympus, and its application to the Vulcan, published in Aeroplane, *September 18, 1953.*

deserved to be, but the company to an extent continued with the problems in terms of production that Metrovick had encountered.[67]

Early in 1944, the Minister of Aircraft Production, Sir Stafford Cripps, met with Bristol Engines chief engineer Norman Rowbotham, and his colleague Frank Owner, and agreed an order for five complete production versions of a proposed 'Propellor-Turbine' engine (turboprop in more modern parlance), together with three sets of spare components. The engine was christened 'Theseus', continuing the Bristol naming theme of classical deities from the large sleeve-valve radial piston engines that were the bread and butter of the firm at the time. The suffix '-us' was added as a sort of turbine 'house-name' from then on. Owner recalled later,

> *"Charles Fraser and I compiled a list of over a hundred such names, divided into four categories: probables, possibles, only if unavoidable, and not on any account!"*[68]

Bristol had settled on a configuration in which the propeller was driven by a free-power turbine of its own, connected via a concentric shaft running through the main gas-turbine power generator shaft. While a further twin-shaft 'compound' engine had been considered, the complexity of three shafts rotating inside each other was understandably viewed as a step too far at the time. Theseus did though create a necessity for the basic development work that could make such a system viable, and when in October 1945 the Project Office found time to look at what might come next, this knowledge was put to good use.

The viable operating envelope of the propeller turbine powerplant was envisaged to have an upper limit in the region of 500mph and 40,000ft. Bristol, like Handley Page, was well aware of the likely future requirements of the Air Ministry for long-range bomber aircraft, and recognised that a pure jet powerplant would be inevitably required to confer the necessary performance, albeit with a fuel efficiency hitherto unachievable.

The starting point was reassuringly 'back of the fag-packet': Owner discussed likely thrust requirements with Bristol Aeroplane company chief designer Archibald Russell and thought them to be in the region of 7,000lb. The engine design team allowed a healthy margin for growth, considering on that basis 9,000lb with

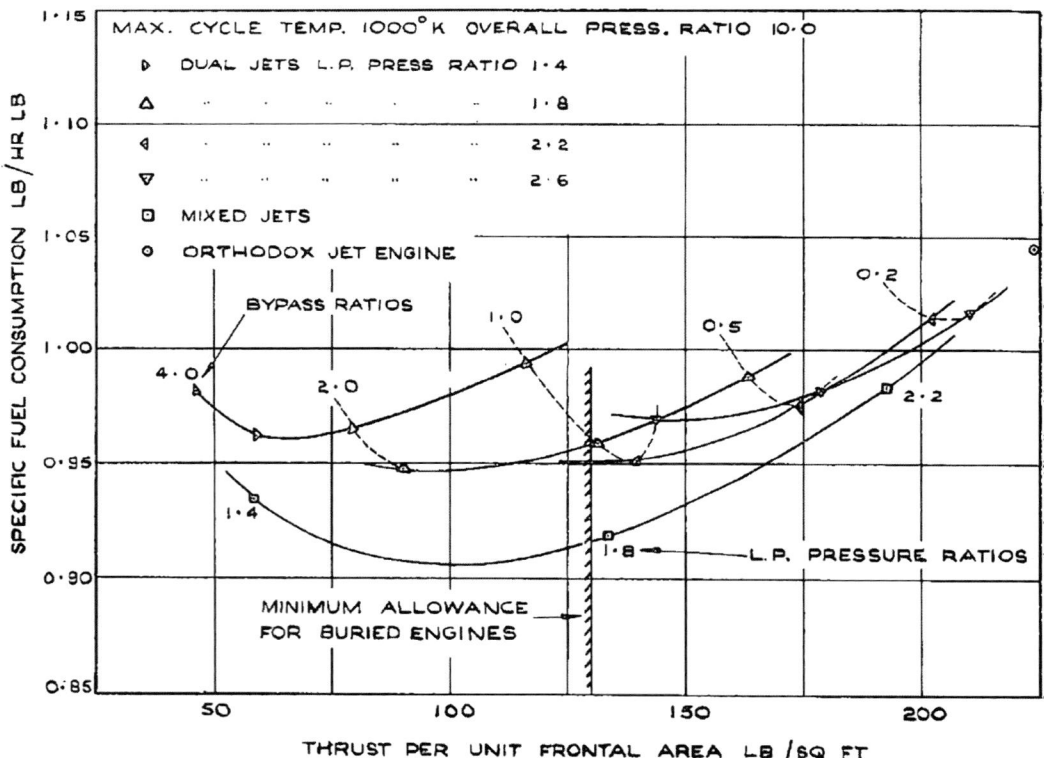

National Gas Turbine Establishment (NGTE) data from around 1948, illustrating that a 'Mixed Jet' or Conway-like design offered the best specific fuel consumption of those considered, at then-sensible pressure ratios. However, this was not fully realisable in a buried installation, pointing to a problem that the V-bombers would encounter in later development.

development to 12,000lb in time. Harking back to similar assumptions made during the Theseus and subsequent Proteus programmes, a maximum diameter of a metre was targeted, together with a maximum viable compressor pressure ratio of about nine. At this point, with the need for a free-power turbine gone, but the capability of engineering a concentric shaft proven, the new engine was sketched as a compound (or two shaft) turbojet, with low- and high-pressure compressors each powered by a single stage turbine and aiming for a pressure ratio of three.

The advantage of the two-shaft axial compressor at off design conditions was described by Sir Stanley Hooker, in his autobiography Not Much of an Engineer. Hooker famously developed the engineering basis to the supercharger technology used at Rolls-Royce for the Merlin. When RR came to be the manufacturer and developer of the Whittle engine, he had been able to put his knowledge of centrifugal compressors to good use. By their nature, these devices were more suitable for relatively high-pressure ratios per stage than an axial compressor, but were not practically scalable in the same manner.

The latter gradually compressed the air by accelerating it through the rotating part of the stage, increasing its total pressure; then subsequently exchanging this velocity for static pressure as it slowed through stators. This process repeated though as many stages as might be required to reach the target pressure ratio, with the channel gradually reducing in size to compensate for the desired roughly constant speed of the air. Therein, sadly, lay the problem: with the inlet and outlet areas fixed to be appropriate for the design pressure ratio, how could the engine operate efficiently away from these conditions? Hooker wrote,

"For example, under starting conditions, where the pressure ratio may be about 2:1, the outlet area should only be 20% less than the inlet area. But because the outlet area is, in fact, only one-quarter of the inlet area, the airflow is severely restricted; putting it crudely, the air taken in at the front cannot get out of the back. This causes the blades at the front of the compressor to stall, vibrate, and not infrequently snap off..."[69]

One route to mitigate this problem, investigated by Armstrong Siddeley in the UK and also, particularly thoroughly, by the NACA as early as 1944,[70] was the use of variable vanes in the compressor to realign the onset flow to an angle the rotor stages were able to accept. Another was the introduction of blow-off valves, which increased the exit area to provide a better match to the inlet as required. Running two independent compressors, dividing the pressure rise between them, had the potential to offer a wider range of efficient design conditions and greater thrust growth potential within the same architecture, while still extracting work from all of the compressor flow.

After some five months of work, a brochure for the proposed engine was submitted to the Ministry of Supply, which in turn issued its own specification (TE1/46) in July 1946, almost precisely mirroring the content of Bristol's proposal. It cannot have been much of a surprise, although it was doubtless a relief nonetheless, when a tender for the production of six examples was accepted, in January 1947. Bristol's radical solution would indeed prove to be future proof in its bomber application; within the same basic dimensions, the nominally 9,000lb design would deliver 20,000lb thrust in the ultimate Vulcans. Such growth was implausible for the single stage Avon and Sapphire powerplants of the Valiant and Victor B.1.

The concentric shaft arrangement of the Olympus allowed better matching of the rotational speed demands for the lowest- and highest-pressure compressor stages, but still directed the entire gas flow through the combustion stage. A jet engine produced thrust by providing an increment in the momentum of a mass of air, proportional to the change in velocity. However, the change in the energy of this mass of air was instead proportional to the square of this velocity delta; ultimately, that cost could only be paid by the calorific value of the kerosene on board.

The resulting high speed (likely sonic) nozzle flow was not necessarily appropriate to the speed of the aircraft. There were propulsive efficiency gains (and hence fuel burn advantages) from accelerating a larger mass of air through a smaller velocity change, to give the same thrust. At the same time, the thermal efficiency of the engine was improved by running at the highest possible turbine inlet temperature, and the whole system at a high pressure ratio.

All of these criteria could be addressed by using some of the energy of the propulsive jet to drive an additional turbine, reducing its velocity but instead driving a second compressor exited directly to its own nozzle, bypassing the combustion and turbine systems. With a lower mass flow through the combustion chambers for a given total engine thrust, less fuel was required to achieve a given temperature too. However, the frontal area of the engine would have to be larger to compensate.

This had been recognised in both Germany and the UK in the war years, with Frank Whittle's Power Jets proposing the aft-fanned W.2/700 and LR.1 both for long range applications and the Miles M.52 supersonic test bed. Theodore von Kármán's first treatise on what he perceived the US reaction to the newly discovered German defence technology ought to be, described what he called a 'turbofan' and suggested it be investigated as a priority.[71]

These ideas were in the forefront of the mind of Griffith, still at Rolls-Royce and now able to expand into the space vacated by Hooker.[72] Interspersed with his contraflow projects, he had considered the concept of a bypass jet around 1940, but with the Rolls-Royce sands shifting in the direction of centrifugal engines, it was not until the availability of the AJ.65 work that a background of axial flow knowledge and rig parts meant development could take place.

A proposal using evidence from a combination of the Avon and Tweed compressors in a two-shaft arrangement was passed to the ARC in the winter of 1946, just prior to the issue of B35/46. Throughout 1947, the concept was both developed and compared to the alternatives. The specific thrust (thrust per unit frontal area) of a bypass engine was always likely to be worse than a turbojet, but it might have a weight advantage due to the reduction in mass flow passing through the correspondingly smaller HP compressor section.

In August 1948 the National Gas Turbine Establishment (NGTE), playing an analogous role in jet engine development to that of the RAE in airframe design, reported on its studies in the comparison between a "double compound jet engine" and a bypass design.[73] Looking at pressure ratios of between 8 and 12, with the latter being identified as the highest considered on paper by industry at that time, it was concluded that a mixed-jet bypass design could yield an improvement approaching 10% in specific fuel consumption. This was for the case of 550mph and 45,000ft, regarded as representative of MoS bomber and transport specifications at the time, together with a continuous cruise maximum cycle temperature of 1,000K. However, for an LP pressure ratio of 2, yielding bypass ratios of between 0.5 and 0.9, the outcome was a reduction in specific thrust of well over 20%.

This showed two key outcomes: the first that the natural development of materials to allow high temperatures would confer a still greater SFC advantage for the bypass engine with time; the second that the limit on minimum specific thrust imposed by the buried engine configuration meant that an optimum could not be reached in the proposed aircraft.

The situation in January 1947 was therefore brimming with potential, but notably the ultimate technology that would be deployed on the Vulcan remained to be demonstrated. A Whittle-type centrifugal-compressor engine would be of such a large diameter as to be impractical for a buried-engine configuration. It was the work of the RAE, Metrovick and eventually the NGTE that was vital in delivering a high pressure-ratio, axial compressor-based engine for the B35/46 competitors. Without this, the project would have been a non-starter.

PRESSURISATION

A contemporary discussion by W. M. Widgery, the chief engineer of Westland subsidiary Normalair, described the necessity for cabin pressurisation in high-altitude aircraft.

"At an altitude in the neighbourhood of 43,000ft the individual is unable, because of the decreasing total pressure in the lungs, to obtain sufficient oxygen even when the local atmosphere consists entirely of oxygen. This sets a strict limit on the operational height without pressure, even for military aircraft in which crew comfort is not the criterion."[74]

Clearly, for an aircraft that would cruise at 50,000ft, this would be essential. However, the desirability of such technology had been considered for most of the previous decade. In June 1941, in a note in which he encouraged rapid deployment of the Mosquito, the former AOC-in-C Bomber Command, Sir Edger Ludlow-Hewitt noted incidentally that,

"The experimental development of the high-altitude bomber is at present being taken care of by Boeing and by the pressure-cabin Wellington. It should be observed, however, that the operational utility of this class will be strictly limited by weather until means of increasing the accuracy of our navigation, out of sight of the ground, can be devised. The solution of this problem—the development of blind bombing—is a necessary condition to the success of the high-altitude bomber."[75]

None other than Frederick Handley Page himself was in the chair at the Institute of Mechanical Engineers on April 21, 1938, when, in his capacity as vice-president of the Royal Aeronautical Society, he introduced Dr John E. Younger as the speaker. Then, as now, the geography of the United States was driving a domestic market for air travel over long ranges. The watershed development of the modern streamlined aeroplane in the early 1930s was typified by the Boeing 247 and Douglas DC-2, providing a relatively safe and comfortable product from which it was plausible for the airlines to make money. The next steps in competitiveness would be to provide worthwhile reductions in journey times and avoid the effects of weather, both of which led to commercial interest in the 'sub-stratospheric liner'. Such an aircraft would fly at 20,000ft to 30,000ft, using cabin pressurisation to provide the necessary comfort for the paying passengers. Younger noted that,

"Two of these sub-stratosphere liners, the Boeing model 307 transport and the Douglas DC-4, will be flown this summer, The most interesting feature of these developments is that they are productions jobs, and not entirely experimental. However, the results of nearly 20 years of research, mostly by the Air Corps at Wright Field, culminating in the Air Corps' Lockheed XC-35 sub-stratosphere aeroplane has been available as the basis for commercial design."[76]

Younger had been one of the co-authors of a vast technical report that described the outcome of the laboratory investigation into practical aircraft cabin pressurisation systems, completed by the Wright Field team in 1936. The next step was to take a representative aeroplane, in this case the Lockheed Electra, to be modified and test the theory. The use of an existing aircraft sidestepped many of the basic engineering problems that a new testbed would have otherwise needed to solve, while ensuring that the results could be seen as applicable in the real world at sensible performance levels.

The Air Corps project team was open to many alternative methods of constructing the pressure cabin, but after discounting corrugated skins and an optimised pressure vessel suspended inside a conventional fuselage, ultimately concluded that a stressed skin structure as seen on the contemporary Boeing and Douglas designs, suitably rated for the pressure loads and carrying these mainly in bending via a circular cross-section design, would be satisfactory. The XC-35 itself was built with a circular cabin but retained the form of the Electra's nose, which added, "considerable weight". This highlighted the importance of designing in pressurisation from the start, even if the structural method was conventional.

Although the readiness level in terms of viable pressure cabin technology was clearly world-leading at the time of Younger's lecture, it would not be correct to think that there was no parallel activity in the UK. On the contrary the Air Staff had, the previous year, issued specification B25/37 for a four-engined high-altitude bomber, while days prior to the lecture the CAS had instructed AMRD

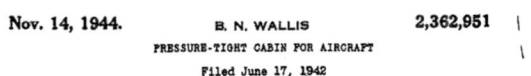
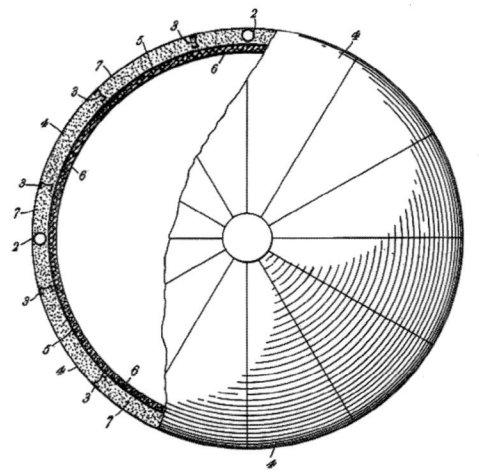
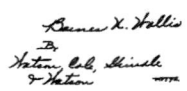

Illustration from a patent by Barnes Wallis, for a pressurised fuselage in a flexible structure. Filed in 1942, this shows the differences between the practical approach used by the big American manufacturers and their stressed skin construction, and the unconventional approach used by Vickers. It was they however that were responsible for the RAF's next heavy bomber to replace the Lancaster, the Windsor.

to study the problem further, eventually resulting in the decision to build a pressure cabin-equipped Wellington at the end of August 1938. The latter would eventually come to fruition, although not as a significant front-line type. As Younger implies, it would be the then neutral United States that would have the bandwidth to take the huge steps into building a modern, grown-up aircraft, able to provide realistic support and living conditions for its crew in the stratosphere.

The contribution of the Westland Welkin's development to British knowledge of the technology required to meet B35/46 did not end with its external aerodynamics. It was the first UK military aircraft designed from the start to be pressurised, and as such exposed some of the challenges this concept entailed. The Rootes blower needed additional silencing, while in contrast to the atmospheric conditions encountered on contemporary bombers, it was cooling rather than heating that was lacking and had to be subsequently provided.

Adequate sealing of the canopy, particularly on the sliding portions, had been thought likely to be difficult, but in practice inflatable rubber gaiters proved successful, using the expertise of the Dunlop company. Such technology would provide useful service during the Second World War, but in de Havilland's Mosquito rather than in heavy bombers. Pressurising the two-crew cockpit enabled the aircraft to carry both offensive loads and essential reconnaissance cameras at high altitudes and speeds. Still in wartime, a similar installation began to be schemed for the Mosquito's eventual replacement, the English Electric A.1. The integration of pressurisation into the accepted philosophy of the heavy bomber, one which would protect itself with distributed defensive armament, would be the ultimate outcome of the USAAC's programme, in the complex and expensive Boeing B-29.

Without a direct equivalent to American super-bomber having been realised, it was the outcome of the Brabazon Committee's requirements in terms of civil aircraft that would provide the experience. Avro was contracted to produce a North Atlantic-ranged passenger version of the Lancaster IV, which would fly safely (in accordance with the then current civil airworthiness requirements) above the weather, while providing its occupants with an appropriate level of comfort that would compete with the experience of travel in the rival American aircraft. In a technology sense, that meant a pressurised cabin in a new, circular section fuselage, and a funded, prioritised route for the company to learn about Normalair's systems.

Concerns about the ability to provide adequate pressurisation, for both crew and systems at 50,000ft, persisted well beyond the issue of B35/46. The RAE's assessment regarded this as one of the key barriers to achieving the specification, arguing that a performance-compromised bomber might result if forced to fly 10,000ft lower, than if designed for that height from the start. Just a few years earlier, the H2S radar system in its earliest form, carried by the Stirling and Halifax, had proven unreliable above 18,000ft.[77]

The cold and low ambient pressure was not a combination that the complex electronics and electrical power systems enjoyed, and effective environmental control and pressurisation was therefore essential to provide the necessary accurate bombing capability. Although it might seem almost innocuous today, this technological combination of aerodynamic, structural and systems engineering was a notable banana-skin in the process of realising the UK nuclear deterrent.

Roy Chadwick CB FRSA FRAeS.

ROY CHADWICK CBE FRSA FRAES (1893-1947)

A fellow member of the Urmston Church Choir would recall the schoolboy Roy Chadwick as being, "…gently reprimanded [for] sketching his Argosies of the skies" rather than paying attention to that day's sermon as he was supposed to.[78] Born in 1893, on leaving school he first took a job as a draughtsman at British Westinghouse. In 1911, he was able to transfer to a similar role at Avro in Manchester, working on products much more to his taste and in time for the advent of the 500 and 504.

Avro's design office transferred to Hamble on Southampton Water in 1916, although production was still undertaken in Manchester, eventually further into Cheshire at Woodford. Perhaps necessity fed the growth of a key characteristic of Chadwick's engineering: the interest in designing for production. Physically separated by hundreds of miles from the factory floor, it was the understanding between Roy Dobson at the factory and Chadwick at Hamble that became recognised as a key part of the success that would follow. This was illustrated when in the late 1920s, Chadwick recognised the potential of Fokker's welded steel technology, and encouraged Avro to invest. When they did, a colleague would comment retrospectively that,

"…there is no doubt that the close collaboration between the two Roys was vital to the success of the Avro welded construction, which is easy for the draughtsman but doomed to failure without the closest collaboration with the production staff."[79]

The true significance of this design for production philosophy would be seen in the much darker times that lay ahead. The spruce and plywood wing technology was then deployed on what was perhaps Chadwick's first clearly successful attempt at a streamlined monoplane, as envisaged by Melvill Jones. The 652 low-wing monoplane, with for the first time main landing gear retracting into the twin, wing-mounted nacelles and clear attention paid to the aerodynamic cleanliness of the graceful fuselage. The main significance of this step was as the basis for a submission to the RAF's tender for a low-cost maritime patrol land plane. The resulting 652A flew in March 1935 and an order for 174 examples, to be named Anson, was received in the July.

But, if the Anson represented the past, no such accusation could be levelled at Chadwick's next major project, an assault on the RAF's multi-role medium bomber requirement. Nothing was left on the table in achieving the exacting performance requirements of the P13/36 specification, but as is well known, it was a fight to counter the unreliable and underpowered Vulture engines that would lead to the Lancaster and ultimate success. It has been described since how Chadwick's team were willing to learn from many sources. Progress was assured by 100 engineers from the design office, or 'Production Office No. 1' as Chadwick's multi-disciplined approach to product delivery had it, being positioned on the shop floor around the prototype aircraft as it developed.[80] He was hands-on and exacting in the standards that he expected from all. William Farren, then of the RAE had cause to work professionally with the Avro technical director during this period.

"Roy used to drive me home, driving over the slippery setts with which Manchester streets were then paved, at a pace that frightened me. He never seemed to take the slightest notice of that, and I gradually lost my nervousness… I remember one night when the sky in the distance was lit with the flashes of bombs, but that didn't seem to disturb him either."[81]

His long-term colleague Sir Roy Dobson described his adaptability as,

"[It has been said] that transonic and supersonic

aeroplanes [call] for a new type of designer. They do not call for a different sort of designer than Roy Chadwick, because I always found that as times changed, [he] changed ahead of them. He did a lot of things that have been perpetuated by other people since... Chadwick will go down as one of England's great designers."

Stuart Duncan Davies CBE FREng FRAeS, c.1970.

STUART DUNCAN DAVIES CBE FRENG FRAES (1906-1995)

In 1946, Avro's chief designer at its Yeadon division, now Leeds Bradford Airport but then responsible for Anson, York and training aircraft, transferred back to the Manchester hub. Stuart Duncan Davies was to become chief designer directly under technical director Chadwick, in charge of the core new projects. Born in London in 1906, he joined Vickers in 1925 and gained a B.Sc degree in engineering while working. He made a major contribution to the Hawker Hurricane programme, joining Sydney Camm's team in the mid-1930s.

He would later describe his role in the small team as a one-man project office, "looking into performance estimation, propeller selection, undercarriage design, engine and radiator installation, as well as structural layout".[82] He would remain at Hawker until the extraordinarily significant eight-gun fighter was comfortably in service, prior to his move to Avro in 1938, as assistant chief designer.

By this time, Avro was churning out the Anson maritime patrol aircraft for the RAF, but Chadwick's team was focused on the much more advanced concept it had submitted for P13/36, the requirement for a medium bomber. Key to its success would be rapid production, over multiple factory sites. Davies would be central to this, creating a series of profoundly modern ideas, strategically grouped as Design for ease of production and Plan to provide rapid response.[83] The ability to rapidly adapt to changing circumstances was illustrated by the metamorphosis of the Manchester, by necessity, into the Lancaster. Production would peak at 293 aircraft per month.

It is clear that by the end of the war, moving into the role of chief designer was entirely appropriate for a man of his experience and talent. Known as 'Cock' Davies, a contraction of 'cockney' given his place of birth, he would work with Chadwick on the Tudor airliner before adopting primary responsibility for the B35/46 jet bomber.

Sir William Scott Farren CB MBE FRS.

SIR WILLIAM SCOTT FARREN CB MBE FRS (1892-1970)

Born in 1892, to a father "relatively unsuccessful in money matters... we were seldom free from anxiety", William Farren was encouraged particularly by his mother to follow a passion for engineering, and earned an entrance scholarship at Trinity College, Cambridge in mechanical science in 1910. After graduating, he spent a short time with British Thompson Houston (where two decades later,

Frank Whittle would demonstrate his first gas turbine rig) before making a jump to the Royal Aircraft Factory at Farnborough. After just a few weeks, he was appointed as head of aerodynamics, and as a measure of how tight the aeronautical world was at that point, he and his colleagues lobbied for flying lessons and were sent to the Royal Flying Corps' Central Flying School.

Farren left Farnborough in 1918 for Armstrong Whitworth, where he played a key role in the Siskin fighter and also lectured in aeronautics at Cambridge. The latter became permanent and he built on his experimental experience from Farnborough in wind tunnels, branching into flow visualisation and laminar flow. Remaining as a consultant to AW, and by implication with wider Siddeley group, he became professionally familiar with Avro as a company and Chadwick and Sir Roy Dobson personally, whom he recalled visiting from 1928 onwards.

By 1937, it was perhaps clear that a position that could offer direct influence and action on the UK's inevitably up-coming struggles was the best use of Farren's considerable, multi-disciplinary talent. He joined the Air Ministry, first as deputy director of scientific research (DDSR), subsequently moving to roles focused on R&D for new aircraft, by 1940 director of technical development at the new Ministry of Aircraft Production.

Under his remit therefore came the new heavy bombers from Handley Page and Avro. But, in 1941 he moved again, this time back to Farnborough and in what must have been a very satisfying appointment, as the first director of the Royal Aircraft Establishment. As a Cambridge academic, he had been involved peripherally in the RAE's plan for a new, 600mph wind tunnel. Now, as director, he focused on championing this project and securing the quarter of a million pounds necessary to build it, at a time when the cost of a military aircraft was measured in thousands of pounds.

The availability of this facility was of paramount importance to the development of the early jet aircraft, and absolutely vital for the B35/46 bombers, which could not have been tested without it. Farnborough's focus on high-speed flight and the aerodynamic technology that would underpin it was linked to the confidence in the UK's progress in the appropriate propulsion; it is notable that in the United States, the NACA emphasised laminar flow development in the same period. Farren's leadership was indeed farsighted.

With the end of the European war, Farren was asked to lead a mission to Germany, with a remit to understand the aeronautical technology of the enemy. He himself was able to interview senior German engineering staff and view the facilities that had stimulated their ideas. His staff included many highly experienced and capable engineers and test pilot—such as Eric 'Winkle' Brown.

Soon after his return, Farren left the RAE for Blackburn Aircraft and, perhaps drawing on the recent memories of the impressive facilities that he had seen in Germany, began a process of improving the technical capability of that company. However, changes at the firm as it merged with General Aircraft left him unsettled, and by 1947 this remarkable engineer was casting his eye around for a new role that would allow him the freedom and authority to make the difference. He would find one at Avro.

Sir Frederick Handley Page, the founder of the eponymous firm that he would lead until his death in 1962.

SIR FREDERICK HANDLEY PAGE CBE FRAES (1885-1962)

Although the first years of his career were spent as an electrical engineer, Frederick Handley Page used the opportunity of an enforced job change to move into aeronautics, an area which had become a passion. In 1909, he started his own eponymous company, which is considered to be the first in the UK to focus exclusively on the design and manufacture of aeroplanes, while also being appointed as a lecturer at the Northampton Institute (now City University, London) in 1910. Two of the Institute's students would play significant future roles in HP's company, the first being George Volkert, engaged as chief designer in 1912 and soon afterwards designing the company's first biplane, aged 22. In the interim, a War Office contract had been received for the manufacture of five BE.2A biplanes, which in part precipitated a move of the company's premises from Barking, to a much larger facility (albeit, a former riding school) in Cricklewood. HP had arrived.

HP and the Admiralty had agreed, by the end of 1914, the specification for what was considered by many to be an unfeasibly large aircraft. It would take almost exactly a year for the prototype of the Type O to make its first flight, on the twelfth anniversary of the Wright brothers' pioneering achievement in the Flyer, which travelled only 20ft further than the new bomber's wingspan on that famous 12 second adventure. In 1917, HP himself lectured at the Royal Aeronautical Society on the engineering of large aircraft, against a background of continuing scepticism. This was a technique he would deploy throughout his career: confidently engaging with a technical and commercial audience to make the substantive, evidence-based case for his products. In 1918, he married Una Thynne, and they would soon have three daughters.

HP's major contribution to aerospace technology, which would also ultimately provide the steady cash flow needed to ride out the lean years of the Great Depression, was the slotted wing. It would take time for the world to come to need the invention, and perhaps it was no coincidence that success followed when Lachmann was engaged full time. Nonetheless, HP's hands-on approach kept him in touch with what could make money in the short term, exemplified by the HP.42 for Imperial Airways, and the sharp end of aerodynamics and structures for the future, demonstrated by Lachmann's Hampden.

By 1938 then, HP had established itself as an advanced technology company and was in the early stages of the major production effort that would characterise the next few years. Tiny orders for six or eight biplanes from Imperial Airways would be usurped by the demand for hundreds of stressed skin, monoplane bombers, made efficient and controllable by their slotted wings. He would serve in that period in influential industrial roles: president of the Society of British Aircraft Constructors (SBAC) in 1938-39, and of the RAeS in 1945-47. He was knighted in 1942 for his contributions to the war effort.

HP's ability to navigate Whitehall would prove important too; the Halifax compared badly to the Lancaster in terms of productivity, and persuasion was needed to continue towards more effective versions. At the same time, a keen ability to extract information on what would be needed next was equally useful. HP had maintained its advanced projects division and—much more than Avro—struck out into new concepts, including tailless aircraft. It was no accident that by the end of the war, HP himself was in a position to accurately speculate on a specification for a future jet bomber. His long experience meant that he had known to whom he should speak.

GODFREY HENRY LEE (1913-1998)

From an aerodynamics perspective, Godfrey Lee is a critical element in the story of the V-bombers, although he did not have overall project responsibility. He was educated at Imperial College in South Kensington, with his undergraduate studies in physics and aeronautics at the constituent Royal College of Science, graduating in 1933 with a B.Sc., prior to further post-graduate study.

After three years with Saunders-Roe on the Isle of Wight, he joined Handley Page in 1937 as a stressman. However, the enforced departure of Lachmann from the company's structure at the end of 1939 created an opportunity for Lee to fulfil the former's wide-ranging role, as head of research. Because the main thrust at the time was towards the Rider-Plane equipped tailless aircraft, for which Lachmann had already generated patents, the most visible outcome of his new function was the HP.75 Manx test machine.

The expertise developed in this time meant that Lee was well placed to sit on the Aeronautical Research Council's committees that were tasked with investigating tailless and swept wing aircraft, while the interest in large, high-performance bombers shown in the second half of the war by the Ministry of Aircraft Production led to his work with chief designer George Volkert on tailless, ultimately jet proposals. By the time it came to address B35/46, his thinking had developed over a period of years and he had the ability and experience to incorporate new ideas into the aerodynamic-dominated philosophy.

In the autumn of 1945, he visited Göttingen and Völkenrode and saw for himself the progress in swept wing design. Perhaps it was his willingness to work closely with the deposed Lachmann in this period, that implies further magnanimity and pragmatism on the part of both men. Between them, it is clear that the relatively advanced concepts of the HP.80 were quickly established, but the practical basis could be seen in Lee's published work presented to the RAeS and others in preceding years.

Dr Gustav Victor Lachmann. VIA NEIL CORBETT

DR GUSTAV VICTOR LACHMANN (1896-1966)

Born in Dresden to father of Austrian origin, Gustav Victor Lachmann had started the Great War as a soldier in the Kaiser's army, but by 1917 had transferred to the Air Corps and was learning (as an officer) to fly as a pilot. His ambition for combat would prove short lived; in August of that year he would stall and spin his aircraft, leading to a crash that he was lucky to survive. His desire to protect against such a situation in the future would lead to his life-long involvement in aircraft design in general, and aerodynamics in particular. His invention of the slotted wing at the same time as Handley Page would dominate the course of his life, importantly leading him to leave Germany. Documenting its development in a doctoral thesis at Aachen gave him the academic credibility to further his advanced engineering career.

After a spell at Albatros, he accepted an invitation at Ishikawajima in Japan. He had married a widowed English woman while working in Tokyo, perhaps giving him another reason to accept an invitation to join Handley Page in 1929, making for the Home Counties accompanied by his two step-daughters. He became an increasingly well-known and recognised figure within both British and German aeronautical circles as the 1930s wore on, subject to approaches and offers of positions within universities and aircraft companies from his country of origin. MI5, openly suspicious of him from the moment of his arrival in the UK,[84] had access to intercepts of Lachmann's correspondence throughout the 1930s.

He was to play a very direct role as a chief designer in the development of the Hampden but is barely mentioned in relation to the subsequent Halifax. This is, in all probability, because Handley Page was forced increasingly to sideline him, away from specific military contracts. Instead, he was appointed as the company's head of research in 1936. He became convinced that tailless aircraft would represent an effective solution for long range, load-carrying problems of the future, and his department kicked off a path that would lead to the HP.75 Manx.

Lachmann was interned immediately after the outbreak of hostilities and spent the war initially in a UK prison. In 1940 he was shipped to another internment camp in Canada. In 1942, he was returned closer to home—the Isle of Man—where he would continue work on behalf of Handley Page although, at his own request, only on projects that he considered were not directly related to the war effort. For the whole period of hostilities, Sir Frederick and others mounted a campaign at levels up to and including the War Cabinet, to have him returned to the 'front line' of design work.

Even when the war in Europe had concluded, the way forward for the former intellectual star of HP was not clear. It would not be until April 27, 1946, that the Ministry of Supply wrote to Sir Frederick, at length and in response to his barrage of letters since the previous November, all on one topic: Lachmann's return to employment. This was finally approved, although subject to four clear conditions. He would—officially—not be working on military aircraft without direct approval of the Minister, while his office would be at his home. Information about military specifications and requirements would not be communicated to him.

It had been more than six years, but Gustav Victor Lachmann was back on the UK mainland, a free man and a consultant to the Handley Page company. What though would be the best application of a German-speaking world expert in aeronautical aerodynamics? As it turned out, there was an ever-increasing pile of technical, aerodynamic and scientific reports to be read, digested and turned to gold, and the majority were written in his native language.

5

FORGED IN ETNA: THE START OF THE VULCAN

THE METAPHORICAL thud on the doormat heralding Avro's receipt of the B35/46 specification coincided with the absence of company technical director Roy Chadwick, unfortunately indisposed with shingles. It therefore fell to chief projects engineer Bob Lindley, just 27 years old at the time, to lead a small team in the preliminary design phase. Stuart Davies wrote,

> "When this problem was first considered by the Avro Company, it was generally known from German sources that the solution to the problem of cruising at very high Mach number without excessive drag rise, required a combination of high sweepback, thickness/chord ratios lower than had been used up to that time (at any rate on bomber aircraft) and lower cruising lift coefficient".[85]

The Avro proposal evolved from a logical starting point in early 1947 as a relatively conventional high aspect ratio, 45° swept and 12% thick wing and tail aircraft, targeting a maximum CL of 0.2 for the cruise. The relatively high sweep angle (compared to subsequent aircraft designed for similar cruise Mach number) was presumably a mitigation for the thickness of the wing, in turn a structural requirement. The perceived impossibility of meeting the weight requirement, together with the inherent longitudinal stability and control capability of the swept wing, led to the investigation of an otherwise similar tailless layout.

This was, indeed, the position from which HP had started their own B35/46 jet bomber development, having studied the tailless configuration for a wartime large piston-engined bomber development. However, from here the paths taken by the two companies would diverge. Both had significant concerns over the ability of the tailless design to meet the specification, and that the development directions necessary would invalidate the strategy. These concerns would include;

* The tailless design would inevitably lack longitudinal damping compared to one with a tailplane. This is important in several ways, but in particular would be associated with poor control of the short period dynamic stability mode.
* The relatively large speed range between cruise and landing implied a requirement for a powerful high lift system. This would inevitably create a large nose down pitching moment, itself exacerbated as the load would be biased inboard and hence (due to the sweep angle) forward. This would be challenging to trim without a tailplane.

to transonic aerodynamic effects.
* The weight benefit of the tailless configuration was still unlikely to allow the specification to be met, without a significant step in weight efficiency of the structure.

The concerns of the firms were mirrored in the opinions of the Royal Aircraft Establishment. Responding to data on tailless bomber designs presented by Godfrey Lee in 1946, the RAE's director William Perring singled out the generation of maximum lift as a stumbling block.

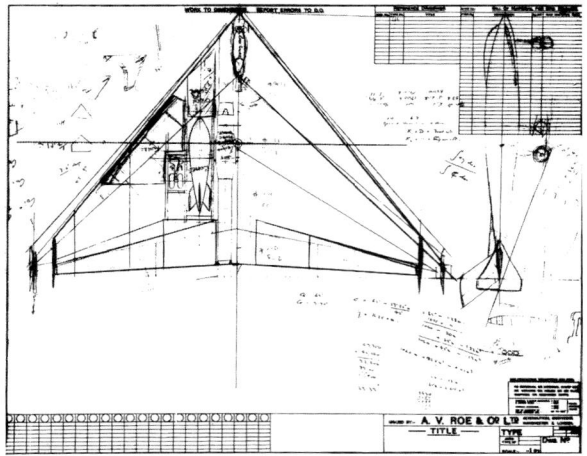

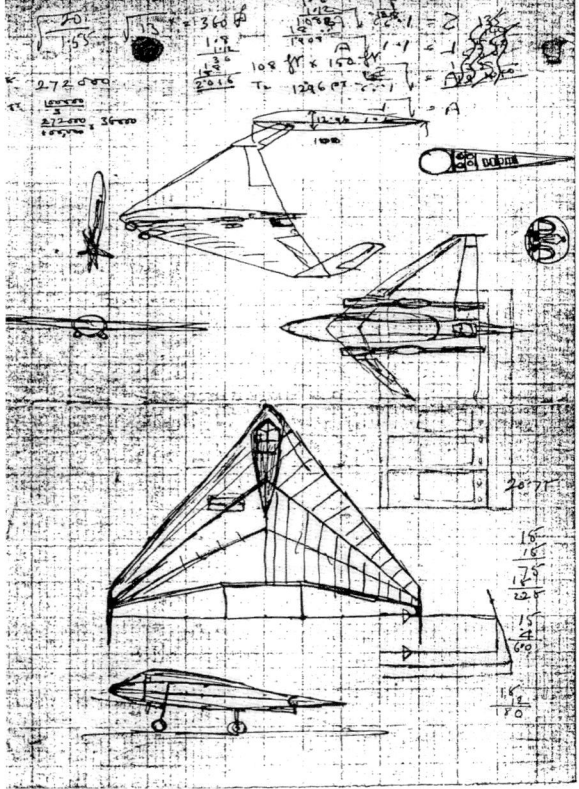

Chadwick's famous sketches from the conceptual design phase of the 698 programme. The annotations suggest a root chord of 65ft, tip chord of about 10ft and span of 90ft. Although the pure flying wing concept would disappear during development, while the engines and bomb bay would swap places, the basic concept was clearly present.

* The effect of tip stall would be exacerbated by the additional sweep angle necessary to provide sufficient moment arm for longitudinal stability and control; it would cause a relatively larger movement forward of the centre of pressure. This was a major issue for an aircraft required to manoeuvre near to its service ceiling and subject

"On the problem of maximum lift at high speeds, the difficulties were still greater. At high speeds, so far as the results known to him were concerned, they had not been able to make use of the maximum lifts that were possible, because on tailless types it had been found impossible to provide sufficient trimming moment; therefore, although the paper was centred on and written around the tailless types, he had a feeling that, for flying at high speed and using swept-back wings, they might be forced to reintroduce a tail in order to provide the trim necessary to utilise the full lift which they knew to be possible. That was quite a serious problem, because when one considered the low maximum lift coefficients at high speeds and remembered the serious gust conditions to be catered for in any design it became almost impossible, within reasonable dimensions, to design an aircraft to meet the requirements."[86]

HP's concern with tip stall led them towards the crescent wing, in which sweep angle was traded for reduced thickness outboard. As shall be described in more detail later, this had the dual effect of reducing the tendency of the tips to load up and also positioning them physically closer to the centre of gravity, reducing the pitching effect of a tip stall. However, the accompanying reduction in the total longitudinal dimension of the wing (root leading edge to tip trailing edge extent on the longitudinal axis) then made a tailplane necessary for stability and control, in turn allowing trim at high lift.

The configuration that was rooted in Lachmann and Lee's studies of a tailless machine balanced by sweep ironically found that the negative knock-on aspects of the flight mechanics were most effectively solved by a tail, just as Perring had thought. Avro's move towards the tailless delta proposal had none of the high-aspect ratio baggage of HP's decade-long pursuit.

Davies made it clear that the initial drive to the delta

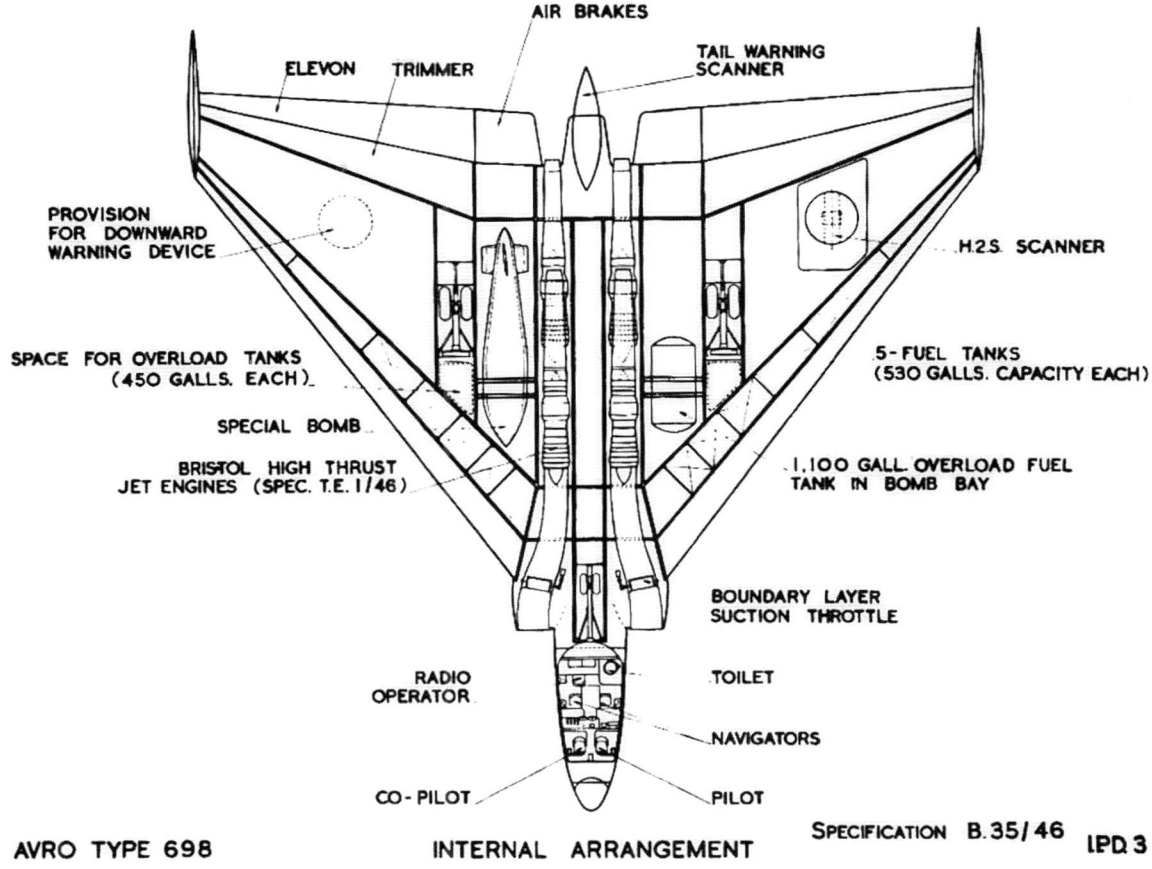

Avro's B35/46 tender proposal, May 1947. Notable are the stacked engines inboard of the bomb bays, together with the tip fins.

planform at Avro was structural; the region behind the wing was simply 'filled in' and aspect ratio reduced to retain the same wing loading as the tailless design. Roy Ewans, Avro's chief aerodynamicist, gave a more detailed description in a contemporary article[87] (pre-dating the first flight of the 698 prototype), highlighting four design aspects as aerodynamic drivers towards the delta: sweep, thinness (sic), low aspect ratio, low wing loading.

Both Ewans and Davies recognised in their writing that by keeping lift coefficient, parasitic drag and structural weight low, the penalty of low aspect ratio (or more correctly, span loading) on induced drag could be paid for. For Avro, this was attractive as it offered a way of using a conventional structural design and construction method. Otherwise, such a method was likely to prove inadequate in providing the stiffness needed by high aspect ratio configurations at acceptable weight.

HP used a spot-welded honeycomb structure for the Victor wing, a pioneering application which required considerable time, effort and at least one restart to make ready, while the inherent stiffness of Avro's projected delta was something of a shortcut that allowed the learning established across 8,000 Manchesters, Lancasters and Lincolns to be deployed. Nonetheless, it is clear that Avro viewed the proposed delta aerodynamically as a low aspect ratio, high taper ratio swept wing; there was no philosophy of exploiting controlled leading edge separation and the slender wing aerodynamics that were still in the future.

Nonetheless, from a purely aerodynamic perspective the physical properties of lower aspect ratio for a given wing loading offered some advantages. A spanwise load distribution optimised for minimum induced drag would be close to elliptical. As such, a shape would approach the necessary requirement of constant spanwise downwash. The Spitfire had provided the obvious example of a wing with a correspondingly elliptical chord distribution across the span, but it could be shown that a tapered wing with the tip chord of, say, 45% of the root approached this ideal very closely, while being much simpler to manufacture.

Sweeping such a wing backwards had the effect of overloading the tip region, and so the optimum swept wing would be one with a greater taper than a straight wing, in order to compensate. Combining the requirements of sweep, area and taper naturally pointed towards the delta shape, while the long, deep root and large internal volume reduced the need for external discontinuities in

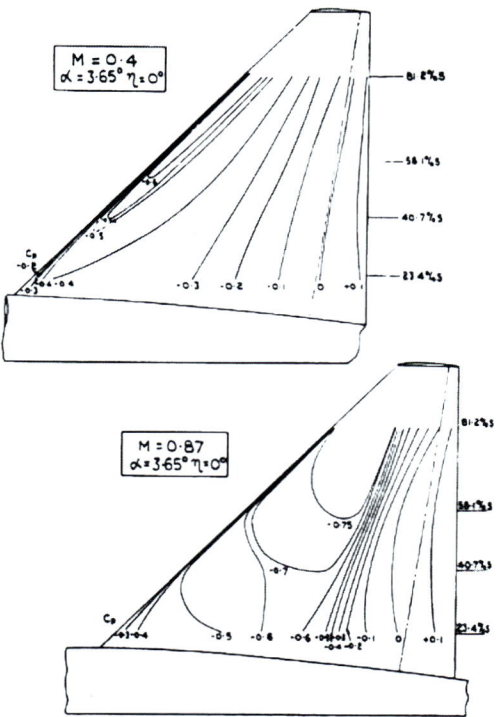

Fig. 7. Upper surface C_p contours. $\eta = 0$ deg.

The challenge of the new regime: These contours are plotted from early British wind tunnel tests on a delta wing. At Mach = 0.4, the pattern is a subsonic one consistent with previous aircraft, showing isobars gradually increasing in suction and then recovering. At the same incidence but at M = 0.87, close to the B35/46 operating condition, the topology is completely different. A shock wave has formed just beyond mid chord, evidenced by the rapid change in pressure of the closely spaced isobars. Ahead, there is a strong pressure gradient towards the tip, with the 'loops' showing the direction of outflow.

the wing, and hence a smooth spanwise load distribution. The concentration of downwash over a shorter span tended to reduce the rate at which lift increased with incidence, known as the lift curve slope. In turn, this suppressed both peak velocity—important to delay shock formation and wave drag—and gust response.

Although it was not necessarily appreciated at this stage by Avro, the long root and blended fuselage could play an advantageous role in the way the wing's rear shock wave would eventually, inevitably form. The streamlines over the upper surface of a swept wing formed a curved shape in planview, which close to the fuselage was constrained by the local geometry. Close to the root, the required compression was achieved by weak Mach waves, which would (depending on the wing planform) coalesce in the outboard region and form a strong shock.

On a delta wing, the trailing edge of the root was positioned well aft and the compression against the fuselage

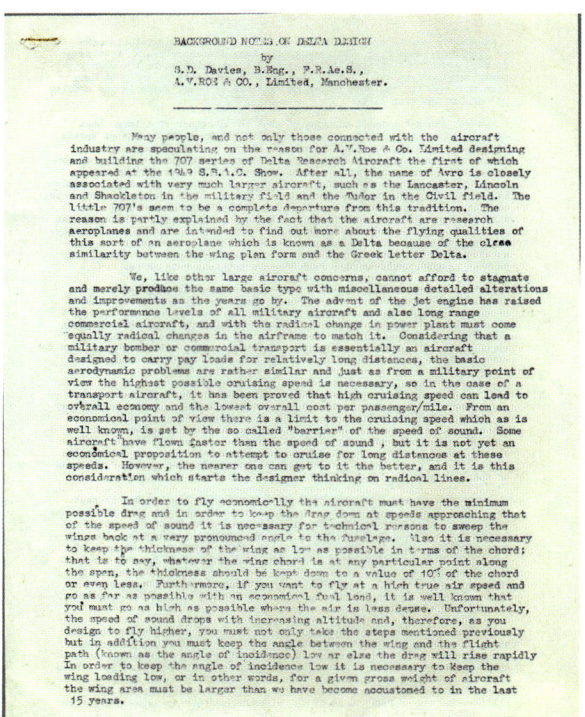

From the horse's mouth: Avro Technical Director Stuart Davies described the rationale for the delta layout in 1949.

able to be managed over a long extent; these factors could be exploited to suppress formation and limit the strength of the rear shock, particularly at very high subsonic Mach numbers or incidences, when a forward shock might also be formed. Overall, the opportunity was there to reduce the influence of wave drag. Finally, the relatively smooth, progressive movement of the aerodynamic centre of 90° apex delta wings had been identified in transonic tests at the RAE. At this early stage, the implication was that stability and control would not be a significant problem in the Mach number ranges being considered for B35/46. RAE wind tunnel specialist Alexander Thom and Perring noted,

"These results, at least up to a Mach number of 0.8, which was the extent of the speed range covered, are rather reassuring... Reducing the aspect ratio has improved the characteristics, the effect of sweepback is also favourable, the curves of aerodynamic centre for swept-back wings having good characteristics up to a Mach number of about 0.85. Above a Mach number of 0.85, on both swept-back wings and delta aerofoils, the aerodynamic centre moves rapidly rearwards and in practice this will result in a large increase in manoeuvre margin. Such an effect need not embarrass the designer, provided that the aircraft can be kept in trim, and the controls remain effective and

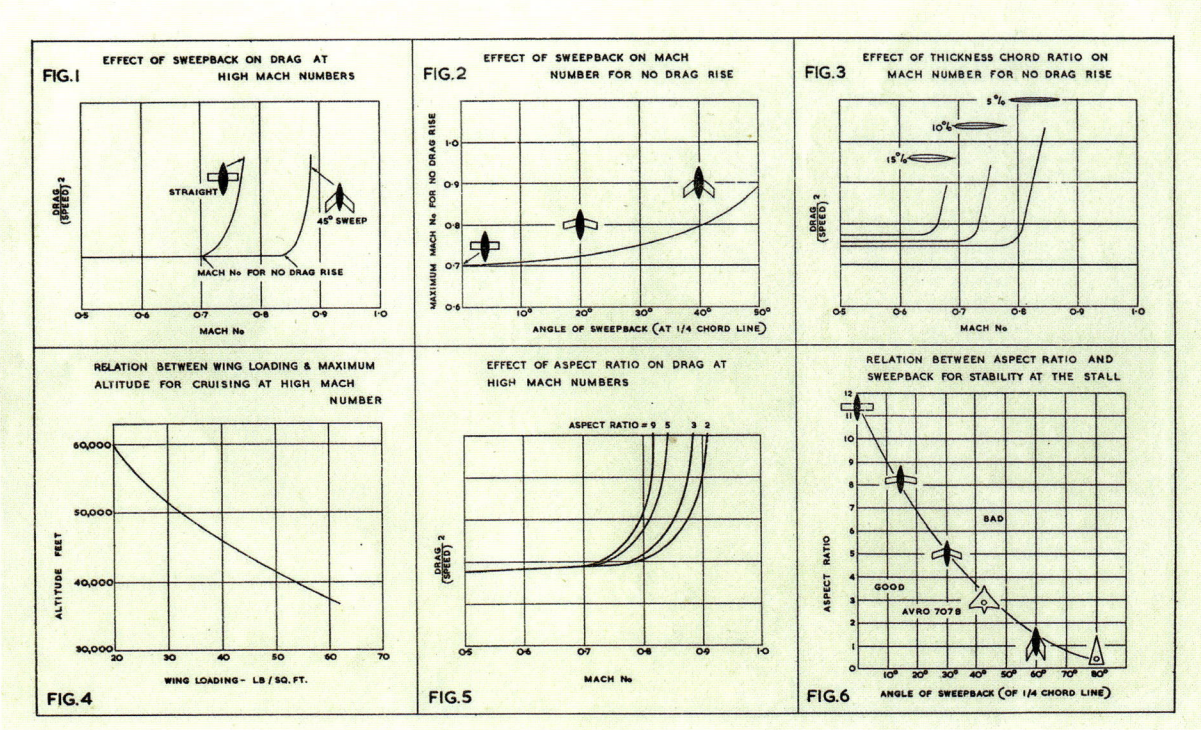

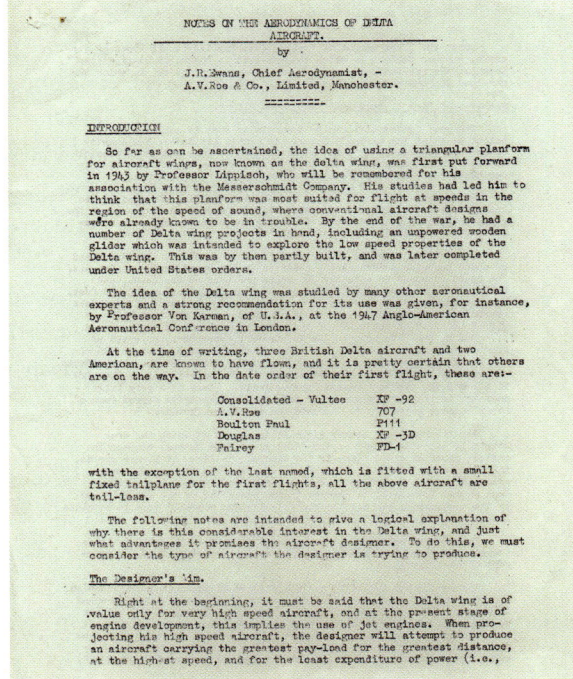

Lecture notes by Avro's chief aerodynamicist Roy Ewans, predating the 698's first flight. This illustrates how the delta became an attractive concept. For example, Fig.4 at bottom left shows how the curve of allowable wing loading became very steep as the required altitude increased, so 'cheap' area was a strong positive. Fig.6 at bottom left showed that the 707 configuration sat on the boundary of acceptable handling at the stall; if an aspect ratio of three was targeted, then it featured the maximum sensible quarter chord sweepback.

powerful enough to overcome this general increase in stability."[88]

The early concepts from both Lindley and as developed by the full team, including the recovered Chadwick, are well known. This must stem from both the widely published sketch by the latter of what looks to be a fundamentally triangular aeroplane, but also the impact on the modern viewer of the realisation of just how different the actual Vulcan was. Lindley's initial studies identified an aspect ratio of 2.4, which when combined with the requirement for sweep clearly pointed towards a delta solution. Avro's first brochure submission offered an indication of the company's thinking, if not a complete justification,

"Our study of the problems produced by the Specification has resulted in the conclusion that a Delta Wing design is the only one which will meet the speed, range, load and gross weight limitations laid down. We have used in our design study all the available information of which we are aware from British and German Reports. We submit that the Delta Wing design provides a neat solution to the requirements of the Specification and whilst admitting that the Delta Wing is at present largely an unknown quantity, we feel confident that it will be less difficult both aerodynamically and structurally to develop the Delta Wing than either a swept wing aeroplane with tail, or the V type tailless aeroplane."[89]

Early tests (1948-49) of the Avro 698 configuration in the Farnborough High Speed tunnel. At this stage, the wing was of constant 0010 section and the cabin blended back along the fuselage.

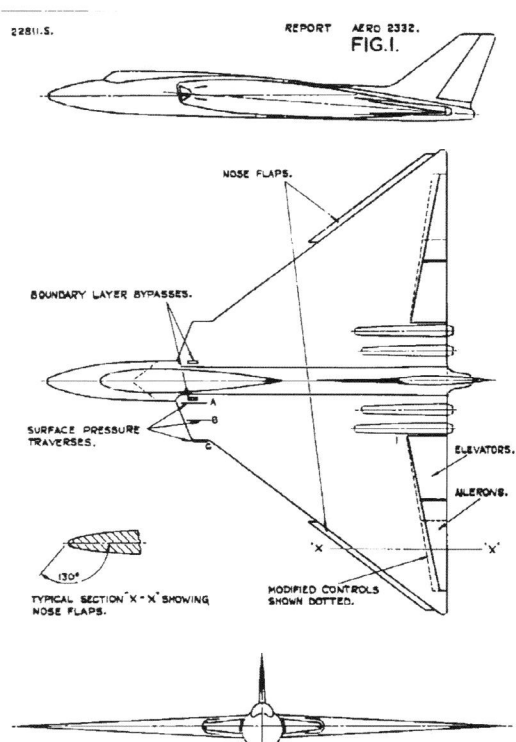

General arrangement drawing of the Avro 698, as tested by the RAE in November 1948. Of note are the pointed tips of the wings, which were of approximately constant 10% thickness NACA section. The engine inlets at this stage were squared-off slots, ahead of the wing. The model incorporated nose flaps, which would not be seen on any of the full-size Avro deltas. RAE TN AERO 2332

From an aerodynamic perspective, the brochure went on to lay down a design assumption that would prove absolutely fundamental to the direction Avro would take: to limit the cruise lift coefficient to a value of 0.2.

"The critical Mach number is a function of the lift coefficient (CL) (decreasing with increase of CL) as well as wing section and sweep back and it will be necessary to use a low value of CL in order to achieve a high Mach number. A mean value of CL of about 0.2 is indicated…"

Studies of a 'conventional' swept wing aircraft, a similar but tailless aircraft, and finally a delta showed that for viable combinations of sweep angle and wing thickness, only the delta could meet the 100,000lb gross weight target. It is fascinating to note that while the Air Ministry might have accepted this premise, Handley Page's fundamental assumption was that a considerably greater lift coefficient would in fact be viable, at greater aspect ratio, using a more aerodynamically complicated wing design.

Was — even at this early stage — the Handley Page design considered higher risk but higher pay off, by the customer? Certainly, it was recognised that HP's proposal could not be physically built using the relatively mature structural philosophy that Avro planned to use; the state of the art of both structural design and manufacturing technology would need to move forward for the HP.80 to fulfil its promise.

Avro's brochure proposal exploited the possibilities of the delta as far as possible. The four engines (based around the Bristol TE1/46) were to be stacked in pairs, fully enclosed and positioned as close to the centreline as packaging constraints would allow. This was driven by the dual aims of utilising the deepest available section of the aircraft and reducing to a minimum the yawing moment that would be created in the event of an engine failure. This would have to be coped with by rudder power, which of course would suggest larger vertical fins and an increase in weight.

Outboard of the engines, an unusual arrangement saw two separate bomb bays provided, each of which was stated to be capable of carrying a "…special bomb". At this time the H2S radar scanner would have been fitted in the port wing and given a 360-degree view via appropriate

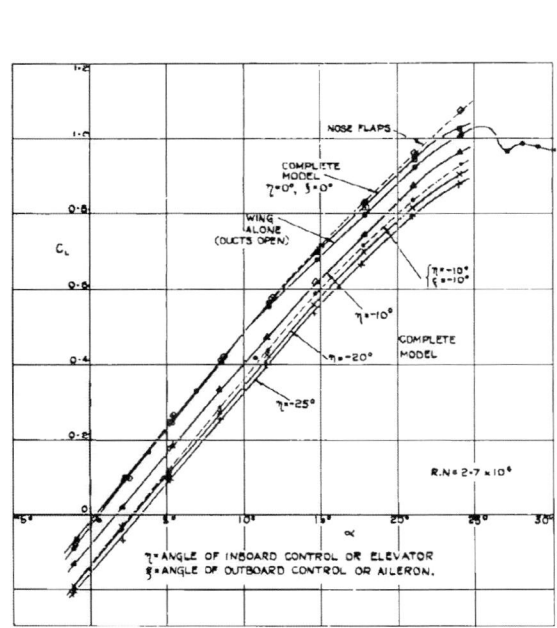

FIG. 3 LIFT CURVES FOR MODEL WITH ORIGINAL CONTROLS, AND FOR WING ALONE.

Curves of lift against angle of attack for the RAE November 1948 tests. The nose flaps are seen to extend the linear range, while progressive upwards application of trailing edge controls reduces it for a given angle of attack, and also the maximum lift coefficient achieved. This was of some concern for the efficiency of a tailless delta, as the achievement of longitudinal stability would require some rearwards download. Nonetheless, the smooth curves well beyond 20° AoA were a new characteristic of the delta design. RAE TN AERO 2332

dielectric panels, then expected to be made of timber. The main undercarriage was fitted further outboard still. The wing leading edges ahead of the front spar included integral fuel tanks and were equipped with both de-icing and suction for laminar flow control, a feature that had been evident in Lindley's proposal.

The nature and intention of the laminar flow control system is obscured to some extent by the brochure description. The statement was made that,

"25% of the wing chord is of laminar flow construction on the lines developed by Sir W. G. Armstrong Whitworth Aircraft Ltd."

This is consistent with other documents, which describe the necessity and relative difficulty of the precision required in the leading-edge surface, as opposed to the

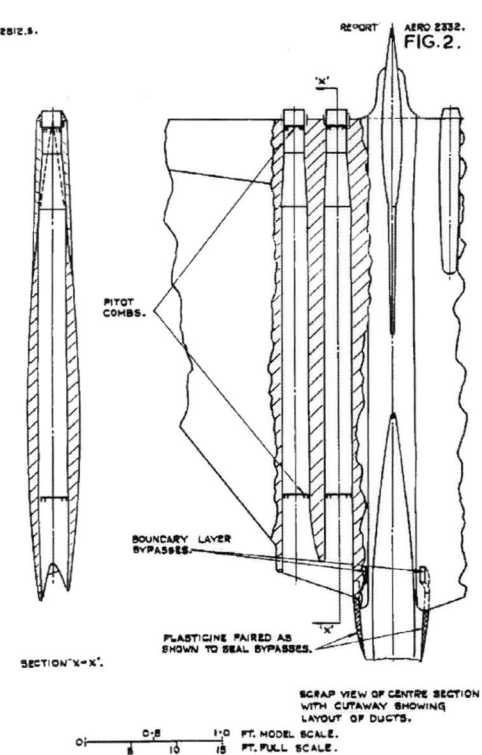

FIG.2. ARRANGEMENT OF DUCTING.

An important role of the RAE tests was in establishing the quality of flow provided to the engines. The model therefore incorporated representative ducting and instrumentation to map the pressures at the compressor face locations, together with boundary layer bleeds. RAE TN AERO 2332

conventional construction that might be used elsewhere. On the other hand, the suction system, as opposed to the surface quality, is noted in a further item:

"Boundary layer suction control is provided over the span covered by the elevons. This suction is obtained by means of a large valve in the air intake just ahead of the point where it bifurcates. The valve will be operated by an electrical actuator and is only in operation during take-off and landing."

This suggests that the suction system might only have been used at high incidence and hence high lift, presumably in order to limit the trailing edge separation due to boundary layer growth. If that were the case, then quite apart from the probability of success it would be an expensive system to incorporate for that limited purpose. One of the key features of the delta configuration that Avro should have been looking to exploit, would have been the ability to avoid any need for a high lift system. It is therefore not completely surprising that the boundary layer suction

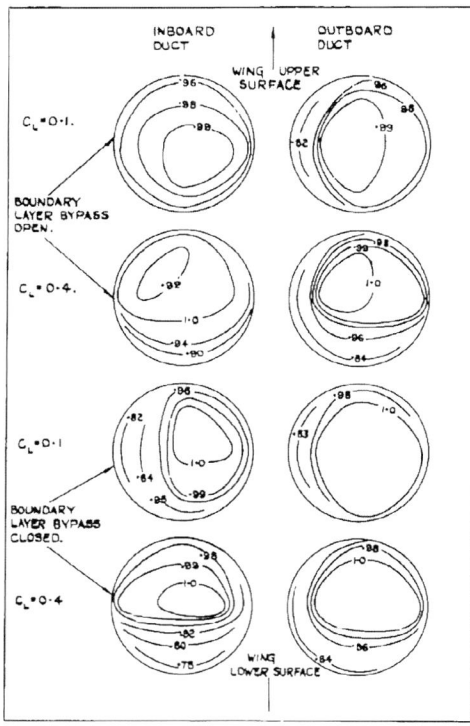

FIG. 19. DUCT TOTAL HEAD CONTOURS.
$P/\tfrac{1}{2}\rho V^2$ AT PLANE OF MEASUREMENT. ENTRY $v/V = 0.58$.

Measurements of total pressure (expressed as a proportion of freestream = 1) at the inboard and outboard compressor faces, at two different lift coefficients. The boundary layer bypass had a very positive effect on the inboard engine as the sluggish fuselage flow was removed. RAE TN AERO 2332

system was quietly dropped during early development.

A relatively new problem to solve was the design of effective air brakes, which in the case of the 698 was compounded by the small body of existing knowledge on the delta wing. There were two major reasons why such devices assumed much greater importance with the early jet aircraft than had hitherto been the case. The first was that the spool-up times of typical turbojet engines of the time were poor compared to modern machines, meaning that a substantial increase in thrust was not instantly available, but subject to the inertia of the engine's rotating components and perhaps not on a timescale compatible with a need to go around from a landing attempt, for example. This was a problem for the Boeing B-47, which operationally was known to stream the drag parachute on the approach, in order to allow the engines to be kept at a higher throttle setting, from which an improved acceleration was available.

The second reason jumped out of elemental flight mechanics: the inverse of the lift-to-drag ratio could be shown to give the aircraft's glide angle (or sink rate), with power off or at idle. Without the cooling requirements of the piston engines that they replaced, the integration of turbojet engines was inherently a cleaner proposition. However, as was found important with the B-29 and the cost in terms of absorbed power of a given drag coefficient as speed increased, the high-subsonic conditions that were the minimum for sensible efficiency to be extracted from the new powerplants were themselves inspiration for designers to attach a renewed emphasis on drag reduction. An outcome for the system as a whole then, was that the exceptional slipperiness of these new aeroplanes, with their unprecedented cruise lift-to-drag ratios, meant they were correspondingly reticent to get themselves on a sensible glideslope to land. The need to provide a means of rapidly varying lift-to-drag ratio was therefore evident.

Answering the question of whether sufficiently powerful air brakes could be designed that would also yield a minimal, controllable, change in trim was one of those subsequently identified as a priority by Stuart Davies. The RAE had produced a resume of the existing knowledge on this topic in 1942, which was thought to still have some relevance at this stage. It would also not be long before Gloster would be seeking to answer a similar question for the F4/48 delta all-weather fighter. In that case though, the use of a tailplane would mean a lower surface flap position close to the trailing edge could be used if desirable. It was thought that this ought to disturb less of the lifting flow over the wing there, but an underside pressurisation so far rearward would cause a strong nose down pitching moment, which in a tailless delta would have to be compensated for by up elevator, hence losing lift.

It is clear that the best approach simply wasn't obvious to Avro. A process of wind tunnel and flight testing would therefore be necessary to confirm the answer to what might otherwise be considered a relatively trivial problem.

Despite the importance of the 698 for the future of Avro, it could not ignore the present, which in the middle of 1947 included the difficult Tudor. The majority of the interest was in the shorter-ranged, but wider and longer fuselage variant, the Tudor II. The prototype of this version had first flown in November 1946, displaying similar aerodynamic problems to the smaller Tudor I and resulting in a similar lack of enthusiasm. On August 23, 1947, Chadwick and Davies themselves joined a test flight piloted, as had been the first flight of the Manchester a few years earlier, by Bill Thorn. Chadwick's daughter Rosemary Lapham documented her perspective of that morning in a memoir,

"Now, he stood at the watershed of his career. He was about to stake his whole reputation on the

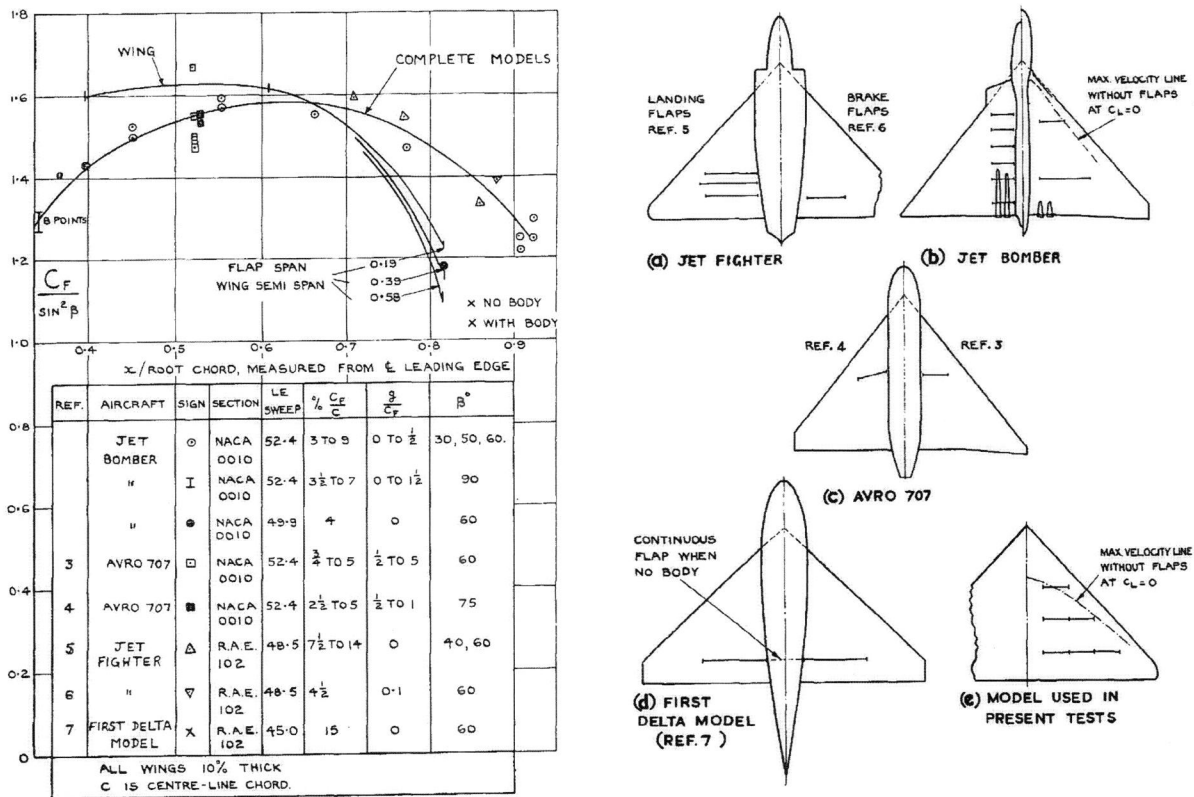

An attempt to correlate the effect of brake flap position with its performance, using the variety of delta models available to the RAE at the time. The 'jet bomber' is the Vulcan, in a number of early states. ARC-R&M-2996

concept of a triangular shape, the future for supersonic flight.

Happy to enjoy the company of his young daughter, he asked simply, 'Would you like to come to the aerodrome with me, today?' But as quickly as the girl's heart bounded with pleasure at being her father's companion, it sank again with responsibility. She replied as simply as he that she had to help her mother.

'Ah yes', he responded, 'you must help your mother.' But moving quickly on, so as not to overly disappoint her, he asked what she would like for her 17th birthday, the following Saturday. Without any deliberation her request was for a watch, but one with small sparkly diamonds around it. He replied solemnly, without hint of amusement at her naivety, that she could certainly have the watch, but he wasn't sure that they could run to the diamonds!

Around noon, she and her mother sat down to a kitchen lunch. During the meal the girl told her mother that she had seen a pretty stainless steel watch in town. A few moments later her mother left the table to telephone her husband. She went out closing the door behind her. She would ask him to call at the jewellers on his way home.

Ros remained at the table. A few seconds later, fate announced itself, piercing the silence by her mother's agonized scream."[90]

During routine maintenance before the flight, the aircraft's aileron control circuit had been incorrectly reassembled. G-AGSU climbed to less than the equivalent of its wingspan above the ground, before lateral control was inevitably lost. The flight crew of Thorn, David Wilson and John Webster, together with Chadwick, lost their lives as the Tudor crashed. Miraculously, Davies and flight engineer Eddie Talbot survived to tell the tale. Davies, the cornerstone of Avro's design and production effort through the war years, was another product of the Northampton Polytechnic, the institution that Frederick Handley Page had taught at and from which he had recruited his own celebrated chief designer, Volkert. In what must have been an extraordinarily difficult time after the tragedy, he would step up to be chief designer and to lead the jet bomber programme.

It was not going to be easy. More than a year after the issue of B35/46 to industry, the RAE remained in no doubt as to the challenge that lay ahead in meeting it.

"Operation at a Mach number of 0.87 and at a height of 50,000ft means that designers are being asked to go right outside the realm of past experience into a region bristling with aerodynamic and structural unknowns".[91]

With the initial designs on the table and aerodynamic data beginning to be gathered, the organisation moved proactively to ensure that any subsequent decisions could be made on a basis of robust technical information. On November 25, 1947, the first meeting of the RAE's Advanced Bomber Project Group (ABPG) was held, consisting of a group of personnel from the Aero and Structures departments, and with a remit of
* Ensuring that the fullest possible use was being made of the existing knowledge and experience,
* Making recommendations for future work on the specific problems thrown up by the advanced bomber.

The group met four times prior to issuing their report, in February 1948. With the intention of informing an audience that needed to make vital decisions, it started at the beginning and pulled no punches when describing the uncertainty.

"It may be said frankly that, with our present knowledge, there is considerable doubt as to whether this postponement of the drag rise to M = 0.87 is in fact attainable on a practicable long-range, high-altitude aircraft. The main means at our disposal… are (i) thinning the wing and (ii) sweepback."

This statement was remarkable in itself: the primary source of independent scientific and engineering advice to the UK government doubted if the specification was even possible to achieve. Could industry meet the challenge?

One reason for this uncertainty was highlighted. While both German data and more recent testing in the United States had agreed that the effect of sweepback was to increase the critical Mach number in proportion to $\sqrt{\text{Sec}(\theta)}$, where θ was the quarter-chord sweep angle, initial tests of the H.P.80 and 698 had not supported this. These tests had been at a relatively low Reynolds number, and it was already suspected that the effect on drag rise and also the induced drag factor (K), were important.

Future wind tunnel tests would indeed show that the rate at which the 'lift-dependent' component of drag, that is anything associated with increasing lift coefficient and so including profile drag rather than simply the three-dimensional effects, would increase at high Mach number was very dependent on Reynolds number.

One reason for this was the impact of tip stall, which effectively registered as an increase in lift dependent drag, but which was itself postponed by increasing Reynolds number towards the full-scale value. Unfortunately, the best data that the RAE had to hand would require a leap of faith in its interpretation. There was a level of confidence expressed though that, if the $\sqrt{\text{Sec}(\theta)}$ relationship was realised, then so too might be the Air Staff's performance targets.

Flight test experience on existing high-speed aircraft, using conventional sections and straight planforms, had indicated where problems might be found. Using the example of the Vampire, the report noted that the maximum achievable lift coefficient fell steadily as Mach number increased, up until M = 0.71. Once this value was exceeded, a kink appeared in the lift curve, with CLmax falling rapidly. This corresponded to the effects of shock-induced separation and resulting buffet taking hold. In a wind tunnel, the lift might follow an extrapolation of the basic curve, but this was not practical in flight. Indeed, when this new curve corresponding to the buffet boundary was followed down to a point where the CL for 1g flight was indicated, the aircraft's maximum speed had been identified. It might have the thrust to go faster, but could no longer generate sufficient lift because of compressibility. The concept of 'usable CL' was introduced: the lift that could be generated without breaching the buffet boundary.

The coupling of aerodynamic and structural effects on the design was recognised as critical in defining the range of viable configurations. The combination of top-level knowledge available on both of these topics was a clear strength of the ABPG, and by definition would direct opinions down the line. The Avro and Handley Page concepts had been contracted to be taken forward, but nothing yet was set in stone in terms of what the RAF would eventually receive. Clearly, the group wielded considerable influence.

As Boeing was in the process of discovering on the XB-47, aileron reversal would play a dominant role in defining the required structural stiffness of the wing. The ABPG calculated that at an aspect ratio of 3.5, a 45° swept wing structure (of taper ratios varying between 5 and 1) would satisfy the strength and aileron reversal requirements equally. Moving in either direction with aspect ratio would require additional structural weight, either to increase stiffness if AR was increased, or strength if reduced. At AR = 6.5, for example, this would amount to an increase of 8% over the AR = 3.5 baseline. That said, the group lacked confidence in the aerodynamic assumptions made for the low AR cases, which they viewed as introducing potential errors of up to 25% in reversal speed.

The significance of these assessments operationally was in the achievement of design diving speeds, which by specification was 415kts EAS. The AR = 6.5 wing designed to be just adequate for strength was calculated to have a flutter speed of 369kts, or 308kts if a sensible 20% margin was incorporated. Redesigning this wing to give an aileron reversal speed equalling the design diving speed moved the flutter safe speed to 466kts. In other words, at this aspect ratio and above, aileron reversal was the first problem to be encountered as speed increased (at 415kts), followed by flutter at 466kts and eventually structural strength.

Similar analysis was conducted for an AR = 2.5 delta, demonstrating ascending requirements of flutter, minimum skin gauge and aileron reversal. The overall conclusion was that strength was unlikely to defeat any of the concepts, but that achieving the necessary stiffness to provide either flutter or aileron reversal protection would be the limiting factor. Both of these relied upon uncertain aerodynamic calculations, but without an adequate level of confidence it would not be possible to choose between them.

A curve ball had been added by the work conducted on the isoclinic wing, in which specific management of the locations of the flexural and inertial axes of the wing could eliminate flutter. Since the low aspect ratio configurations hit flutter as their initial limit, incorporating this thinking into the design might give a relative weight advantage to the delta over the swept configurations, as the need for some additional stiffness would disappear. However, the assessment of the state of the art was realistic: "…how this would affect the comparison of plan-forms cannot of course at this moment be even guessed". Nonetheless, with the APBG report incidentally providing a basis of an understanding of what form a Soviet bomber might take, it is both understandable and reassuring that no stone was being left unturned. Conversely, the need to not jeopardise the urgent timescale of the advanced bomber project was also clearly in the forefront of minds.

An interesting aspect of the APBG report is the way in which it challenges some of the requirements of B35/46 itself, together with discussion on the impact of operating an advanced bomber expensively designed to meet the specification, outside of the planned mission. One of these points is truly prophetic: an investigation into the effect of flying a Hi-Lo-Hi mission, which the V-Force would indeed adopt in the mid-1960s.

It was found that the AR = 2.5 delta was by far the better prospect, compared to the AR = 5 swept design, for a mission in which the cruise was at 50,000ft but followed by a descent to sea level for the target. This was basically attributable to the high dive speed inherent in the delta design. Was the RAE already thinking of 'under-the-radar' attacks being necessary, as early as 1948? If nothing else, it would have offered some support to the idea that the low aspect ratio concept, at odds with the conventional thinking for a long range, load carrying aircraft, was a viable and to some extent future-proofed avenue.

The final conclusion—the scientific advice presented to the government and the Royal Air Force—was that it was impossible to draw one with the present state of knowledge. Nonetheless, the search needed to be systematic and the RAE emphasised the need to capture the widest spread of plausible options. If both HP and Avro had proposed deltas, then the implication of the advice was that one would have been chosen. However, the two favoured contenders mapped onto the candidates that the RAE had in mind.

> "It is quite apparent that, with the present uncertainty in basic information, we cannot put all our eggs into one basket. Several designs must be chosen in order to spread the risk. Assuming 500 knots true airspeed in the neighbourhood of 50,000ft is the aim, and that the greater the height attainable over the target the better the aircraft, the project analysis … leads us suggest that at least three designs should be chosen.
>
> Firstly, a tailed aircraft of aspect ratio between 5 and 6, and with about 45° of sweepback…. The tailplane is considered essential to minimise and control …[aeroelastic] distortion effects… This aircraft will potentially be the best of those suggested, in terms of height over the target, but will suffer greater risks from a small margin between cruising CL and maximum useable CL and from distortion effects.
>
> Secondly, a tailless delta aircraft of aspect ratio 3. An aspect ratio of less than 3 is undesirable, owing to large sacrifice of height over target. This aircraft will be the safest of the three in terms of lift reserve when cruising, and distortion effects should be manageable. It may prove a very attractive engineering proposition if satisfactory means can be found of stabilising large areas of thin skin…
>
> Thirdly, we would suggest a design intermediate between the above two extremes—namely a tailed aircraft of aspect ratio 4 and with only 35° sweepback, the smaller sweepback being compensated by

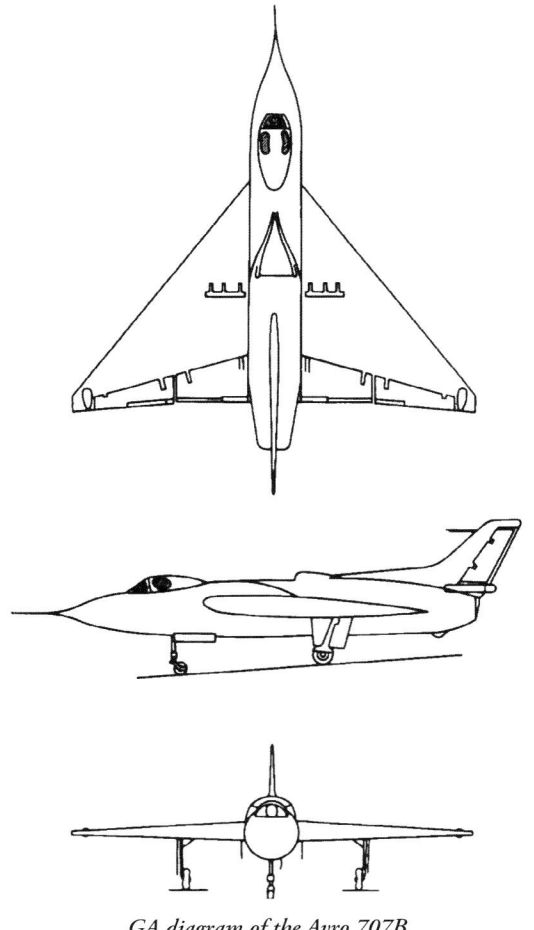

GA diagram of the Avro 707B.

thinner wings (say 9%-10% thickness/chord ratio as compared with the 13.5% thick wings of the 45° swept aircraft)."

On the table, the Avro 698 met the basic requirements expressed in the ABPG's second option, although the aspect ratio at the time lay below the minimum that that they deemed desirable. The HP.80 lay somewhere between the first option and the third 'all out' ambitious proposal. The 45° sweepback said to appropriate, was significantly above the average of the HP proposal, which fits with the uncertainty about its effectiveness in the low Reynolds number tests. On the other hand, the HP wing was already thinner on average than the third proposal, and the aspect ratio of about five which was desirable for cruise efficiency was made possible by the reduced outboard sweep. It seems then that the RAE did not necessarily buy into the crescent wing at the time as being one of the best strategies available. Despite the challenges, they were clearly sold on the delta and this provided the technical underpinning of the early decision to commit to Avro.

The lack of experience with delta-winged aircraft in the real world had been highlighted by the RAE's ABPG report as a major uncertainty in delivering B35/46. It was a problem potentially solved only by flying real aeroplanes. While a wind tunnel should provide accurate engineering information about the forces and moments generated by these novel configurations, if this could only be done at low Reynolds numbers, then the conclusions remained suspect and open to interpretation. Assuming such a machine could be made effective, efficient and controllable, were there still hidden dangers to be found its operation?

Avro's outline plan to tackle this was a progressive development programme of flying models. The first stage would be low speed trials, with a small single seat, single engine aircraft, at about one-third scale to the final bomber design. Proving the concept at higher speeds where Mach effects became relevant would require a larger and more powerful machine, for which a twin engine, half scale aircraft would be needed. Finally, the prototype of the 698 itself would be simply a 'flying shell', potentially lacking systems but able to prove the flying qualities at full scale. The one-third scale demonstrator was designated by Avro as the 707, and the half scale as 710.

The 710 was never to reach fruition. Powered by two Avon engines, the level of engineering effort would have been very significant, diverting precious resources and funding from the 698, while likely delaying the first prototype. It was essentially axed on that basis, following a meeting between Avro and the Ministry of Supply on August 30, 1948. However, to meet the requirements for high Mach number flight investigation, a developed version of the 707 was proposed instead, stressed to both higher dynamic pressure and M = 0.95. This was approved as the 707A.

Construction of the two low-speed 707 aircraft began soon after the MoS decision; the geometry of the wing was of a correspondingly early standard. Advice from the high-speed wind tunnel research at the Royal Aircraft Establishment during and immediately after the war[92] suggested that thin, symmetrical sections were likely to be useful at high-subsonic speeds, as they minimised the change in pitching moment resulting from shockwave formation. Avro settled on the NACA 0010 (10% thick, zero camber) section, from the same family as that used by the Spitfire, and the result of NACA's research back in the early 1930s that as we have seen, could ultimately be traced back to Max Munk and WW1-era Germany.

The 707 was therefore designed with full span NACA 0010 section and the pointed tips of the initial planform. To reduce the engineering overhead, appropriate components from existing aircraft were used. The main

Details of the structure of the early, low speed 707 designs.

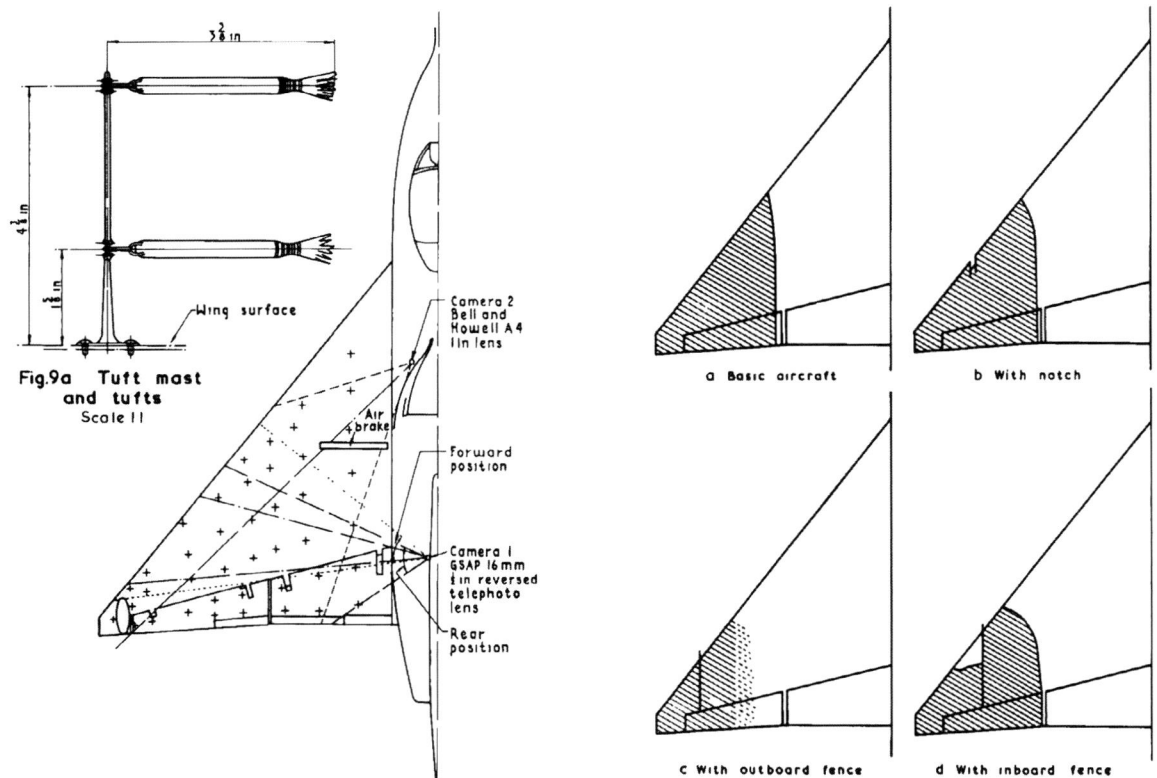

The 707B was instrumented for delta wing flow investigations with tufts on masts over the wing. This enabled the position of the separated region at high angle of attack to be recorded by movie camera. Subsequently the effectiveness of passive flow control devices, including a notch and fences, could be checked.

undercarriage came from the Avro Athena, while the cockpit canopy and nose undercarriage was adapted from that of the Gloster Meteor. The powerplant was the proven Derwent 8, also in production for the Meteor. Harking back to the Fokker experience, some consideration was given to using a wooden wing as a method of expediting things further, but ultimately a simple metal structure of two spars and pressed ribs was adopted.

The first 707, VX784, was ready for taxying trials at Woodford in August 1949, but it had been decided that the long runway and test facilities available at the A&AEE Boscombe Down, were more appropriate for the first flight. VX784 was therefore dismantled and moved by road, being ready again for flight on an auspicious tenth anniversary: September 3, 1949. Unfortunately, an excessive crosswind meant that Avro deputy chief test pilot Eric Esler could not risk a step into the unknown that day and it appeared the out-of-limit conditions would frustratingly rule out the following day too.

Nonetheless, high-speed taxi tests were conducted and these highlighted excessive elevator force being required to rotate the nose up, at sensible take-off speeds. Adjusted overnight, the longitudinal control was still too heavy in tests the following morning. With full nose-up trim, VX784 abruptly rotated at 100mph and became fully airborne at 103mph; Esler quickly pushed forward and braked heavily to complete the 'taxi'. This necessitated a change of brakes and tyres, but after bedding in with further taxying and another brief hop, he felt confident to fly. By the evening of September 4, the crosswind had abated somewhat, and the dice were rolled. Esler and VX784 took off at 19:50, squeezing into the day the first flight of a UK-designed tailless delta aircraft.

The impetus for the evening flight was the looming presence of the SBAC Airshow at Farnborough, to which Esler flew VX784 on the 6th. While it would remain on static display, the mere sight of the unusual and futuristic triangular jet must have been captivating to those present.

The specification for the 707 did not require an ejection seat; the maximum permitted speed was restricted to 300kts EAS. For the high speed 707A, this speed would by definition be exceeded, hence a new monocoque nose design incorporating a Martin-Baker seat was a necessity. Design and production work for this proceeded in parallel, but to a longer schedule, than that of the baseline low speed design.

On September 30, less than a month after the first flight, Esler was piloting VX784 when it stalled and crashed. It

FORGED IN ETNA: THE START OF THE VULCAN

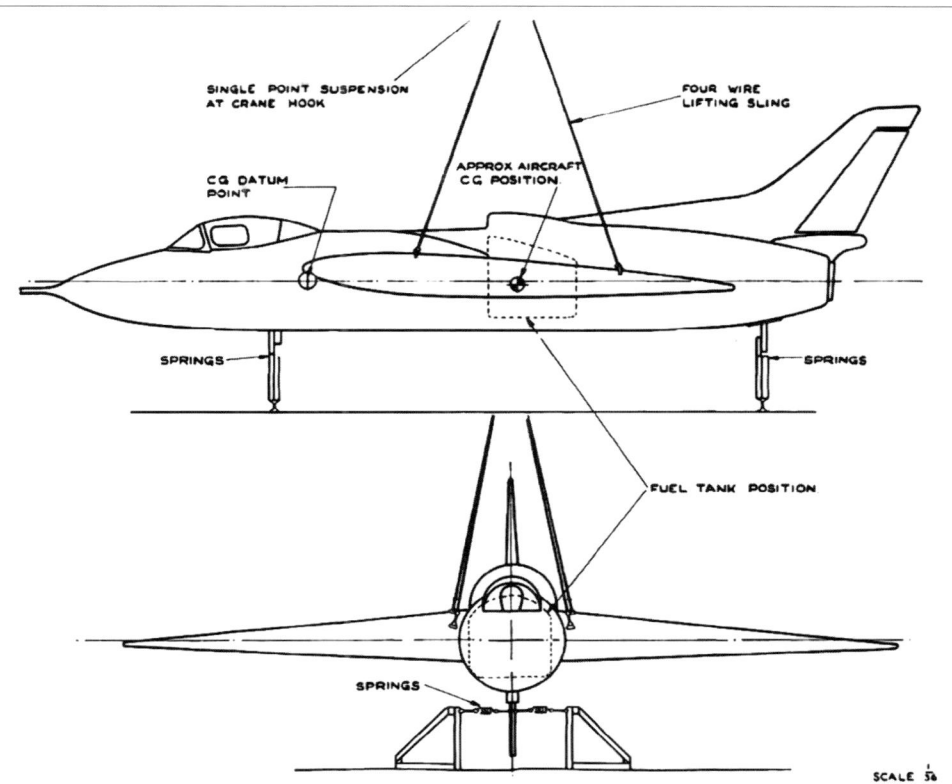

To understand the flight dynamic behaviour of an aircraft, it was necessary to know the CG position and moments of inertia about the three axes. The Avro 707B was subject to such treatment by the RAE.

was subsequently determined that the lower airbrakes had deployed fully due to an electrical problem, which had in turn reduced elevator effectiveness to the point where longitudinal control was rendered impossible. Without adequate means to escape, the 31-year-old Esler was killed instantly; another brave casualty of the relentless quest for high performance military aircraft at this time. It wasn't just a blow to life and limb however, as the aircraft had very recently been fitted with a data recorder (or automatic observer, as it was known at the time), and apart from the qualitative evidence that the delta wing was satisfactory, had barely produced any quantitative data as part of the test programme.

It would be almost exactly a year between the first flight of VX784 and that of the second low speed 707, VX790. Although intended in the contract to be identical, the accident that befell the first led to significant engineering changes in the second, covered by amendments to the specification. An ejection seat was now considered essential, but was incompatible with the nose design. A pragmatic solution was to borrow the nose design of the high-speed 707A, then deep in design, which gave a much sleeker and more aggressive aspect to the 707's appearance. In keeping with the 'rapid' philosophy, the new nose again borrowed an undercarriage design, but this time from another Hawker Siddeley group product: the Hawker P.1052. Designated 707B in order to differentiate the revised design and painted in bright blue, VX790 took to the skies on September 6, 1950. Like the earlier aircraft, after a brief series of flights, it was deemed fit to go on show at Farnborough, albeit again in the static park only.

Much has made in the intervening years of the 707 programme's apparent lack of contribution to the parent 698 aircraft, on the basis that the former only flew after the design of the latter was finalised. However, it must be recognised that the sum of knowledge of delta wing aerodynamics was small, and early validation of concepts was powerful. Even if subsequent redesign of the 698 itself was required as a result of the 707 tests, the latter at the very least accelerated the embodiment of such changes into the production bomber.

The 707B was unrepresentative of the Vulcan in at least two significant ways. Firstly, the design of the wing had moved on, in that the original constant 0010 section that it used had been superseded by important modifications at both root and tip. These were intended to improve aerodynamic performance at high Mach numbers, so while less relevant to the low-speed tests for which the 707s were designed, would nonetheless have had an effect. The second was the influence of the dorsal intake. The report noted that,

"The effect of the engine thrust on stability is complicated. At low lift [high speed] coefficients in the cruise configurations the stability improves slightly with thrust … At higher lift coefficients the effect of thrust is generally to delay the reduction in stability margin … This may be explained by the local air flow over the body and inboard sections of the wing being particularly sensitive to intake conditions at the higher incidences."[93]

This result showed that, much like as in wind tunnel testing, the results even of flight tests had to be interpreted and could not be taken as immediately transferable to the full-scale bomber.[94] Some of the key results were to demonstrate the effectiveness of transfer from the different tunnels in use, for example between the Avro low speed, RAE high speed and NPL CAT facilities, all of which were to some extent involved.

The 707B demonstrated the behaviour that would become familiar with relatively low speed delta wings: the relative increase in outboard load would lead to separation, but unlike the slender shapes of the future, this was not in the form of controlled stable vortex. It showed that the design was fundamentally stable up to a lift coefficient of 0.5, roughly double that which would be required by the full-scale bomber over the target. It highlighted the rearwards limits of the centre of gravity range and how the effect of wind tunnel testing at lower Reynolds number was to underestimate the pitching moment and the linearity of its slope. Reynolds number has a considerable influence on both boundary layer development and separation, and so the fact that the outboard separation-dominated pitching behaviour was not well predicted by low RE tests is unsurprising, in retrospect.

A feature of the 707B was the provision for asymmetric ballast weights to be carried in the wings, together with wingtip parachutes. These could be used to apply known rolling and yawing moments in flight, for the assessment of control power. The RAE observed,

"The lateral behaviour of aircraft at approach airspeeds has become increasingly important in recent years … whilst the changes in aerodynamic configuration required for high performance have led to a general deterioration in aircraft lateral characteristics."[95]

The 707B's rolling performance was found to be similar to that of the Meteor, which although obviously not the most up to date machine by 1953, was in widespread use and shared a similar approach speed and rolling moment of inertia. The approach speed of the delta was limited by both longitudinal and lateral control limitations, rather than absolute lift; it could have landed more slowly and hence reduced field length if control power and stability could be improved. Below 115kts IAS, "problems of speed and flight path control due to the drag characteristics of the aircraft…" were limiting, while aileron snatching related to the encroaching flow separation at the wing tips crept in below 118kts. Consequently, the approach was flown at 120-125kts, touching down at 110kts.

The 707B therefore indicated, among other things, that the future bomber needed to improve on the aerodynamic limiting factor of tip stall. This was, as shall be seen, already in hand, but the small, cheap test aircraft proved that the work was both necessary and on the right track. In the case of the lateral tests, the costs and risks associated with yawing parachutes and asymmetric loading were less appropriate for the vital full-size prototype.

The dorsal intake was not only atypical of the ultimate bomber design, but a source of project work in itself, showing in a limited way just how much capacity would have consumed by the complex 710 had it been proceeded with. The initial intake consisted of little more than a lip and vertical splitter plate. With the cockpit ahead and the long fuselage boundary layer to be ingested, it demonstrated poor ram efficiency, recovering little more than 20% of the dynamic pressure at cruise speeds.[96] Because the 707B soon showed that achievable level speed limited the possible Mach number of its testing regime, a new intake design was sought that would reduce the losses and give the equivalent of a higher thrust engine.

The eventual solution developed and tested at Rolls-Royce Hucknall was the use of a NACA-style submerged ramp, with vortices generated by the curved side edges working to sweep the boundary layer outboard. This style of inlet was considered in the UK to be poor in terms of drag (largely due to the blockage and skin friction caused by the vortices), but on the 707 it increased ram efficiency to 70% and fundamentally solved the speed problem. This may have been due to the unusually long boundary layer and hence the relative unimportance of the drag aspect compared to the pressure recovery and improved thrust. The chief development engineer at Hucknall, F. B. Greatrix, noted that this was equivalent to a 25% increase in thrust at 500mph.

What effort would have been needed to make the 710's twin Avon installation all that it needed to be, with little if any of it being applicable directly to the 698? Clearly, there was a delicate balance to be struck. As it was, the 707B received its new intake and increased Mach number capability in early 1951, a year and a half prior to the first flight of the 698.

6

EMERGING KNOWLEDGE: THE MAJOR REDESIGN OF 1949

THE APPOINTED date for design freeze of the 698 was September 1949 and by that stage the design had moved on considerably from the 'flying wing' of Chadwick's early sketches. The concept for detailed design handed over by project staff to the engineering departments was now firmly recognisable as an early form of the aircraft with which we are familiar.

Perhaps the most obvious differences to what would be built were associated with the engines and wing tips. The power plants were fed from wing root slots, but this was a feature of a forward, lower-sweep projection of wing, while the jet pipes were roughly centrally mounted such that similar volumes projected above and below the wing. The tips were pointed, forming a 50° swept triangle and there was a single central tail fin, finally moving away from the fins close to the tips that had seemed attractive previously.

The new design fitted into the implied Avro philosophy of not making life harder than it needed to be; the potential effects of the outboard fins on tip stall were avoided, while a horizontal tailplane could be mounted if—in extremis—this was the only viable longitudinal stability and control strategy left. Nonetheless, a curve ball was about to be thrown, thanks to the emerging work on transonic aircraft aerodynamics at the Royal Aircraft Establishment, that would make both company and customer commit to a major redesign even at this late stage.

High speed wind tunnel tests of a half model of the 698 (which allowed a greater Reynolds number to be simulated than a full model) had been conducted earlier in the year at the RAE. At the time, this technique was in its infancy and a satisfactory method of sealing the root to the half fuselage or wind tunnel wall had not been established. Consequently, drag measurements were treated sceptically and those from necessarily smaller full models were used. At around the same time, a series of full model tests on swept and delta configurations in the same tunnel began to trigger alarm bells.

There were hints that the all-important drag divergence Mach number might in fact be lower for a delta than an equivalent swept wing, and that when applying these results to the 698 configuration, drag rise might occur considerably below the specified M = 0.87 cruise. This was clearly a major risk to the success of the programme, especially so given the alternative swept design available from Handley Page. A conclusion from these tests was that modifications to the wing root were likely to give the best improvement, but these were some of the most disruptive imaginable. Why was this so?

As the Avro 698 approached its original design freeze towards the end of 1949, the aerofoil for the full span had settled on the four-digit NACA symmetrical profile at a thickness of 10% (NACA 0010). The rationale of this

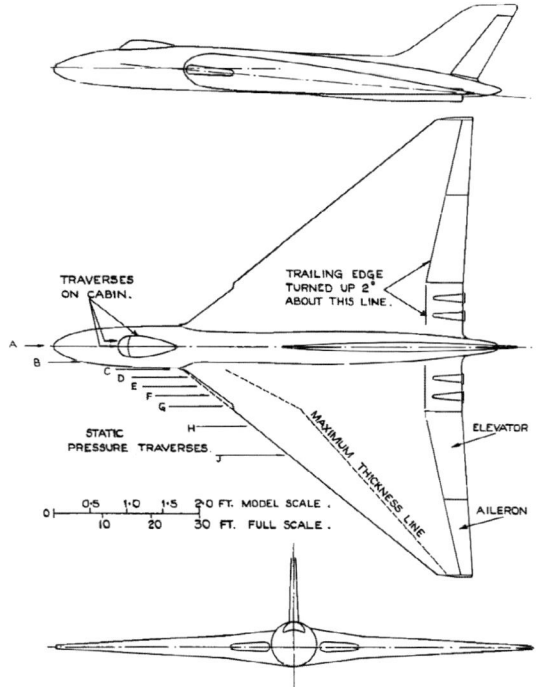
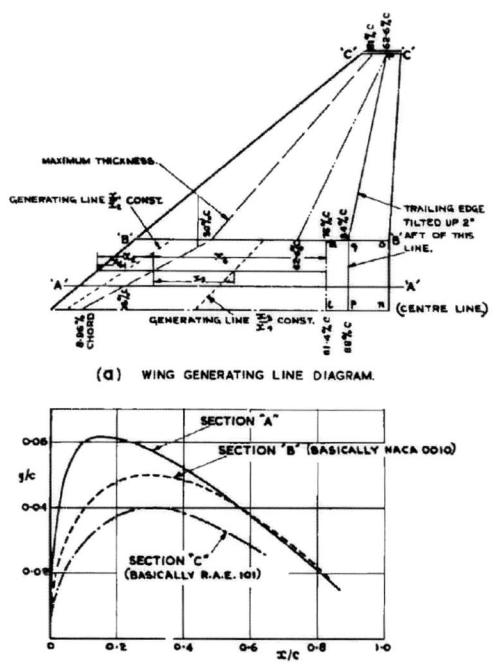

General Arrangement drawing of the Avro 698 Prototype configuration, as wind tunnel tested by the Royal Aircraft Establishment. Note the indication of the line of maximum thickness on the wing, and how this sweeps forward towards the centreline. This was due to the revised inboard profile adopted for the 1949-50 redesign, moving the suction peak (and aerodynamic sweep) correspondingly forward. RAE TM 2558

The redesigned wing of 1949-50. On the lower diagram, one half of the symmetrical shape of the baseline NACA0010 profile is shown as a dashed line. The maximum half thickness (5%) occurs at about 30% of the chord. The tip shape is now 8% thick RAE 101 shape, which has the maximum thickness slightly rearwards, while inboard the bespoke section blends back towards the 0010 towards the trailing edge, but reaches 12.3% thickness very close to the leading edge, at around 15%. These sections correspond to B-B, C-C and A-A respectively, in the detailed planform drawing at the top. By blending between these three defined shapes, the three-dimensional surfaces of the wing were defined. RAE TM 2558

choice, as opposed to the more recently designed British high-speed aerofoils of the Squire series (for example), is less immediately obvious. These latter geometries were designed to give a gradual, favourable pressure gradient; the relatively low magnitude suction extended over a long region of the chord towards a rearwards peak, giving rise to the 'rooftop' description of the pressure distribution. The Squire A to E series were eventually redesignated as RAE 100 to 104 respectively. Each described a thickness distribution, but would be scaled to a required thickness and could have camber applied.

Both the NACA 0010 and RAE 101 sections had maximum thickness at around 30% chord, but the peak suction on the RAE 101 was intentionally further aft, with the pressure recovery commencing at around 30% chord compared to 17% chord for the 0010. Here the three-dimensional effect of the Vulcan's delta wing came into play. The geometric sweep (i.e. the lines formed by plotting constant percentage chord) gradually reduced from ~50° at the leading edge to less than 10° at the trailing edge. Assuming that the isobars (lines of constant pressure) would approximately follow the geometric chord, the further aft peak suction of the RAE 101 section would generate isobars that were less swept for a given suction level, hence analogous to operating at an increased Mach number.

At the low lift coefficient levels typical of cruise for the Vulcan, this might be expected to manifest itself as causing the drag rise to occur at a lower Mach number, which a comparison between similar planforms using the two sections did in fact show in the wind tunnel. This difference reduced as lift coefficient was increased and the peak suction naturally moved forward, such that at CL = 0.2 there was little to choose. The RAE observed that in a general case (i.e. an aircraft with a design point at a higher lift coefficient or with a wider intended operating/ manoeuvre range) the RAE 101 would be the better choice. As shall be seen, this trade had been made on the similarly

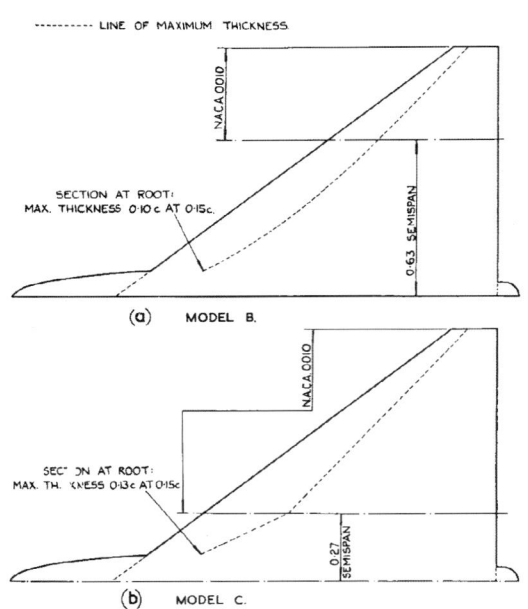
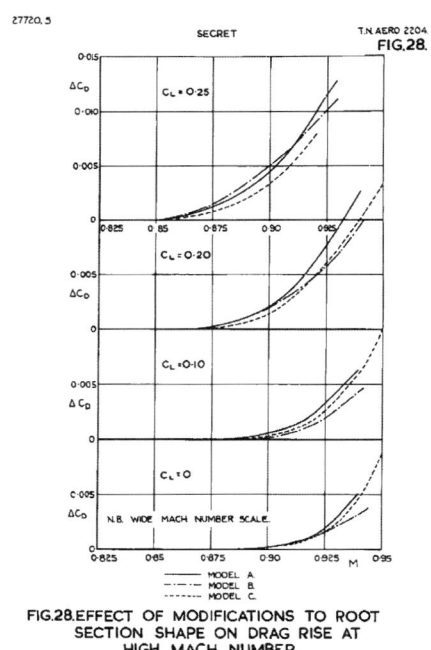

Stages of development in the inboard wing shape. The maximum thickness line was moved forward, then the section increased in thickness. As lift increased to cruise levels, the thicker wing improved drag divergence. RAE TN AERO 2204

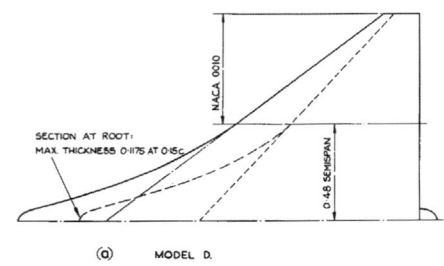

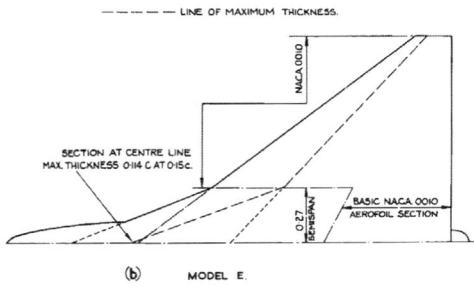

An alternative strategy was a lengthened inboard fillet, connected to a longer nose. Both this (model E) and an artificially blended fuselage (model F) were tested as possibilities in the RAE campaign. RAE TN AERO 2204

shaped Gloster Javelin, which in its fighter role had to pull greater 'g' and over a wider Mach number range than the Avro bomber.

The reduced level of peak suction of the RAE 101 was primarily of interest for another reason however: protecting the outboard wing from tip stalling. Because the sectional lift coefficient outboard would naturally be higher, due to the effects of sweep and planform taper, the British design would more often be operating above the crossover threshold in drag divergence with the NACA 0010, allowing it to play to its strengths. The final design incorporated this section therefore, at 8% thickness at the tip.

Across the central plane of the aircraft (or indeed wing), the geometric sweep angle must pass through zero. The important effects of this on the spanwise isobar formation and seeding of the rear shock was a hot topic and Avro heeded advice (usually attributed to A. B. Haines of RAE) on sweeping the maximum thickness well forward in the inboard region.[97] The strategy also required taper of the thickness/chord ratio over the span of the wing, with the rate being highest near the root.[98]

Ultimately, this manifested itself as a 12.3% thick bespoke inboard section, with maximum thickness line displaced well forward to about 15% chord. The t/c taper was achieved by blending back to the existing 10% section outboard of the engine bays, then a very subtle taper down to a tip thickness of 8%. The decision to proceed with this change was made in late 1949 and the 707A high speed test vehicle was delayed in order to ensure this feature was incorporated.

The inboard spanwise sweep was aided further by data from RAE high speed wind tunnel tests on alternative configurations of the engine intakes. As the design coalesced in 1948-49, the inlets evolved from the large single pitot design of the original proposal to a squared off plan view projection ahead of the wing root and rectangular slots, each feeding a pair of engines. Finally, the plan-view projection was removed and the inlet slot followed the underlying wing leading edge, as is now familiar.

Propulsion integration was certainly a discipline in which the Royal Aircraft Establishment could boast expertise, both organically developed and imported from Germany. Dietrich Küchemann and Johanna Weber had literally written the book on this, based on their extensive wartime research at AVA. Following their monograph for the British, it would be published by McGraw Hill as Aerodynamics of Propulsion in 1953 and remains relevant today. It would be incorrect however to think that they were speaking to an ignorant audience. Among many other things, Küchemann was an accomplished cellist, to the extent that he considered making this his career after the war. His biographical memoir as a Fellow of the Royal Society would record,

"He arrived in England on October 16, 1946 and took part in his first concert [at Farnborough Town Hall] on December 1. The person responsible... was Dr John Seddon."

Seddon was probably the closest equivalent to Küchemann's aerodynamic expertise in the field of inlet aerodynamics, having spent much of the war working to improve the cooling systems of piston-engined aircraft. He was enthusiastic about the "remarkable" installation of radiators in the Mosquito, ahead of the main spar and fed from leading edge inlets. He would write that,

"The installation, one of the first of its kind, was so successful that it was reported to give zero overall drag, or even a small net thrust, at a time when cooling drags were accounting for anything up to 10%. of the engine brake horsepower. Since that time the wing leading edge intake has never been without adherents in this country, and during the gas turbine era the number has grown."[99]

Seddon published an extremely relevant monograph on the integration of leading edge inlets into delta wings in 1950. This document articulated the more aerodynamically subtle impact of buried engines, when compared to the obvious reduction in skin-friction drag. Far from hindering the wing pressure distribution, at the root of a swept wing it might actually be made more efficient and able to penetrate deeper into the transonic regime.

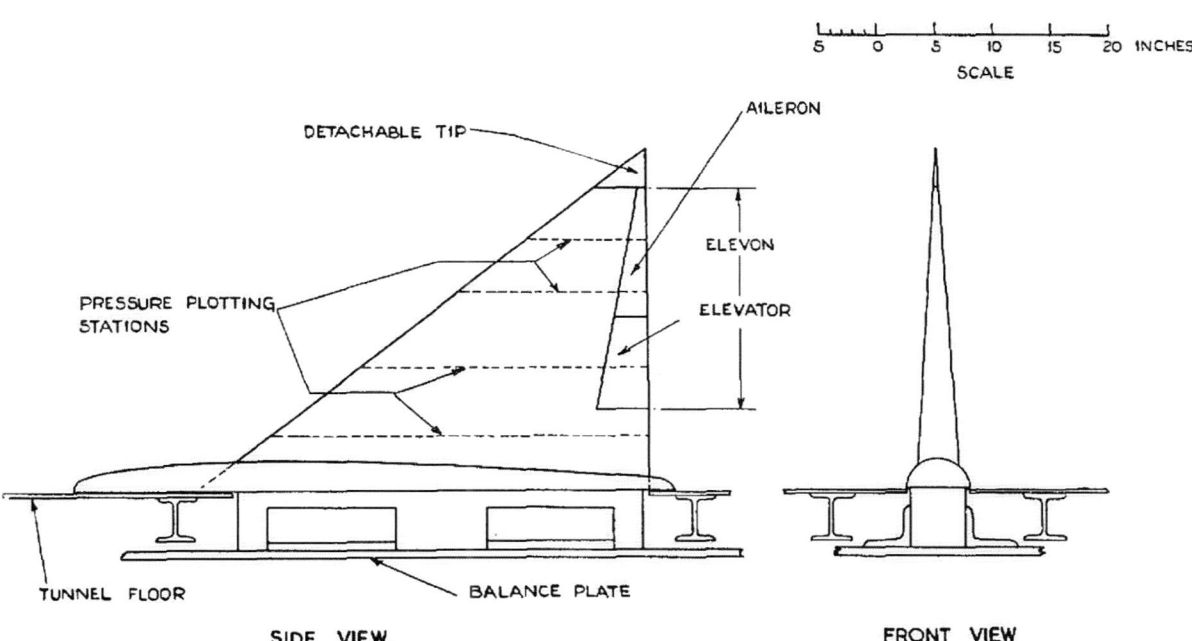

FIG. 1. Installation model in R.A.E. 10-ft × 7-ft High Speed Wind Tunnel.

Model used by Ken Newby on RAE 10ft x 7ft high speed wind tunnel, to investigate the effect of control deflections. The model was equipped with a removable wing tip, outboard of the aileron. This proved important, as introducing a finite tip chord helped to reduce the local load and alleviate tip stall. These tests therefore pushed the 698 planform towards a cropped tip. NEWBY, ARC-R&M-2999

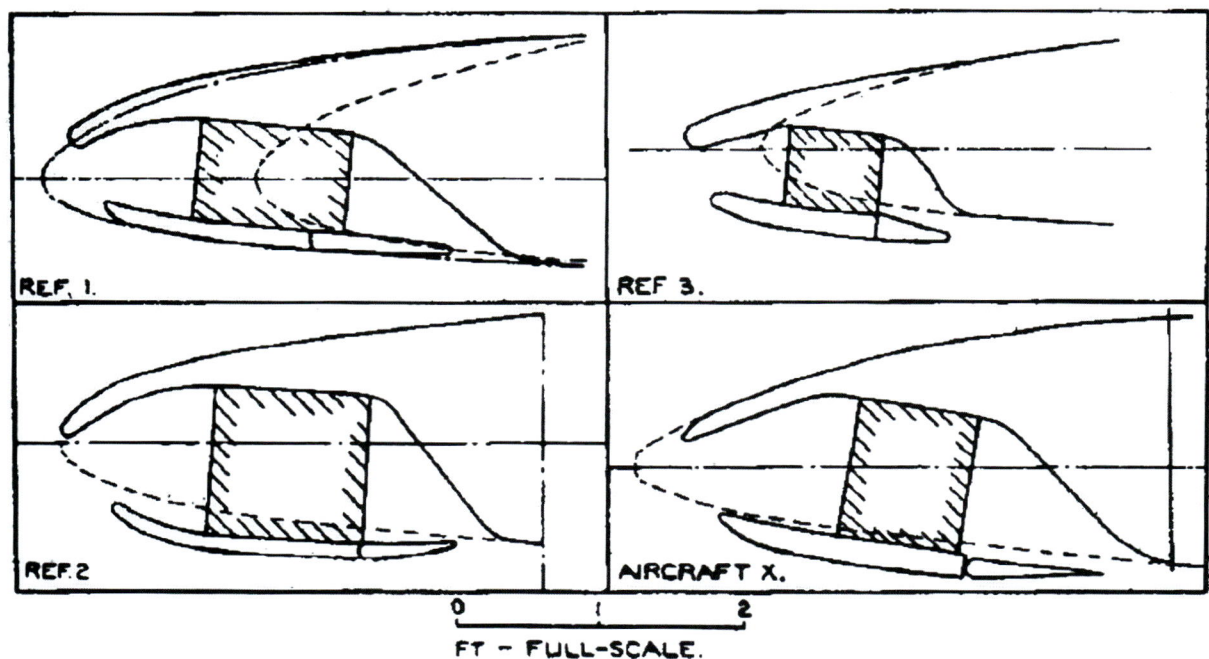

Cross-section through a selection of wing leading-edge radiator installations, from a contemporary Royal Aircraft Establishment report. The radiator core itself is the hatched block, with the exit area controlled by the lower surface flap. This is typical of the research of Seddon and his colleagues. Clockwise from top left: Mosquito, Tempest I, 'Aircraft X' and Hornet. ARC-R&M-2498

De Havilland Mosquito B.35 TJ138 at the RAF Museum, Hendon. Like the V-bombers, an aircraft conceptualised to use altitude and speed to avoid enemy defences, part of its remarkable performance was attributable to efficient engine cooling conferred by the inboard leading edge radiator installation. John Seddon and others were to link this with the development of gas turbine installations in the immediate post-war years.

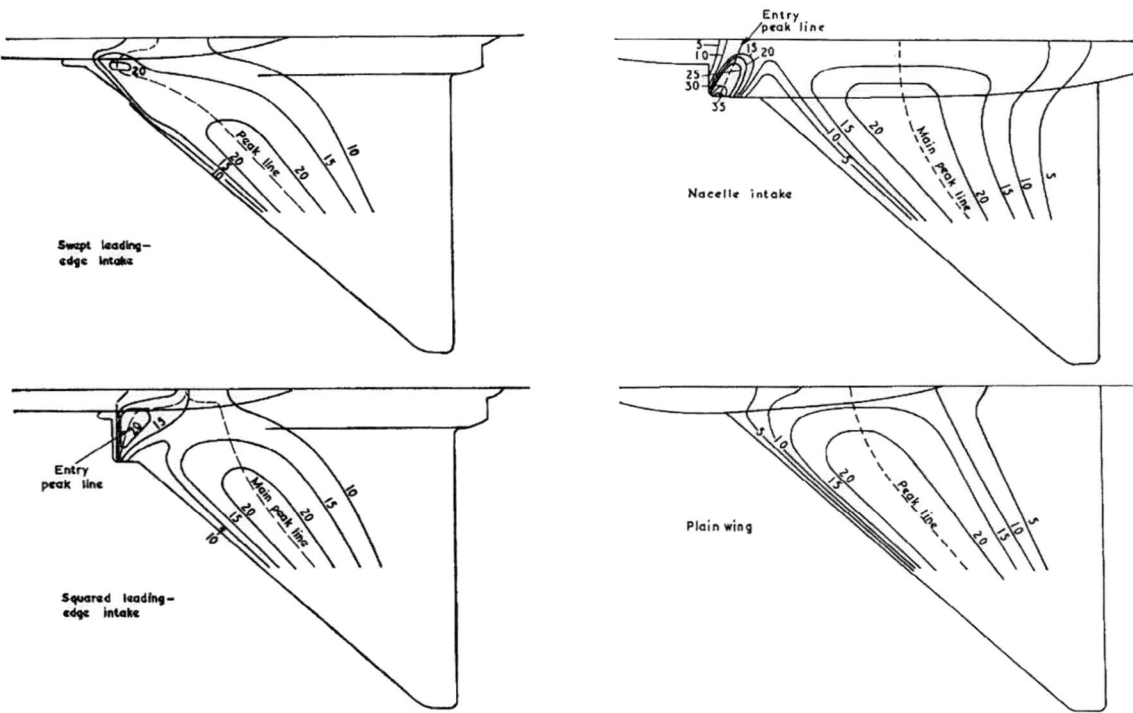

The effect of different inlet configurations on the pressure distribution of a delta wing was investigated by Seddon and Kettle of the RAE. The model was an aircraft of 50° leading edge sweepback, Aspect Ratio of 3 and full-scale span of 52ft, all of which correspond to the basic form of the Gloster Javelin, but are representative of the Vulcan prototype too. The striking result is for the swept leading edge inlet. This configuration bends the peak suction line forwards towards the aircraft centreline, which is favourable for maintaining aerodynamics sweep. The other configurations have the opposite result. ARC-R&M-3353

"In the design of swept-wing aircraft for flight at high subsonic speeds it is important to minimise the proportion of wing covered by the body and nacelles, because of a loss of effective sweepback which normally occurs on these items and in their junctions with the wing. One way in which this can be done is by the use, in suitable cases, of leading-edge air intakes with buried ducts. Wing-root intakes, in particular, provide an opportunity not only to reduce the amount of wing covered, but by modifying the root sections, to correct the body-junction effect also."

In the relevant RAE tests, compared to fuselage side nacelles (which the pitot inlets were essentially a form of) the squared-off inlets were found to provide an effective sweepback improvement equivalent to 10° over the inboard 10%-30% semi-span, while a similar arrangement retaining the leading edge sweep of the wing for intake lip increased this increment to 22°. The result was an increase in critical Mach number from 0.85 for the squared off inlet, i.e. below the specified cruise M, to 0.94 for the swept leading-edge inlet. This was not for free; the ram pressure recovery was reduced by nearly 5%, but it was also shown that ducting the fuselage boundary layer away from the inboard wall recovered most of the deficit. At the point of design freeze therefore, the swept leading-edge inlets had been adopted and would incorporate a boundary layer splitter and inboard bleed duct.[100]

Later RAE work specifically related to the developed Vulcan configuration noted concerns with the quality of flow in the outboard ducts in particular, given that,

"tests on engines with axial flow compressors have shown that it is important to obtain a fairly uniform velocity distribution at the compressor inlet."[101]

The wind tunnel tests involved keeping the flow in one duct roughly constant, while varying that in the other. When similar flow to both ducts was targeted, as would be the case in the vast majority of conditions other than engine starting, it was found that the outboard end of the inlet left much to be desired. Flow separation was seen, with a consequent distorting effect on the pressures in the plane of the compressor face.

Attempts to cure this by using an extended central splitter simply reproduced this problem in the inboard

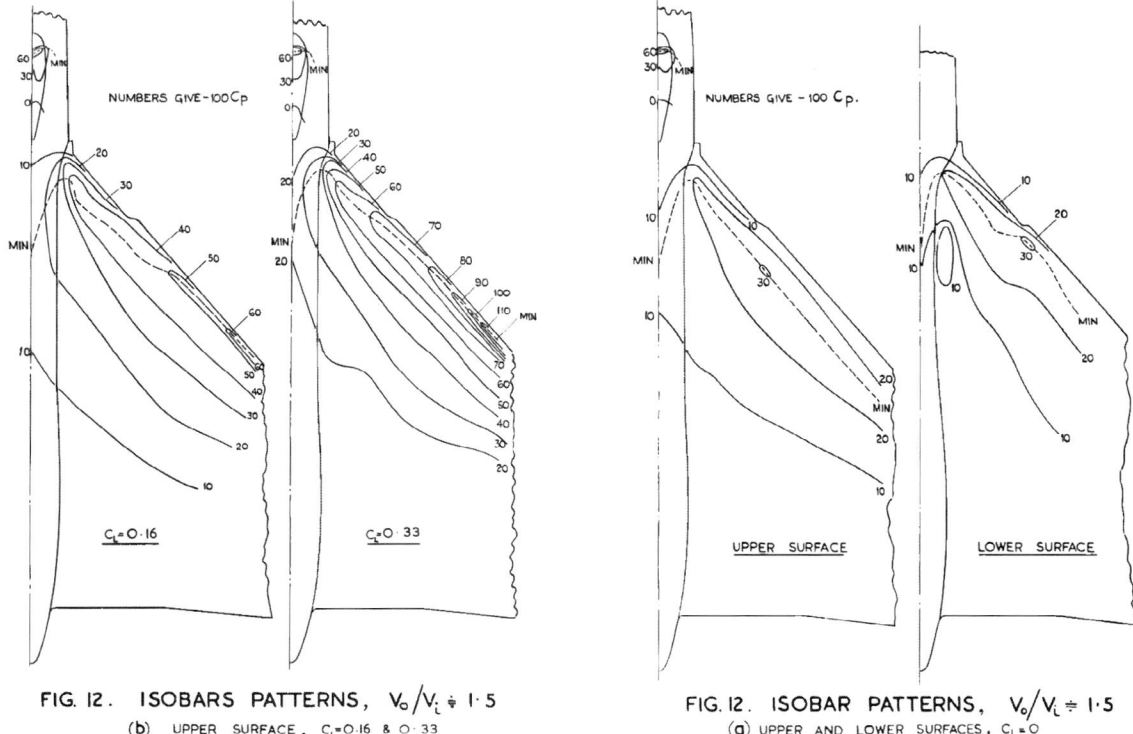

Surface isobar patterns obtained at the RAE through pressure plotting analysis, on the 698 prototype configuration. The upper and lower surfaces are shown at zero lift, while CL = 0.16 and CL = 0.33, which straddle the intended cruise condition, are shown for the upper surface only. All plots show just how effective the revised, thickened inboard wing design was in carrying the isobar sweep all the way into the fuselage. It can also be seen that the intakes were a positive influence, reducing the peak pressure locally. RAE TN AERO 2416

duct too; a more positive effect was induced by cutting an auxiliary inlet slot immediately aft of the outboard inlet. Varying the flow velocity between the ducts to replicate starting showed that the outboard flow was improved when the inboard duct flow was reduced; this was attributed to the exaggerated inclination (and so loss through turning back onto the duct axis) caused by the latter. The RAE was able to conclude that starting an outboard engine first was likely to be more successful, but it is interesting that these results also provided an indication that the loss of an engine on one side was likely to result in at least a temporary reduction in thrust from the other. This was fundamentally the experience and assumption of Vulcan crews throughout its service life.

Control effectiveness through the flight Mach number range was also investigated in the high-speed wind tunnel, leading to an incidental planform change. Early studies show pointed tips, i.e. a true delta planform with tip chord reducing to zero. The RAE tests showed the benefit of finite tip chord (with a taper ratio of 0.115), particularly in the situation of control deflection. For a given lift coefficient, the local lift at the tip was reduced, alleviating tip stall and improving the linearity of the pitching moment characteristics. This modification was an improvement under cruise conditions and was therefore carried forward.[102]

The question of the airbrake layout was also assessed at Farnborough, concluding that a rear upper surface position would be far from optimal, causing,

"...substantial changes in lift and pitching moment, and appreciable losses of lateral stability and elevator power... Increases in the drag are obtained if (a) the flaps are moved forward on the wing and (b) there is a gap between the flap and wing."[103]

This was quite possibly welcome news, as the greater depth available further forward on the wing was much more conducive to stowage and packaging of an airbrake design, particularly if it could no longer be a simple hinged flap. The latter point was taken full advantage of as the design progressed, with a concept in which the flat surface of the airbrake was held aligned with the freestream flow, while it extended on struts. It could then be rotated to allow a variable drag increment, but crucially it was held far enough away from the wing surface to not immediately disrupt the boundary layer, leading to lift reduction and a change in trim.

The aligned state during deployment was important too: a trim change when the airbrake was in its final position was one thing, but the potential for rapid changes as the device extended was perhaps more worrying, given how quickly it would need to be dealt with by the crew. The final solution was therefore elegant and effective.

The changes in the configuration between the proposal of April 1948, by which time the Avro 698 had moved away from its initial flying wing configuration to a thinner delta wing and fuselage, to the final design were drastic. The decision to pause work on the 707A to ensure that it was representative can be seen as a logical decision, even if by definition it brought the test aircraft's first flight closer in time to that of the 698. However, events would show just how important this call would turn out to be.

As the high-speed test aircraft, the 707A needed to be able to fly at Mach numbers beyond the planned cruise speed of the bomber, and so was stressed to M = 0.95 and an indicated air speed of 415kts, compared with the 300kts of the original 707. As well as the highly developed (and final) wing shape, it would use the leading-edge inlets designed for the 698, albeit with internal ducting to suit the single Derwent VIII. In keeping with the attempt to faithfully replicate the 698 shape, the control surfaces adopted the final design, meaning that the elevators in particular were reduced in span. In small inboard region that would be occupied by the jet pipes of the 698, an emergency dive recovery flap was fitted; a nod to the aircraft's role in opening up the unknown parts of the flight envelope. As described earlier, the nose featured an ejection seat and had already flown on the modified 707B.

The 707A would demonstrate the existence of another problem that had been anticipated, but that would certainly need to be solved for the 698. In the words of Tony Blackman, Avro chief test pilot from 1956, "the aircraft exhibited an uncontrollable, although fortunately not divergent short period pitching oscillation at high Mach numbers."

What did this mean? In addition to its longitudinal static stability, which will cause the aircraft to return to its original incidence when disturbed (and hence by maintaining constant lift, drag, weight and thrust, continue at the same speed and altitude), it must also show dynamic stability. Static stability requires what might be thought of as stiffness in pitch, as if the aircraft were a mass and spring system. When moved from its equilibrium position, so compressing or stretching the 'spring', the stiffness of the latter will return it. However, the mass-spring system will potentially overshoot the target, setting up an oscillation, in this case in the pitch angle of the aircraft. What is also needed is reaction to the rate of pitch oscillation that the spring has caused; the system needs damping as well as stiffness, to be statically and dynamically stable. Static stability is assured by satisfying the two conditions of having the aerodynamic centre of the aircraft behind the centre of gravity, together with a positive pitching moment at zero lift. Dynamic stability is, unsurprisingly, rather more complex.

The reaction to a disturbance, whether from a gust or pilot input, could be described mathematically and to some extent predicted. The maths predicted two responses. The first took place over a relatively long period, tens of seconds or even minutes. Imagine a gust caused a small increase in the aircraft's airspeed. This would cause in turn an increase in lift, such that it was slightly greater than weight and the aircraft would climb. However, the increase in lift would result in greater lift-dependent drag, slowing the aircraft, reducing lift and consequently reducing altitude. This so-called Phugoid Mode is typically not well damped, but is easily controlled by the pilot. In its own studies for a B35/46 aircraft, the RAE estimated a period to half amplitude of the order of 400 seconds at the required cruise altitudes and speeds.

In contrast to the Phugoid and its change in pitch angle at roughly constant incidence, the short-term response to a pilot elevator input is a movement of the nose and so incidence, but also an accompanying reaction due to the aircraft's natural weathercocking into the current wind. There is therefore a second response that, due to its nature as a high frequency response, is known as the Short Period Mode.

Unlike the Phugoid, the response is usually well damped by the aerodynamics of the aircraft. Its frequency might be too rapid for the pilot to control without causing their own Pilot Induced Oscillation (PIO) effect and rapid divergence, but because it could be understood at the design phase, it was possible to provide sufficient damping to avoid this. The inherent damping of the Short Period Mode in earlier aircraft was so well expressed that the RAE would comment,

> "Until aircraft flew transonically [it] was normally so good that the derivatives governing it had been studied but little. Recent flight experience revealed a serious loss of damping in pitch of the short period oscillation at transonic speeds, and resulted in increased interest [in them]."[104]

For a given flight condition, the damping was a function of vertical velocity (heave), pitch rate and inertia. What became apparent through empirical testing by the early 1950s, was that for a select group of wing geometries, as the aircraft became transonic the damping due to pitch rate term would change sign. The conclusion was that for, "wings of moderate aspect ratios and leading-edge sweepback of the order of 55° or less…" the term became destabilising, but the addition of a tailplane suppressed this effect.

For lower aspect ratios and greater sweep angles, the loss of damping did not occur, meaning that the 60° sweep of the English Electric P.1 for example demonstrated good behaviour though the transonic region, even when tested without a tail in the wind tunnel. The Avro 707 (and of course 698) had landed firmly in this troublesome region and a solution had to be found. A tailplane might have been one (and in fact was conceptually designed as a contingency for the bomber), but would have negated some of the aircraft's weight and drag advantages.

The resolution that Avro settled on was as radical as the delta concept itself—automatic stabilisation would be used in both pitch and yaw axes, to provide the necessary damping. Louis Newmark Aviation Ltd, an offshoot of a watch company that had moved into the field aircraft avionics systems, would eventually provide the pitch damping equipment in the production Vulcan.

High speed wind tunnel tests had a much better chance of capturing the static stability behaviour of a transonic aircraft, and the RAE's in-depth assessment of the redesigned Vulcan configuration had much to say about this too. It was found that the now uprigged trailing edge (achieved by redatuming the full span control surfaces) provided a nose up pitching moment that successfully counteracted the otherwise nose-down moment that the wing would generate at zero lift.

For a pair of sensible centre of gravity positions, the elevator angle to trim varied little up to $M = 0.9$, but then an increasingly nose up trim was required as the aerodynamic centre moved rearwards. This was the opposite of the usual control orientation, in that the pilot would need to pull nose up to trim at a higher speed, rather than pushing the stick forwards. Consequently, as Ken Newby's report observed,

> "At higher Mach numbers however, though the aircraft is stable with respect to changes of CL at constant Mach number… it is unstable with respect to changes of Mach number… hence an increase of Mach number will cause the aircraft to pitch nose-down unless the pilot applies more up-elevator."[105]

The report went on to describe succinctly why this characteristic might prove more than an inconvenience to the 698's crew:

> "The variation of drag with Mach number above $M = 0.87$ might have some limiting effect on this instability, but for an aircraft such as the Vulcan, the component of the aircraft weight, acting along the flight path in a shallow dive, could be of the same order as the thrust available from the engines at high altitude.
> The main disadvantages of this instability are:
> (i) Pilot fatigue when cruising for long periods at high Mach numbers.
> (ii) Difficulty of maintaining a constant altitude and speed during a bombing run at high Mach number.
> (iii) The possibility of exceeding the design Mach number of the aircraft should the pilot's attention be distracted."

The bomber's mission implied a need to cruise for some hours at a Mach number close to where this longitudinal instability might be found. Even for an autopilot, it would be necessary to have reliable, predictable and ideally stable behaviour to make the control system viable. The comments on the aircraft's ability to perform as an accurate bombing platform were surely not made lightly, given the significance of the statement and the respect in which the institution making them was held. That said, the problem was glaring: wind tunnel results that might be at worst slightly pessimistic, strongly suggested that at a few units of Mach above the aircraft's design cruise point, longitudinal static instability would be encountered. Avro had already proposed a solution though, which explains the lack of panic in the narrative.

> "One method which has been suggested by the firm for overcoming this instability, is the use of an automatic pilot unit to increase the elevator deflection (independently of the control column movement) by say 1/2° for every 0.01 increase in Mach number above about $M = 0.87$: this would stabilise the aircraft, and make the control column move forward continuously with increase of Mach number for trimmed flight at constant altitude".

Consequently, the presence of both static and dynamic longitudinal instabilities close to the required cruise conditions were to be addressed by the use of autostabilisation

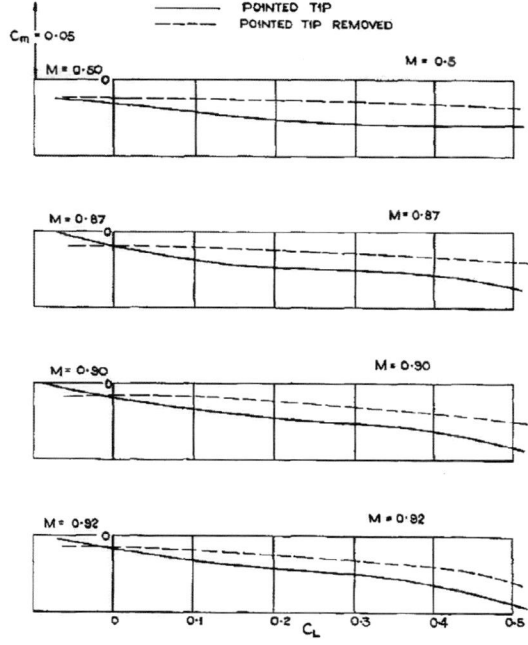

The effect on pitching moment of the cropped tip, as lift changes for a range of Mach numbers. A negative pitching moment pushes the aircraft nose down; for a trimmed condition this is balanced by elevator deflection. Lift in the cruise required a Cl of about 0.25; in all cases the dashed line of the cropped tip trace is flatter, being almost invariant at M = 0.5 and close to linear at the design cruise Mach number of 0.87.

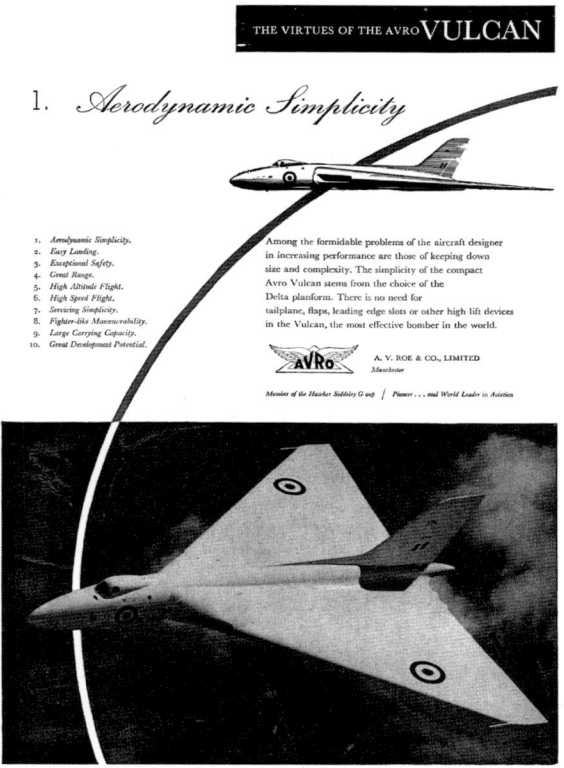

Avro advertisement of 1953, extolling the virtues of the delta wing and featuring the prototype Vulcan, VX770.

systems: a Mach trimmer to apply steadily increasing up elevator for the former, and a pitch damper that would control the elevator to react to a sensed pitch rate for the latter. Had Avro misunderstood the requirements of the delta wing design, to find themselves in such a predicament? There are several ways of answering this question, and it is fair to say that autostabilisation was not an intrinsic feature nor an expectation from the start of the programme.

On the other hand, it was abundantly clear that there would be unexpected problems to solve along the development path, and the simplicity of the delta layout was a way of avoiding a large tranche of known problems, leaving capacity to deal with the unknowns as they arose. In the case of the longitudinal stability issues, Avro deployed a new and advanced technology, exactly as they should have done. Inadequate longitudinal damping and consequent stability issues was a feature of the Northrop flying wing designs (the B-35 and YB-49) in the United States, and would almost certainly have been found by Handley Page if Lachmann and Lee's proposals had been taken up. They too would have taken the advantages, such as they were, and been forced to tackle the problems as they found them.

The first phase of test flying with the prototype VX770 showed that the short period mode was indeed felt on the full-scale aircraft at altitude, at an Indicated Mach Number (IMN) of 0.96 and above. The 698 had a large and increasing error in IMN with Mach; a peculiarity of the wing-tip mounted pitot probes, it would largely disappear when these were moved to the nose later. The IMN was equivalent to a True Mach Number (TMN) of about 0.89, so within the cruise Mach requirement.

The vital clearance of the autostabilisation systems was a major undertaking for the second prototype Vulcan, which in the winter of 1955-56 would fly more than 100 hours of appropriate tests. Five decades later, the presence of such systems in the control loop was one of the reasons that the Vulcan was designated as a "complex" ex-military type by the UK Civil Aviation Authority, leading to the need for the original equipment manufacturers to be involved in the successful effort by the Vulcan to the Sky Trust to return XH558 to the air.

The major redesign which commenced at the end of 1949 incorporated three substantial changes, using the understanding of three-dimensional transonic wing design that was being concurrently generated. Notably, these all featured inputs to varying degrees from the RAE into Avro's design team. The variation of thickness inboard

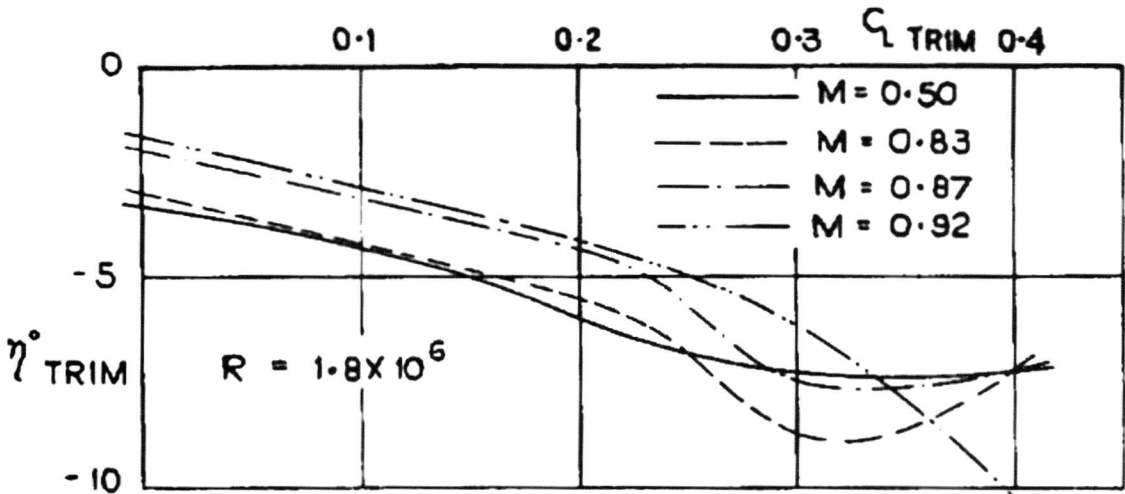

FIG. 20a. Elevator angle to trim.

Elevator angle to trim, against aircraft lift coefficient. Each curve is for a fixed Mach number, with increasing lift (via angle of attack) from left to right. For longitudinal stability, the curve must slope downwards, indicating that to increase the angle of attack, more up elevator is required to trim. If disturbed and pitched up, the elevator angle will be insufficient to hold, and so the aircraft will pitch down again, automatically maintaining lift and speed. However, as the Mach number increases, the curves become flat (neutrally stable) or positive in gradient (unstable). For the Vulcan's M = 0.87 cruise, a maximum stable trimmed lift coefficient of about 0.3 was indicated, which was insufficient for an aircraft expected to need 0.2 to 0.3 for 1G flight. The result for M = 0.93 appears anomalous, as the reversal of the curve does not occur; this does instead hint at what is happening. The tests found that tip stalling was helped by up-elevator, while as Mach number increased, the effect of shock control by the elevator become dominant. It was postulated that the M = 0.93 curve would be typical of flight Reynolds numbers in the cruise; rather than becoming unstable, it would require increasing up elevator to trim, eventually run out of authority as Mach increased beyond a limit. NEWBY, ARC R&M 2999

Looking inboard from above the wing of Vulcan B.Mk.2 XH558. The very forward position of maximum thickness as the wing blends to the fuselage is clear, caused by the bespoke section in that region. The Inboard region of the B.Mk.2 wing was geometrically similar to the B.Mk.1, although the lower lip of the inlet was deepened for increased mass flow.

should not be considered as simply a wing modification; instead, it was an example of the new paradigm of planform effects of the whole aircraft becoming critical.

No longer would it be possible, as had happened in the case of the Lancaster, to consider the aerodynamics of the wing and fuselage as essentially separate. The impact of the interfering pressure distributions of the swept wing and fuselage, together with the geometric unsweeping across the central plane, was very important. There was little use in carefully designing the wing to postpone drag rise if the critical Mach number of the central region swamped the gains. Secondly, the remarkable integration of the engine inlet region in supporting this function, while at the same time delivering high total pressure flow with little compressor face distortion to the buried engines, was a major success.

Finally, the investigation of the effect of the wing tip geometry on control power directed attention to the pitfalls of relying upon the pointed tip. Its short chord and high load gave it a propensity to stall. If that were symmetrical—an unlikely best case—then the sudden loss of lift at the rear would pitch up the nose, into a worsened stall for the whole aircraft perhaps. More likely was an asymmetric event, with uncommanded pitching and rolling moments leading to spinning. It was a natural conclusion that the wing area required must be provided with a finite tip chord, while Avro would not be the only B35/46 manufacturer that would shudder at the potential loss of directional control through stalling of fins near the tips.

The revised conceptual design was issued in February 1950, with detailed component drawings starting to be released by a team of up to 190 draftsmen in the following May. This was a precursor to the start of systems test rig and mock up construction that would feed the ultimate building of the two prototype aircraft ordered by the Ministry of Supply in June 1948: VX770 and VX777.

THE IMPERFECT DELTA: MAKING THE VULCAN WORK

THE WHITE-PAINTED prototype of the Avro 698 (VX770) first flew under the famously pin-stripe suited command of chief test pilot Roly Falk on August 30, 1952. Susan Kennet, daughter of chief designer Stuart Davies, recalled it as,

> "…a beautiful summer's day and we assembled in front of the Club House at Woodford to watch this historic event—namely, the first flight of 698 as she was called. She did look rather lovely as we waited for her to move."[106]

Falk was the sole pilot (indeed occupant) of the vital prototype that day. His career to that point had been remarkable in its unorthodoxy. A pilot prior to the Second World War, he had flown charters for the press covering the conflict in Abyssinia (the announcement of the opening of which had interrupted Busemann's lecture on swept wings at the Volta Congress) and then the Spanish Civil War. Becoming an Air Registration Board test pilot in 1938, his skills were of course in high demand with the coming of war. By 1943 and after serving in the RAF, he was the chief test pilot at RAE Farnborough.

A look at his logbook from around VE Day in 1945 gives a fascinating hint of the variety of tasks he would have found himself involved with. On April 30, three flights in a Spitfire with "tests of large tailplane", which may relate to the efforts to improve the longitudinal stability of the Griffon-powered final variants of the legendary but by then venerable fighter. Two further Spitfire sorties were flown on May 7, with a flight in the Sikorsky XR-4, the first production Allied helicopter in between. On May 16, with the war over and the need to learn as much as possible from the aeronautical technology of the defeated enemy, he flew a Liberator on a return trip to Schleswig. Two days later, he flew back in a Hudson in order to collect a prize of great potential: a Messerschmitt Me 262 B-1a/U1, which he flew to Farnborough via Gilze-Rijen in the Netherlands.[107] This machine would provide important data on high-speed flight, which of course was in short supply in the UK at the time.

David Thirlby, an apprentice in the Avro Flight Research Unit (FRU), was witness to Falk's confidence.

> "[He] said to me, 'The 698 will take off like nobody's business, and if you position yourself at the ring road crossing I should think it would unstick there.'"

After a single high-speed taxy to confirm her readiness, VX770 accelerated down Woodford's runway and, five years after the contract had been signed, became an

Pages from Roly Falk's log book in April–July 1945. Among the wide variety of types flown, the German Messerschmitt Me 262, Heinkel He 162 and Arado Ar 234 jet aircraft are notable, but perhaps one of the rarest types in any British log book must be the Alexander Lippisch-designed Me 163 Komet. COPYRIGHT JOHN FALK, USED BY PERMISSION

The Avro 698 prototype VX770, at Farnborough in 1952. As can be seen here, this was the only Vulcan not to be fitted with the visual bomb-aiming position underneath the cabin. At the time, it fitted with Rolls-Royce Avon engines. JOE BARR COLLECTION

aeroplane. Thirlby noted that she took off where Falk had predicted.

Falk flew VX770 for about 30 minutes before making the 698's first landing, back at Woodford in front of those who had designed and built her. The flight would have been largely uneventful—a significant result in itself—were it not for the loss of both main undercarriage bay rear doors, which were seen to fall away close to the airfield. Just days later in early September, the aircraft appeared in public for the first time at Farnborough.

For one week annually the Hampshire airfield was the venue for the SBAC airshow, which in the early 1950s was a celebrated showcase of the prowess of British (and only British) technology. On a more permanent basis of course, it housed the Royal Aircraft Establishment. The 698 was far from the only new and futuristic British aeroplane to be demonstrated that year but its dramatic triangular shape, sheer size and startling performance made it exceptionally newsworthy. The *Aeroplane* magazine recorded,

> *"In view of the 698's vast size and unorthodoxy, it was remarkable to learn that Roly Falk was the sole occupant of the new bomber which, when viewed in the circuit, had the appearance of a species associated with submarine life. It resembled, in fact, a giant stingray so common in tropical waters.*

> *Although the 698 had done less than three hours' flying when it first appeared at Farnborough, Falk handled it with what may be termed abandon."*[108]

The limits of this technology were ever present though; the week is also remembered for the breakup of the de Havilland DH.110 prototype and the loss of its crew, together with 29 spectators on the ground. These would remain the most recent spectator fatalities at a UK airshow until the 2015 Shoreham Airshow crash, 11 people on the ground being killed when a Hawker Hunter T7 crashed. This was, coincidentally, the year of the final display appearances by a Vulcan. The 1952 crash highlighted the profound difficulty in building aircraft that could be operated safely through their design operational envelope using the tools available at the time.

John Derry, the pilot for the DH.110 on that fateful flight along with observer Tony Richards, had ironically survived through probably the most dangerous period of test flying at DH: the development of the DH.108 Swallow tailless jet, in which he may have exceeded Mach 1 in 1948. Although considered an important high speed test bed, not least for the DH.110 itself, the three examples built would all crash fatally for various reasons.

Development of the 698 airframe had run ahead of that of the definitive Bristol Olympus engines around

THE IMPERFECT DELTA: MAKING THE VULCAN WORK

The clean lines of the Vulcan prototype, in its purest form.

which the aircraft was planned. VX770 therefore flew initially with Rolls-Royce RA.3 Avon power, providing only around 70% of the static thrust anticipated for the initial production aircraft. This obviously limited the extent to which the flight envelope could be explored, but by July 1953 VX770 had graduated to Armstrong Siddeley Sapphire engines (which were also the design powerplant for the rival Victor), to be followed rapidly in the September by the second aircraft (VX777), equipped from the start with Bristol Olympus 100s. The latter began the process of finding the limits at altitude and high Mach numbers, unfortunately halted by a heavy landing in July 1954. It fell to the first prototype, catching up in thrust, to confirm that the buffet boundary at cruise was inadequate. Stuart Davies commented with some understatement,

> "With the straight leading-edge of the prototype it was apparent that the performance boundary of the production Vulcan aircraft with later versions of the Olympus engines would be uncomfortably near the high Mach number buffet threshold. It was considered that modifications could be made to the wing planform in the region of the outer portion of the wing."[109]

In fact, this was severe enough to prevent the aircraft achieving a combination of lift coefficient and Mach number that would allow the mission requirements to be met, in terms of speed and height over target, let alone manoeuvrability. Although the issue was not identified during the wind tunnel test programme, it had become apparent in the early testing of the high-speed version of the delta test aircraft, the 707A. In the slightly stronger words of Tony Blackman,

> "...unless this buffet could be cured, the aircraft would be of no use as a bomber."[110]

This was of course a major problem for the programme. The Vulcan was now in production and a solution needed to be found that could be retrofitted to existing aircraft, as well as being incorporated into new ones. While the delivery of a B35/46 bomber to the RAF was critical to national security, the availability of the Handley Page alternative meant that it did not have to be Avro's design, whether or not the first examples were quickly coming together at Woodford. In reality, the buffet boundary was extended sufficiently by the incorporation of a new outboard wing leading edge. This was aerodynamically designed by Newby of the RAE at Farnborough.

The apparent simplicity of his solution, which was certainly true from a mechanical perspective and enabled the rapid implementation of the substantial modification into the production line, perhaps masks its aerodynamic complexity. The resolution is specifically referenced by Küchemann in his (albeit, posthumous) manifesto, The

The sole two-seat 707C, based on the 707A design, was WZ744.

The first of the two high-speed 707A test machines to be built was WD280. Showing here its original straight leading-edge planform, it would play a vital role in the Vulcan's wing development programme.

Second of the 707As was WZ736, emphasising in this taxying image the nose-up attitude that had been found necessary for rotation during testing of the low-speed versions.

Aerodynamic Design of Aircraft.[111] Discussing the challenge of transonic swept wing aircraft,

> "…. no improvements in numerical and experimental design tools are ever likely to dispose of the need for physical insight. On the other hand, a good understanding of the flow phenomena involved has led to successful designs even in the early days when the available design tools were still rather poor. An example of this kind is the modification of the wing of the Avro Vulcan aircraft by K. W. Newby (1955), which effectively converted the original delta wing into a wing where sweep effects were successfully exploited."

It might have been justifiable for the earlier RAE contributors to the aircraft's aerodynamics to feel aggrieved at the implication that Newby alone had made the Vulcan work, while Küchemann himself had been deeply involved in the design of the Victor. A former wartime Blackburn Aircraft apprentice, Newby was a contemporary of two others who would subsequently forge very notable careers in the British military aircraft industry: Alec Atkin and Ralph Hooper, known respectively for their work on the Lightning and Harrier in particular. Hooper would say of their time at University College Hull,

> "I think really Ken Newby set a particularly high standard and the rest of us had to try and be competitive. He did go into the scientific civil service, waste of a good man but there we are."[112]

In order to define the issue encountered by the Vulcan, it seems reasonable to refer to some approximately contemporary sources. Mabey reported in 1964 on wind tunnel studies investigating buffet on several wing planforms, stating that:

> "Wing buffet means the wing response, mainly at its fundamental bending frequency, to the random excitation from pressure fluctuations in separated flow. Wing buffet can limit aircraft performance by producing unpleasant vibrations for the aircrew and/or passengers, by disturbing sensitive equipment or even by endangering the structural integrity of the aircraft. Wing buffet often occurs before stalling or longitudinal instability and hence buffet boundaries are as important as stability boundaries."[113]

Buffet could be encountered by any aircraft in which regions of separated flow existed due to its fundamentally unsteady nature. However, the significance to the then-new transonic aircraft being developed was the existence of a new mechanism by which separation was triggered. The local regions of supersonic flow on the wing inevitably terminated in a shock wave; the pressure rise across this would in turn result in boundary layer separation when

Avro 707A WZ736 was assembled at Bracebridge Heath, near Lincoln, and carefully towed down the A15 to RAF Waddington for flight testing, in February 1953. This aircraft was intended primarily for general research work on behalf of the Ministry of Supply, and would play a role in automatic landing system development with the RAE.

The celebrated SBAC Farnborough Airshow formation of 1953, in which the two Vulcan prototypes were joined by four 707s. From left to right, 707A WZ736, the sole 707B VX790 is identifiable by its pointed wingtips of the pre-1949 design, the two seat 707C WZ744 and the first 707A, WD280. The second Vulcan, VX777 leads the first VX770.

The wing-mounted, conventionally hinged and vented airbrakes on the Vickers Valiant. This concept inevitably caused a flow separation behind, which had the potential to trigger buffet either locally or on the rear parts of the aircraft in the wake's trajectory. This situation was avoided on the Vulcan by moving the airbrake surface away from the wing, and on the Victor by mounting them on the extreme rear fuselage.

it became sufficiently large. This was alluded to by Seal in a report published in 1959.

> "Two types of flow separation give rise to buffeting, firstly, the actual flow separation over the wing itself and, secondly, premature flow breakaway, which may occur at the wing-fuselage or tailplane-fin junction, for instance. Although it is unlikely that flow separation can be completely eliminated it may be postponed and alleviated by good design … Decreases in thickness-chord ratio and aspect ratio and increase in sweepback all tend to alleviate buffeting at high Mach numbers and moderate CL where compressibility triggers off the vibratory flow. In this region sweepback can reduce the buffet amplitudes by as much as 50% to 70% as compared with a similar straight wing aircraft and it may also delay the onset of buffeting to a higher Mach number."[114]

While Avro had encountered problems with buffet before, notably in the development of the Tudor airliner, these had typically been related to the second type of separation highlighted by Seal. The new challenge was the mitigation of flow separation specifically associated with shock-induced flow separation on the wing—an example of Seal's first type, but with a novel cause.

The high-speed buffet problem involved the interaction of a flow separation and the structural dynamics of the aircraft. Irrespective of how these might have been modelled during design, it would have required both of these factors to have been resolved correctly in order to either avoid (in the case of the aerodynamic issue) or bound them (applicable to either positioning the structural modes or ensuring the aerodynamic issue was shifted out of the planned flight envelope). In 1950 when the Vulcan design was frozen, this was simply impractical and helps to explain the significance of the scaled flight test programme that was instigated for both the Avro and Handley Page proposals. Indeed, nearly a decade later an RAE report would note that no method had yet been developed that could accurately predict the onset of buffet or the magnitude of the loads generated.

Vortex generators on the fin and tailplane of Vickers Valiant XD818, the only surviving complete example. Now preserved by the RAF Museum, in 1957 this aircraft dropped the first British H-Bomb, near Christmas Island as part of Operation Grapple.

In terms of detecting the flow separation in the wind tunnel, the only facility that was available to provide the required Mach number was the 10ft x 7ft high speed tunnel at RAE Farnborough. This could have shown the presence of a steady shock-induced separation, providing a likely upper limit on Mach number before buffet might occur. The models built to survive in this environment were not necessarily dynamically representative, so could not show how the real aircraft would statically deform and dynamically move, changing both shape and incidence in a time variant manner. Also, it could not simultaneously provide Reynolds number similarity, which was important for a boundary layer effect. There was a level of confidence that increasing the Reynolds number would postpone problems to speeds beyond the cruise condition.

Some calibration of the Vulcan's development issues against the state of the art can be made by considering the contemporary development of the other two V-bombers. The Valiant had the benefit of a less rigorous specification, but the associated challenge of a compressed development timescale. G. R. (later Sir George) Edwards described it as, "far and away the hardest aeroplane that I ever did", making that further point that it was, "… given no mercy because we could not build flying scale models; it had to be right first time".[115]

Buffeting of the tailplane was encountered on the Valiant prototype, when the wing-mounted airbrakes were extended. Clearly this was the effect of wake impingement rather than local shock-induced separation, but nonetheless was unexpected despite the extensive wind tunnel testing that had already taken place. It was temporarily eliminated by deletion of the upper surface airbrakes, which was not a long-term solution due to the reduction in drag increment and increase in nose-down pitching moment caused by the circulation from the lower surface pressurisation.

Tests of alternative arrangements to reinstate the upper surface devices were undertaken at Farnborough, with the report noting,

"No reliable guide was available in the model tests for indicating whether the brakes would cause buffeting … "[116]

At high M (>0.8), the effect of the upper surface air brakes was perceived as likely to cause shock induced

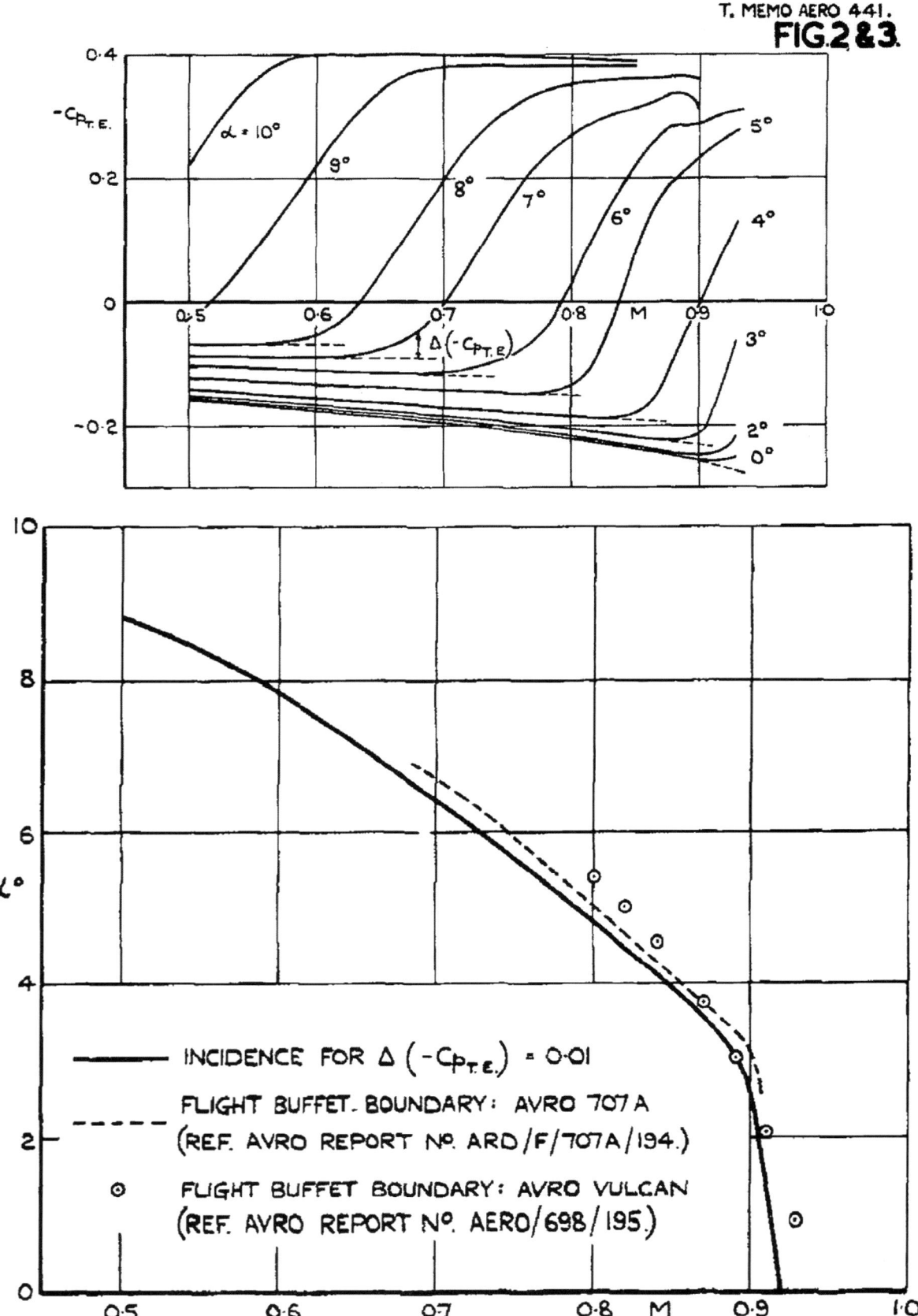

Corresponding data for the Vulcan baseline configuration: a) mapping of separation boundary from WT tests and b) comparison of WT Separation Boundary with inflight measured Buffet Boundaries for baseline Avro 707A (WD280) and Vulcan prototype (VX770). NEWBY, 1955

separation and consequently wing buffet, due to the exacerbated adverse pressure gradient. Again though, this could only be inferred, "in the absence of any measurements of buffeting..." It would seem that others solving similar problems, contemporaneously with Newby's work on the Vulcan leading edge (early 1955), also looked to improve a surrogate parameter for buffet in the wind tunnel.

Flight test of the first production Valiant aircraft revealed an unrelated buffeting of the tailplane and fin at high M and EAS conditions, cured by improved sealing and fitting of a vortex generator array. Although these fixes were relatively simple to incorporate, it does still emphasise how far through the programme the issue had remained hidden. More concerning perhaps was suspected buffet-induced fatigue failure of an aileron control rod at high Mach and with applied G-loading, indicative of manoeuvring under typical cruise conditions. This was addressed both aerodynamically and structurally; a strengthened rod and another row of VGs ahead of the aileron.[117]

Wind tunnel tests on the closely related V-1000 wing had shown that a separation boundary associated with outboard shock formation existed for moderate incidences at cruise M, which was correlated well to changes in the lift and pitching moment slopes. These were, however, steady measurements on a rigid model, again providing a local upper bound (assuming Re similarity) but without quantifying the dynamics of the separation and assessing the structural response.[118]

Although not an issue at high speed, the Victor suffered from an analogous low speed aeroelastic problem, which would manifest itself with fatal results. The aircraft's characteristic fin and large dihedral tailplane evolved at a similar time to the final Vulcan configuration, in 1949. The Head of Aeroelastics at HP would report that,

"At the time, before analogue or digital computers were available, anything like comprehensive flutter calculation was out of the question".[119]

The approach for validating the design was the manufacture of a dynamically representative low-speed wind tunnel model, with iterations being made between it and completed full scale parts to gradually close in on the expected structural performance. The final inputs to the flutter calculations were from ground resonance tests of the completed aircraft, to identify the modal frequencies.

The timescales for this work are striking: the low speed wind tunnel model began running in 1952, produced useful results and influenced the requirements for elevator mass balancing and the control circuit stiffness. However, the overall Victor programme was advanced to the stage that the first prototype flew in December 1952; clearly, the wind tunnel model was driving retrospective updates rather than the original design. The results of the flutter calculations based on the aircraft ground resonance results were not available until early 1954, over a year after the prototype had flown.

Although it was known that simplifying assumptions had been made in these calculations, including neglecting the tailplane dihedral, the agreement between the results and the wind tunnel model gave confidence that the appropriate trends had been captured. However, in July 1954 and while performing low altitude position error work at Cranfield, the prototype suddenly dived into the ground, killing instantly pilot Squadron Leader Ronald Ecclestone and three flight test observers.

The urgency with which the introduction of the V-bombers was viewed at the time is illustrated by a parliamentary question asked of Duncan Sandys (Minster of Supply) a few days later by Frank Beswick MP. After being asked to make a brief statement on the accident, it is clear from Beswick's follow up where his concern lay;

"While appreciating that expression of sympathy and wishing to associate my hon. friends with it, may I ask whether this unfortunate tragedy will have the effect of greatly delaying the production of this aircraft?"[120]

The 'what' of the accident (failure of the tailplane) was rapidly obvious, given the eyewitness accounts and its relatively intact although detached status. The 'why' took considerably longer; several months and the availability of a complete full-scale tail unit for testing. These tests eventually revealed a level of flexibility about the fin to tailplane junction that had not been modelled, while the aerodynamic effect of the dihedral had not been applied to the ground resonance-based calculations. The proximate cause was the induced fatigue failure of the tailplane attachment bolts.

The accident to Victor WB771 was not caused by high Mach number effects but does serve to illustrate the profound difficulties at the time in analysing aerodynamic-structural interactions of any kind in a timely manner during the design cycle. Although the Vulcan Phase 2 leading edge would be one of the most visible 'fixes' applied to a V-bomber, it is nonetheless apparent that the requirement for such a retrospective modification

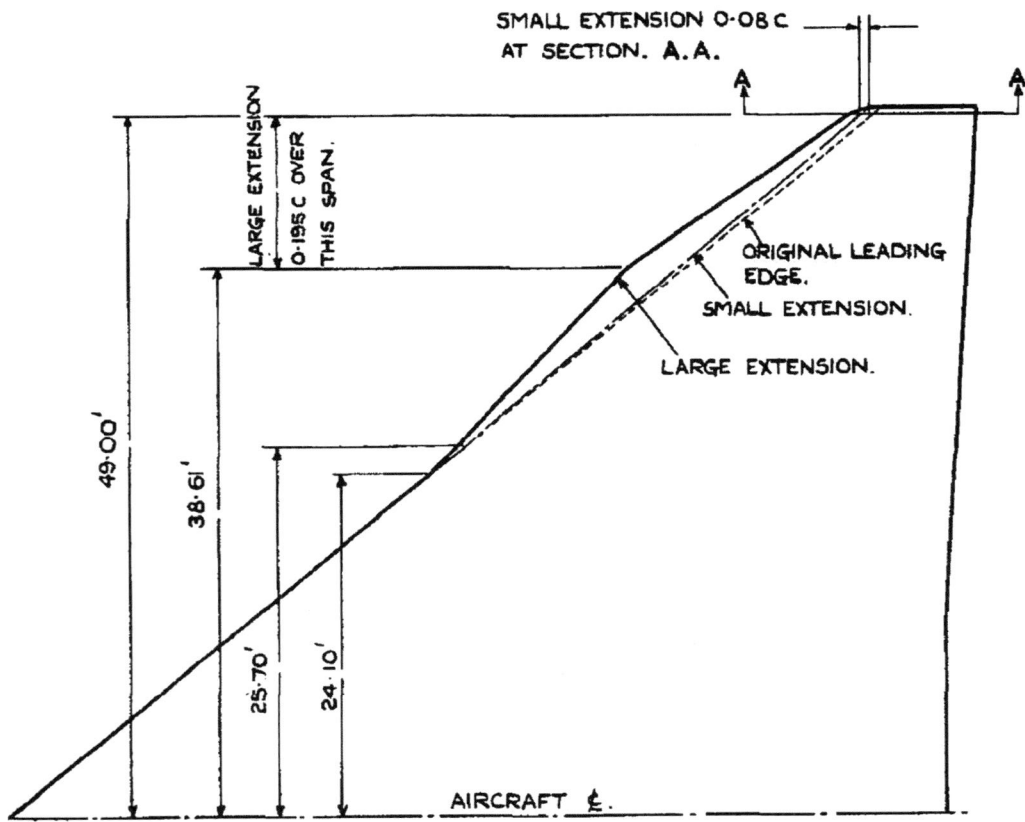

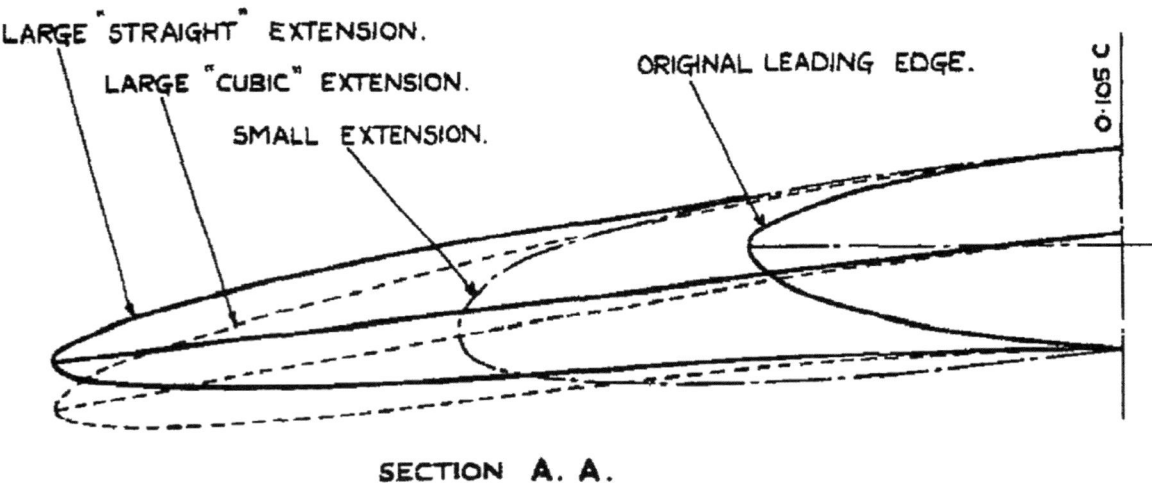

Vulcan wing leading edge modifications as designed and tested by K. W. Newby at RAE Farnborough. NEWBY, 1955

was consistent with the contemporary analysis capability.

In the absence of such analysis, a substantial rig test was essential, and the earliest available 'laboratory' was the first Avro 707A, WD280, flown over a year after the 698 prototype design freeze, in June 1951. The thoughts of Sir George Edwards are pertinent; he specifically highlighted the risk imposed on an equivalent programme at the time by not using scaled flight test aircraft. The known cost of this accepted in advance was an inability to meet B35/46.

In the words of Avro technical director Sir William Farren,

> "We had plenty of warnings, but then we had plenty of warnings of other problems as well. If you set off on such a venture with nothing to help you but a list of warnings, you are never quite likely to start at all."[121]

The initial attempt to solve the problem was through an

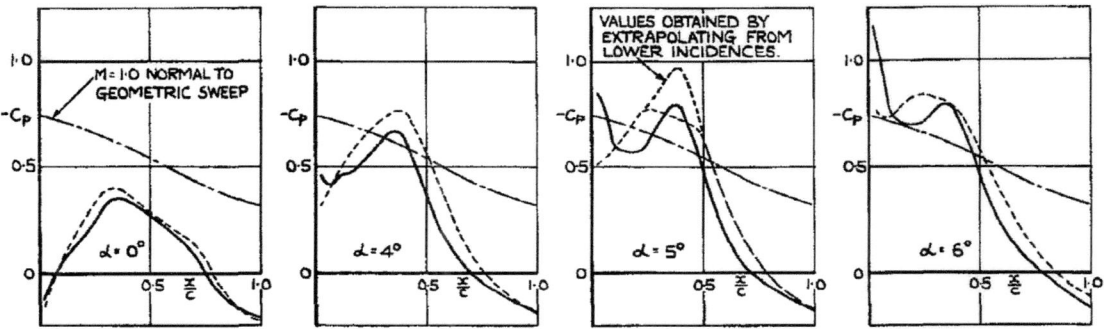

Upper surface pressure coefficient distribution at M = 0.85 with straight (solid line) and cubic (dashed line) extensions. NEWBY, 1955

array of vortex generators on the 707A itself, at 35% chord ahead of the ailerons. This proved ineffective in solving the buffet, although successfully improved shock-induced separation above M = 0.9. However, this did bound the problem: at the M = 0.87 cruise condition, the shock was both ahead of the array and moving rapidly forward. The aerodynamic problem was close to the leading edge and the solution would be found there too.[122]

At the RAE, Newby first had to identify the indicative problem to solve in the wind tunnel where, as discussed previously, the nominally rigid model was not of course subject to buffet. The assumption was made that buffeting was likely when the flow separated at some point on the wing and failed to reattach ahead of the trailing edge. Because this would mean that static pressure, relatively simply measured by a pressure tap and manometer, would not recover to 'attached' level, this steady phenomenon could be detected practically in the wind tunnel. By sweeping through an incidence-Mach space, a boundary could be plotted between conditions potentially corresponding to buffet-free and buffeting at full scale.

Looking back on this period after six decades had elapsed, Pearcey wrote:

"We were using our small-scale wind tunnels to explore the details of the transonic flows that were limiting the flight envelopes of the aircraft then being developed. The opportunity to interact with the full-scale test flying was motivating and provided the all-important confirmation that what we were doing was relevant to the real problems. For example, it was clear that the buffeting that was being experienced was caused by the shock- wave induced separation of the boundary layers and that the flight conditions for its onset could be predicted from observations at model scale of a divergence in the variation of pressure at the trailing edge."[123]

The development of this criteria for relating wind tunnel measurements to the flight buffet boundary was described in a number of reports by Pearcey and his NPL High Speed Aerodynamics section colleagues, roughly concurrently with Newby's work on the Vulcan leading edge. It is not explicit where the idea originated, but given the depth of understanding being generated in terms of shock wave/boundary layer interaction, it is tempting to think that the NPL aerofoil work was the source, filtering in the direction of the RAE. In a paper presented at an NPL symposium in early 1955 and in the text later published as an ARC report, Pearcey stated:

"The onset of the effects of separation on the overall flow for a two-dimensional aerofoil, or for a given spanwise station of a three-dimensional wing, can be detected by observations of the divergence of the trailing-edge pressure from its normal variation, or of that of the static pressure at a point just upstream of the trailing edge on the suction surface."

These conclusions were drawn from previous work at NPL, certainly predating the publication of Newby's report (May 1955), if not the tests described. Farren stated that he believed the technique to have been first described, by the NPL team, in a paper of 1954. Speaking at his RAeS Wilbur Wright Memorial lecture of 1956, he used almost identical data to that published in Newby's Vulcan report, to illustrate the technique. In any case, the combined Avro/RAE date clearly demonstrated at least three relevant conclusions,

* The Vulcan's baseline buffet boundary was very close to the typical α ~ 3.5°/M = 0.875 cruise condition.
* Newby's criteria was indeed a useful surrogate for the flight buffet boundary.
* Flight tests on the 707A would be expected to

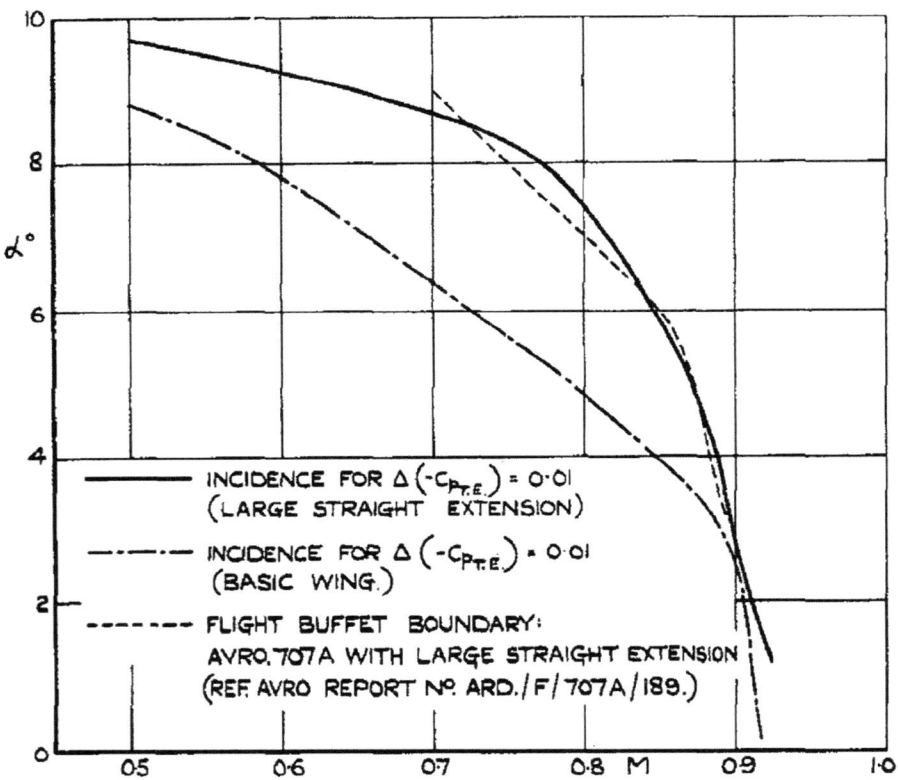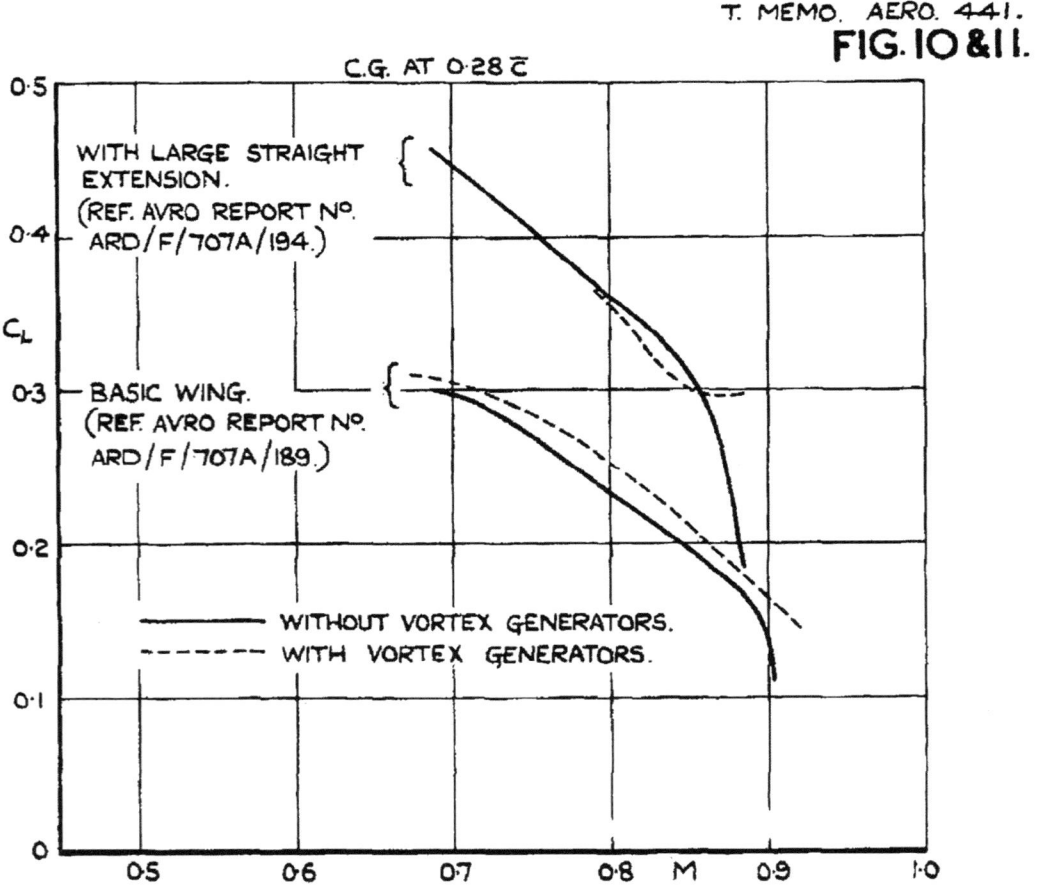

Final configuration flight test results for Avro 707A. Pictured above is a comparison of wind tunnel separation boundaries with flight buffet boundaries against (below) flight buffet boundaries as altered by the addition of vortex generators. NEWBY, 1955

yield results transferable to the Vulcan, allowing a more rapid and cheaper validation.

Any solution was constrained by the need to be retrofittable to the existing primary structure of the Vulcan's wing, which in practice meant a new leading-edge shape that faired into the baseline geometry at approximately 11% chord. This provided the opportunity to tailor the spanwise camber of the new leading edge, improving alignment to the local flow direction and hence reducing the suction peak.

The implied cruise Mach number from the B35/46 specification made localised supersonic flow an inevitability for practical combinations of wing thickness and sweep. Acceleration of the flow close to the leading edge would give rise to a closed "sonic boundary" because, at some point ahead of the trailing edge, compression back to the external subsonic conditions would have to occur. In a purely supersonic flow, this would have happened at the trailing edge itself via a shock wave—a simpler and more predictable state of affairs. However, in transonic flow the presence of the shock at some intermediate position along the chord, variable with Mach number and incidence, would certainly result in wave drag due to the increase in entropy and loss of total pressure across it, but also potentially in flow separation behind it. Referring again to Ramaswamy,

> "However, Pearcey in the UK showed experimentally that nearly shock-free flow over aerofoils could be developed, and significantly, he showed besides, that small perturbations either in geometry or in free stream Mach number would result only in small changes in the flow field. He also gave a physical picture of how the development of supersonic flow depended on the curvature distribution of the aerofoil in that region and consequently how the contour could be modified to support a shock-free flow. This was a crucial and important development which gave a tremendous fillip to the revival of transonic aerodynamic research throughout the world."[124]

The realisation of Pearcey and his colleagues at NPL was that the sonic boundary might offer a method to partially mitigate the effects of the shock wave, by a combination of increasing leading edge thrust through rapid supersonic expansion, and, subsequently, controlled isentropic recompression to reduce the local Mach number and hence the strength of the shock when it occurred.

The stagnation point at the leading edge of an aerofoil in transonic flow (at which the flow is brought to rest and divides to pass around the upper and lower surfaces), would be somewhere on the lower side assuming a lifting configuration. This would give rise to an accelerating flow around the highly curved leading edge to a supersonic suction peak. From this point, further acceleration takes place through weak expansion waves while the surface curvature remains high. It is the reflection of these waves from the sonic boundary, resulting in compression waves, that is the critical mechanism.

By abruptly reducing the surface curvature, the supersonic expansion is halted, limiting both peak velocity and the chordwise extent of the supersonic region. Instead, the low surface curvature causes the relative strength of the compression waves to be dominant and, so long as they do not coalesce, a region of isentropic compression can exist. To quote a report on the NPL work from 1966,

> "The exact form of the curvature distribution, at and after the curvature change, is very important in controlling the rate of the compression and the isentropic nature of the flow. Even when a shock wave does form, its strength can be minimised by a well-designed curvature distribution."[125]

Clearly, while this was true, in the general sense and for the purpose of understanding the technical thread described here, the initial descriptions of a suitable aerofoil geometry from Pearcey's work emphasised a very highly curved (potentially small radius circular arc) leading edge, changing abruptly to an upper surface of minimal curvature. This was a very different profile to either the early NACA symmetrical four-digit aerofoils or the newer high-speed aerofoils of the RAE 10x (Squire) or NACA 6 series.

Two planforms were tested, the larger of which also had two different sections. The small extension was found ineffective at the Vulcan's cruise Mach number in flight tests on the 707A, and so attention was switched the larger extension, which maintained a 19.5% chord increase from the tip to the 78% semi-span point, thereby capturing the 83% peak load position. The new planform then kinked to blend back to the original by about 49% semi-span. In terms of section, the 'straight' extension was based on the leading 30% chord of a 5% thick RAE 101 symmetrical aerofoil, with the axis drooped at 7° from the chord line of the basic wing. The 'cubic' extension cambered the chord line of the extension such that the curvature continuity was achieved with the upper surface of the basic wing at 10.5% chord, which reduced the local incidence but generated a longer region of curvature. Although the source of the information is not discussed, the Newby's report states,

> *"The drag results for most thin wings (t/c = 0.04-0.06) at incidence, show a sudden decrease in CD at about M = 0.8-0.9 when plotted at constant incidence; this suggested that forward movement of the main shock might be prevented when supersonic expansion of the flow round the leading edge and onto the upper surface occurred."*[126]

This convincingly explains the aerodynamic mechanism that Newby was targeting in his design and the reason for the incorporation of the already thin leading-edge profile as a forward extension, as opposed to simply reducing the nose radius of the existing wing. It also implies a reason for the disappointing high Mach results of the small extension. As Pearcey and his colleagues had theorised, the supersonic expansion relied on a significant curvature over a small chordwise extent, rapidly changing to a region of very low curvature behind to allow the isentropic recompression.

Clearly, Newby did not stumble across the 'peaky' pressure distribution concept but was familiar with its application. Simply drooping the leading edge of the existing profiles, which was essentially the approach of the small extension, would not provide a suitable geometry for the desired aerodynamic mechanism.

The comparison between the two 'large' extensions yields further information on this exploitation. Newby illustrated upper surface pressure distributions at M = 0.85 and angles of attack of α = 0° (close to zero lift), α = 4° (slightly greater than cruise) and α = 6° (well above cruise lift). The key was the relationship between local pressure and that which would occur at M = 1, above which the flow would decelerate nominally through a shock wave. Note also that the pressure coefficient at which M = 1 is attained reduces along the chord, as the geometric sweep does due to the almost unswept trailing edge.

At very low lift, the margin to sonic flow was large at all points on the chord, which illustrated in general terms the attractiveness of the strategy of using a large area and low wing loading (allowing operation at low α) for an aerodynamic perspective, as previously exemplified by the Canberra. At a lift coefficient comparable to cruise, it was shown that there was a gradual increase in suction on both designs towards approximately 40% chord, as would be expected ahead of the eventual shock, but that the straight extension has consistently lower velocities (using Cp as the indicator).

The paucity of pressure tappings available on the wind tunnel model meant that the rapid changes in Cp around the leading edge were not completely captured, but what was clear particularly for angles of attack slightly greater than for cruise, was the significant suction peak at the leading edge, followed by a subsequent recompression towards sonic velocity and then an acceleration again. This was much more significant for the straight extension; indicative of 'peaky' behaviour at relevant incidence.

The straight extension was just beginning to demonstrate a Cp.te rise at this α-M combination, whereas data in the report shows that the conic extension was beyond the ΔCp.te threshold and hence the flight buffet boundary. Newby's conclusion was that more rapid leading-edge acceleration (and hence strongly favourable pressure gradient) caused by the continuous curvature of the cubic extension was responsible for a laminar boundary layer being maintained to the shock position, whereas the straight extension was turbulent at the same station. Although evidence was not offered for this, the point is made that the laminar boundary layer would separate more readily at the shock position. This was due to the lack of turbulent mixing and implied reduction in ability to maintain the equilibrium with the shock pressure rise in the subsonic boundary layer flow adjacent to the wing surface.

The straight extension was the clear winner in the wind tunnel test, but the impact of boundary layer transition and knowledge that the Reynolds number was far below that of flight must have been a warning flag. The perils of scale effect at transonic velocities would catch out the engineers involved in several aircraft programmes for some time after this work; one of the more notable examples being the Lockheed C-141 Starlifter.

It was normal RAE practice at this time to allow natural transition in the wind tunnel, rather than forcing at a fixed location using a trip strip, although there was growing evidence that at the Reynolds numbers achievable in the 10ft x 7ft wind tunnel with complete aircraft models, extrapolation to flight scales was far from assured. Newby's own tests earlier in the 698 programme had highlighted the danger of concluding that changes in a laminar boundary layer found in the wind tunnel, but highly unlikely in flight, could lead to very misleading conclusions. In this case, was it sheer luck that the geometric requirements that promoted the 'peaky' supersonic expansion-isentropic compression phenomenon also favoured early transition, thus placing the wing in an aerodynamic regime that was more representative of flight?

NPL's sectional work remained a valuable input to industrially (and strategically) important aircraft design programmes as the decade continued. The 'peaky' pressure distribution was now a flight-validated technology for the mitigation of wave drag in transonic aircraft. New

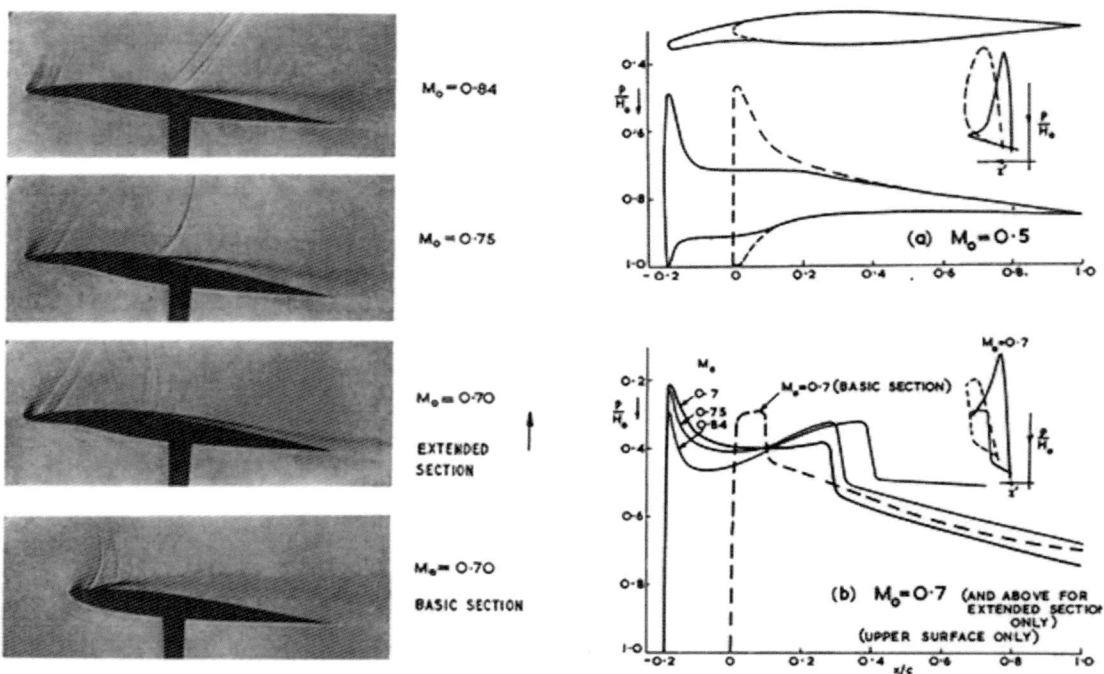

Schlieren images of the baseline NACA 0009.5 and extended chord version (left) at α = 7°, comparative upper surface pressure distributions (right lower) for these conditions and additionally at M = 0.5 (right upper). PEARCEY, IN LACHMANN, 1961

aeroplanes would of course have properly designed aerofoils which inherently exploited this mechanism, but it is clear from the available literature that the Newby implementation was being used to tell the story among the aerodynamic community. This provides a second source of data that shows what Newby inferred but could not observe: that supersonic expansion and rapid isentropic compression certainly did occur in two dimensional experiments on the same sections.

In 1960, for example, Pearcey discussed the conceptual basis of the technique at a conference. The original article makes no direct reference to the Vulcan but uses as a case study an aerofoil of NACA 0009 ½ (sic) section. This non-standard thickness seems an odd choice for a nominally generic study, until it is noted that the Vulcan baseline wing design used a 9.4% thick NACA 00XX section at the spanwise station corresponding to the eventual kink location.

This is surely beyond coincidence, but the lack of an explicit statement was likely the result of hands tied by security considerations. A year later however, he "was able to show some of these results in a contribution in boundary layer and flow control" as part of Boundary Layer and Flow Control edited by Lachmann, a very significant and influential work that has been referenced frequently in the intervening decades. In this work, Pearcey did imply the application, by describing it as work by Avro and Newby—enough evidence for those in the know.[127]

The critical points in this data from the perspective of the current work are the direct comparisons between baseline and modified aerofoil at α = 7°, M = 0.7. The pressure distribution for the baseline case showed an acceleration to supersonic flow at the leading edge (peaking, according to Pearcey's data at M ~1.4), and a short plateau ahead of a rapid compression (the shock wave), which is well forward on the chord. In contrast, the modified aerofoil showed a much more significant leading edge peak, with the flow accelerating to M ~ 1.7 before rapidly compressing back to M ~ 1.24.

The local Mach number ahead of the eventual shock wave was therefore reduced, in turn reducing the strength of the shock itself and the local adverse pressure gradient imposed on the boundary layer. The continuous nature of the pressure recovery ahead of the shock on the modified aerofoil was the key; it cannot have been caused by the discontinuous mechanism of a shock wave as seen on the baseline aerofoil, hence implying the only plausible alternative of compression waves being reflected from the sonic boundary. The peaky mechanism could be visualised in action.

In contrast to the balance measurements and limited pressure data available to Newby, the NPL team were able to add the final piece to the puzzle using flow visualisation through a Schlieren system. The corroborating data was reassuringly clear: the baseline aerofoil showed the sharply defined, curved line of the shock just aft of the leading edge, with the corresponding dark boundary in the flow initiating at the foot, extending tangentially to the local flow and marking the separated boundary layer.

On the modified aerofoil under the same conditions, the shock was only weakly visible towards the mid-chord, with the thickening of the attached boundary layer a more obvious marker. The Mach number could be increased markedly to M = 0.84 (causing a movement of the stagnation point towards the upper surface and reducing the supersonic expansion), until separation to a similar extent as on the unmodified aerofoil at M = 0.7 was seen.

It should be noted that the Mach numbers quoted in these two-dimensional experiments were equivalent to those normal to the leading edge of the three-dimensional wing in Newby's tests. This accounts for the Mach number at which shock induced separation occurred being much less than the cruise Mach number for the Vulcan; its leading edge was of course swept for this purpose. As described above, this configuration was also improved in the final Phase 2 configuration using vortex generators, which was not the case here.

Pearcey also referred to the work at NPL on vortex generators and their role in the suppression of shock-induced boundary layer separation.

"Their flight tests also confirmed that we were justified in recommending that vortex generators scaled up from our tests on small aerofoils (5in chord) would be successful in delaying the onset of the buffeting."[128]

The use of vortex generators for this purpose, seeming so innocuous today, was in its infancy at the time. It would rapidly become commonplace on transonic aircraft, not least on the Vulcan's delta fighter contemporary, the Gloster Javelin, and its bomber competitor, the Victor.

The second prototype VX777 was flown in an evaluation by the A&AEE in May 1955. While the subsequent report recorded the appreciation of the unit for the assistance provided by Avro, particularly in servicing and the briefing given by the company's test pilots, they were scathing of the standard of the Vulcan itself, as presented. It was noted that development had advanced and would go further, but also that making a judgement based on the prototype was difficult.

"Initial production Vulcan aircraft will be fitted with a drooped leading-edge and vortex generators on the outer wing, Olympus 101 engines rated at 11,000lb take-off thrust, and will have an operational take-off weight of about 165,000lb. The absence of the all-important wing modifications together with the non-representative engine thrust (10,000lb) and take-off weight (130,000lb maximum) on VX777 prevented a full evaluation of the Vulcan's flying qualities and operational potential, but certain serious deficiencies adversely affecting the acceptability of the aircraft for service use were shown up by the tests."[129]

The three major issues already identified were highlighted, namely the change in static stability at high Mach number, the reduction in pitch damping, and finally the poor buffet boundary. These were all recognised as having attention focused on them and clearly great store was being held in the benefit of the new leading edge.

"The expected cruising Mach number is 0.87 M (500 knots TAS) and the design Mach number is 0.95 M. Above 0.86 M a nose down change of trim occurred, which became pronounced with increase of Mach number towards the limit, making the aircraft difficult to fly accurately and requiring great care on the part of the pilot to avoid exceeding the maximum permitted Mach number. This characteristic is unacceptable; the Firm propose to eliminate it in production aircraft by the introduction of an artificial stability device.

With increase of Mach number above 0.89 M the damping in pitch decreased to an unacceptably low level, particularly near the maximum permitted. Mach number and the aircraft was difficult to fly steadily. The Firm propose installing a pitch damper in production aircraft.

As tested the Mach number/buffet characteristics were unacceptable for a high-altitude bomber, but considerable improvement is hoped for with the drooped loading edge and vortex generators.

Associated with the buffet were oscillating aileron hinge moments which in these tests imposed severe manoeuvre limitations from considerations of structural safety. These conditions should be alleviated by the aforementioned wing modifications and strengthening of the aileron circuit now in hand by the firm."

By the time of the leading edge programme, the urgency of advancing the V-bomber programme and the nuclear deterrent was palpable. On February 9, 1955, the Under-Secretary of State for Air, George Ward, was challenged on the subject in the House of Commons by Woodrow Wyatt.

The start of Vulcan production at Woodford. XA890 in the foreground was the second production B.Mk.1, and in the 'as built' condition had the original straight leading edge.

"Is not the Minister aware that he said in the debate on the last air estimates that the Valiant will be in squadron service in 1954, and it is now 1955? Is it not the case that we have as much delay and inefficiency in the production of bombers as we have in the production of fighter aircraft?[130]

Is it not the case that we now have no modern fighter aircraft to defend us and we have only one squadron of Valiants as medium bombers with which to launch an attack?"

Ward was able to respond that,

"As I have said, the Valiants are coming in now and we hope to have Victors and Vulcans next year. Meanwhile, we have excellent aeroplanes in the Lincoln and the Canberra."[131]

Surely, he could not have been comfortable placing an emphasis on the Lincoln, an aircraft so comprehensively outclassed by the Washington (B-29) taken second-hand from the USAF. Despite the speed of the Canberra, as a high-level nuclear bomber it could play no role at the time, and it lacked the requisite radar bombing capability. His colleague Air Commodore Harvey may have recognised Ward's uncomfortable predicament and attempted to interject helpfully.

"Will my hon. friend not agree that if the party opposite had not lost their heads when the Korean War started and ordered so many Canberra bombers, half of which I believe had to be cancelled, many of these other types would now be in use?"[132]

However, Labour's George Wigg, who had been PPS to the Secretary of State for War under Attlee and thus implicated, countered with,

"Whether the Opposition lost their heads or not has nothing to do with the fact that the tail dropped off a Victor last year."[133]

Ward was in with half a chance of delivering on his promise, although there is no record as to whether he would ever thank Ken Newby and Herbert Pearcey. The first production Vulcan, XA889, had flown for the first time just five days earlier. During the following October it was returned to the factory for installation of the Phase 2 leading edge, which would be the required production standard for the RAF. After being delivered to Boscombe Down in March 1956, the aircraft was worked through a programme by 'B' Squadron of the A&AEE, towards CA Release for the Vulcan B.Mk.1 and subsequently approval for service.

This was granted on May 29, 1956, a situation that must have seemed very distant a year earlier, when the A&AEE had reported on their opinions of VX777. On July 20, 1956, Vulcan B.Mk.1 XA897 landed at RAF Waddington in front of the eager core personnel of 230 Operational Conversion Unit, with whom it would only briefly stay prior to being replaced by XA895 and returning to Woodford for additional work. Nevertheless, nine and a half years after its issue, B35/46 had finally been fulfilled by the delivery of an operational aircraft.

Early production Vulcan B.1 XA896, with modified Phase 2 wing leading edge.

On display at Avro's Woodford factory on August 13, 1955, this early Vulcan B.Mk.1 is thought to be XA891, the third production aircraft. It first flew in the following September and spent its life on trials, before crashing as a result of electrical failure in 1959. ELAINE HOOKE, VIA GRAHAM HOOKE

Vulcan B.Mk1 XH502 in the USA, 1958.

The sixth production Vulcan B.1 XA894 spent its life on trials work, most notably as the flight test bed for the Olympus 22R, destined for TSR2. It was proposed to use the turbine from this engine, which operated at higher temperature, as an upgrade for the Olympus 301 in the Vulcan B.2. Like TSR2, this plan came to nothing and XA894 was destroyed on the ground when its test engine exploded. ALISDAIR MACDONALD

8

THE VICTOR

Following a rather unhappy war, Gustav Victor Lachmann was eventually able to once again take up a full role at Handley Page—setting about translating into English of the many newly-acquired documents on aerodynamics in his native language in furtherance of a bold new cause. In his words, expressed many years later,

> "The Victor represents technologically the most challenging project undertaken by the company. The high subsonic cruising speed alone represented a great leap forward having regard to the state-of-the-art at that time. It involved exacting demands in aerodynamic and structural engineering in meeting the criteria for strength and flutter free operation, together with a very high standard of performance and reliability demanded in electrical, hydraulic and other systems."

Handley Page and Lachmann had based their careers and prosperity on innovations concerned with safety in the air. The worries of tip stall, then subsequent loss of control, led them to adopt conceptually and develop to its potential the crescent wing, in which sweep angle was traded for reduced thickness outboard. From a longitudinal stability perspective, this had the dual effect of reducing the tendency of the tips to load up and also positioning them physically closer to the centre of gravity, reducing the moment arm and consequently the pitching effect of a tip stall. However, the accompanying reduction in the total longitudinal dimension of the wing (root leading edge to tip trailing edge extent on the longitudinal axis) then made a tailplane necessary for stability and control, in turn allowing trim at high lift.

The crescent planform was of course a variety of swept wing, and it has been described already how HP came to conceive of a swept wing bomber in the class of the eventual B35/46 requirement, but in which the sweep was incorporated entirely for reasons of trim and stability. When were the favourable transonic effects of this swept planform realised and an attempt made to exploit them? It is tempting to wonder if Lachmann had been aware of the German work through his extensive contacts there in the late 1930s, but if he was, this is not apparent in the available written record.

It is clear that on the Allied side of the enforced information divide, R. T. Jones at NACA in the United States was independently considering the use of sweepback by March 1945, whereas this does not appear to have been the case in the UK. The latter is somewhat ironic given the lead it enjoyed in gas turbine propulsion development at the time, a technology that needed to be linked with effective transonic airframe technology to yield its full benefit. Brinkworth noted in his study of the Gloster E28/39, that sweepback had been suggested at a meeting between the manufacturer and the RAE in October 1939. However, it is clear from the manner in which the paper

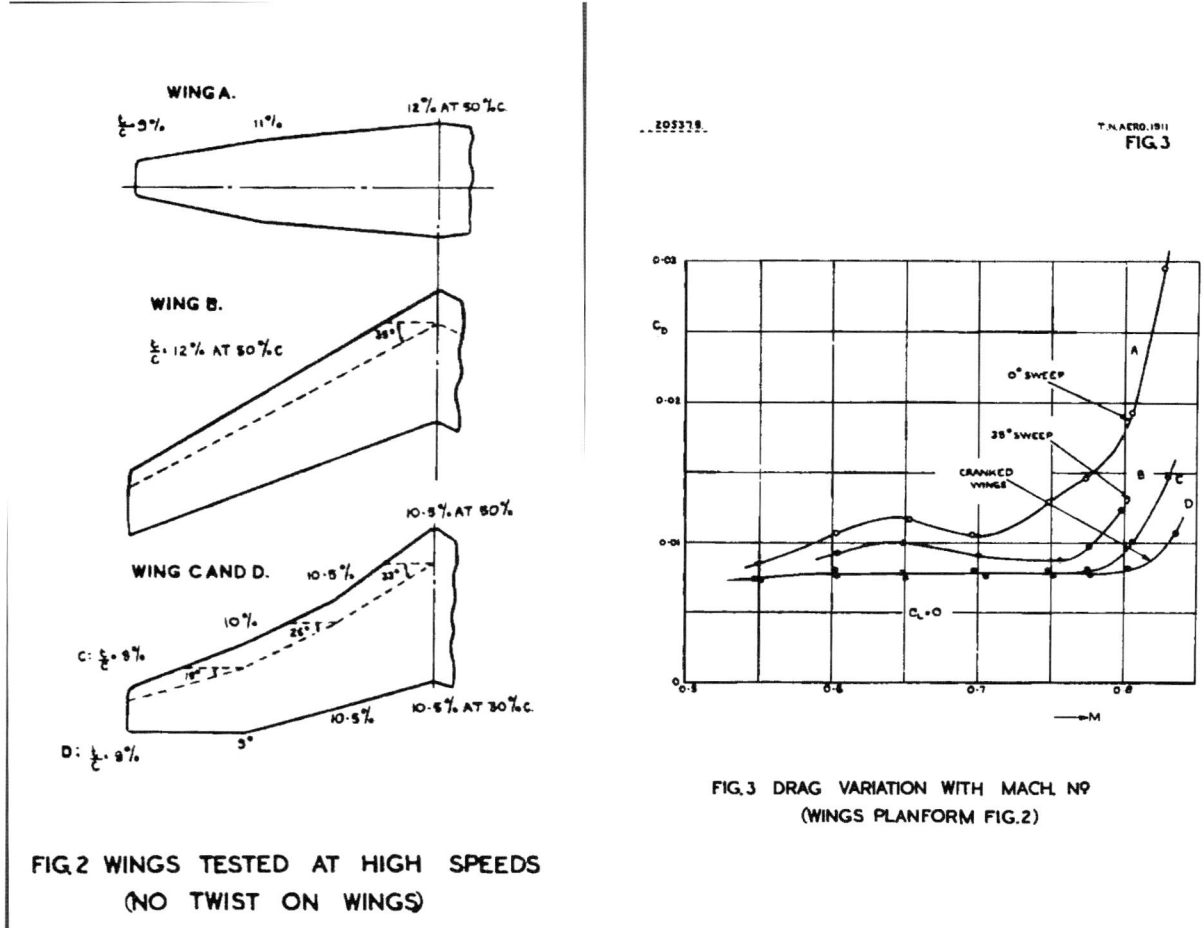

Planforms tested by Ludwieg at Göttingen, and subsequently reported by Hill and Küchemann in RAE TN 1911. This data showed some of the potential for high speed performance of the 'cranked' wing.

addresses this, that he was also unconvinced that this was related to high Mach number performance.

More than a year earlier than the initial findings of Jones, the advanced state of swept wing research in Germany was being discussed in front of the assembled aeronautical body by the LFA's Theodor Zobel. In describing the theoretical basis behind the technology, he was able to call on the mass of data that had by then been collected in high-speed wind tunnels, dating back to immediately prior to the war and the work of Betz and Ludwieg at AVA Göttingen. In August 1944, the development prototype of a large, swept (forward) wing, turbojet-powered bomber, the Junkers Ju 287, made its first flight. Although far too late to have entered service prior to VE-Day and in fact purely a test 'mule' incorporating the fuselage of a Heinkel He 177, the physical existence of such a machine serves as an illustration of the practical state of the art in Germany.

A consistent theme of the Allied reports from the German research establishments, is that only around half of the theoretical benefit of the sweep effect would be achieved in reality. To understand why this might be so, we can refer to work produced somewhat later at the RAE, by Küchemann and Weber and describing ideas that had started to develop among the German researchers.

Their analysis began by dividing the wing into three regions: centre (root), tip and the intermediate region characterised by 'sheared wing' flow. The latter could be considered the 'target' flow pattern; in planview, the streamlines were curved due to the spanwise pressure gradient established by the swept isobars. For an infinite wingspan, there is nothing to bound (or interfere with) this flow pattern. However, on almost all practical aeroplanes, with finite wings, symmetry is established at the centreline. This must mean that the sweep angle is reversed from one side to the other, and in turn at the centreline the sweep angle must pass through zero. For some distance on either side of the central plane, there will be a region in which the isobars feature little or no sweep, behaving as if they were part of a straight wing.

Similarly, at the tip, there is no mechanism to maintain the spanwise pressure gradient and streamline curvature beyond the extent of the wing, so the pattern must again tend to a straight streamline. Towards both centre and tip therefore, the isobars will gradually 'unsweep' with respect to the apparent geometrical sweep. Without specific attention to avoid this effect, these regions of the wing will not operate as aerodynamically swept, with consequent loss of suppression of the suction peaks, increase in adverse pressure gradient in recovery and a reduction in local critical Mach number.

The existence of designs, data, wind tunnel models and partially complete test versions of the Arado Ar 234 jet aircraft in Germany in 1945 should mean that there is no doubt that the cranked or crescent planform had been invented there. Ralph Denning, later to become a significant presence at Bristol Siddeley in the Olympus era, was part of the Operation Sturgeon effort at Völkenrode. Sixty years later, in a friendly exchange of views with former Victor aerodynamicist Harry Fraser-Mitchell, he quoted the view of Arado designer Rüdiger Kosin himself, as expressed in a book on the subject of German jet development by Wolfgang Wagner.

"When, in 1942, Arado considered the application of the swept wing effect, three wind tunnel models with different constant sweep angles were tested in the low-speed tunnel of the DVL... At higher angles of sweep the models showed massive flow detachment at the wing tips and, in fact, increasingly so at increased sweep which, with even greater growth, would have led to nose-up pitch, stall and nose dive. Experiments fitting leading edge slats, similar to those used on the Me 109, led to nothing so they hit on the idea of sweeping the wing more steeply on the inboard section than on the outboard section. Arado patented this idea in 1942."[134]

Several sources state (perhaps based on each other) that Kosin and Walter Lehman patented a design for a wing with a stepped reduction in sweep angle, on December 9, 1942. The promise shown by the Ar 234 as it began its flight test programme led to interest in a larger bomber and reconnaissance aircraft of similar philosophy. On July 30, 1943, the German Air Ministry issued a requirement for a two-seater powered by either four Jumo 004 C or four HeS 011 jet engines and able to carry 2,000kg of bombs for 2,200km. Arado's response, detailed in a description of February 10, 1944, was the E-395. While the E-395 was rejected in favour of the Ju 287, the description was among papers collected by the Allies after the war and was subsequently translated by Lockheed.

It detailed two alternative wing planforms; the first was a straight planform similar to that of the earlier jet while the second followed Arado's patented progressive sweep concept. The use of such a planform was a tangible idea by this time; at least five variations were wind tunnel tested and a wooden wing constructed to be fitted to the Ar 234 V16 prototype. This was apparently destroyed before it could be flight-tested, as the British Army captured its airfield at Dedelsdorf.

Back in the UK, at a conference held by the Royal Aeronautical Society in November 1946 and chaired by Sir Frederick Handley Page, Godfrey Lee discussed in detail the engineering problems associated with tailless aircraft. Discussing the thickening of the boundary layer near the tip and the consequent reduction in margin to stall, he observed,

"Another approach is to try and reduce outflow at the tips by reducing the obliquity of the isobars for the outer part of the wing, and this clearly suggests a progressive reduction in sweepback from the centre line to tip. Such a wing would be awkward structurally, and at the tips the gain in critical Mach number from sweepback would be reduced, but for moderately high speeds a reasonable thickness/chord ratio could still be retained outboard while full advantage could still be taken of sweep at the structurally much more important centre section."[135]

Lee offered several references to German reports, in both their native and translated forms in his paper, but was silent about the origin of this particular concept. As he implied, perhaps it was because none were needed—the idea followed so naturally from observation, that it had occurred independently in both Warnemünde and Cricklewood. Elsewhere, German development in this field was briefly summarised as part of the ongoing analysis of the data acquired by Operation Sturgeon and subsequent work. In May 1947, Hill and Küchemann reported,

"The idea seems to have been first suggested by R. Kosin of Arado, who claimed it would reduce the tendency to tip stalling, give smaller values of lv [rolling moment due to velocity?] at high CLs and also have a higher critical M since the largest sweepback would be at the most critical region at the centre of the wing."[136]

It was noted that the lowest critical Mach number region of the wing would be close to the centreline, as the isobars unswept themselves towards the symmetry plane (with zero sweep, by definition). This could be addressed by exaggerating the inboard sweep and carrying this as far into the fuselage as possible, which is what is inferred. This is, however, a different concept to the constant spanwise critical Mach number wing, combining appropriate values of sweep and thickness together.

Perhaps the most relevant evidence examined within Hills and Küchemann's study, was the report on a series of Arado Ar 234 specific tests at Göttingen conducted by Hubert Ludwieg. These described the baseline straight wing, an initial swept wing and two cranked wings, the latter differing in t/c and position of maximum thickness. The low Reynolds number achieved on the 400mm span models (Re = 0.9 x 106) was noted as a potential source of error in terms of identifying the true critical Mach numbers of the respective designs, but it was suggested that the trends seen, "may be more reliable".

Because the wings were not of identical thickness distribution or t/c ratio, some interpretation was required. The 'straight' swept wing was of 35° sweep at the quarter chord and 12% thickness. The cranked wings were of tapering thickness, but with a maximum of 10.5% thickness on the inboard region. The report noted that,

"…results do suggest that the cranked planform has at least as high a critical M as a straight [swept] wing, since the difference of 0.02 on critical M between wing B and wing C is the same as might be expected for a difference in t/c of from 12% to 10.5%…"

In other words, had a 10.5% thick swept wing of identical planform been available to test, it was considered that the critical Mach number would have been improved to the same level as that achieved by the cranked wing. The latter, however, achieved this with a combination of reduced sweep and thickness outboard, so offered potential advantages in boundary layer and tip stall behaviour. The geometric parameters for 'wing C' are very close to those described by Fraser-Mitchell in his comparison with the Victor, in his discussion with Denning.

"The Ar 234 wing tested in the wind tunnel had sweeps of approx. 32, 24 and 17 deg from the apex outwards and corresponding thickness:chord ratios of 10.5, 10.5, 10 and 9 %. The Victor wing in its earliest form had sweeps of 50, 40 and 30 deg and t:c ratios of 16, 12, 10 and 9% linked to this to keep the Mcrit constant across the span. I do not think that the Arado wing had this linkage and I believe this constitutes a major difference between the two layouts".[137]

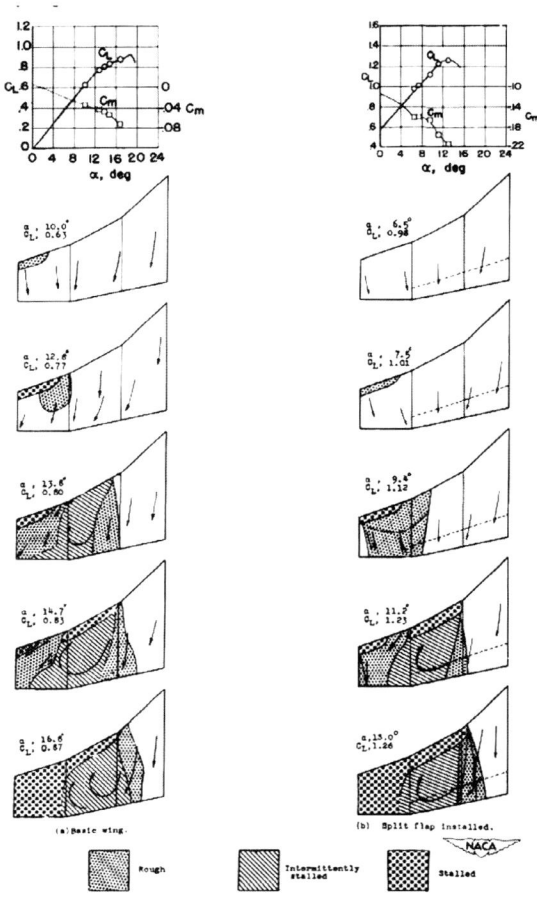

Flow patterns observed by Barnett and Lange on their cranked wing at high lift. Notably, flow separation started at the outboard leading edge and was exacerbated by the local loading at the cranks. This helps to explain why the nose flaps used by HP were important.

Fraser-Mitchell was adamant that it was this tailoring of the combined sweep and thickness properties of the wing across the span that was the radical new feature of the Handley Page wing, and that this had never been considered in Germany. Denning remained less convinced, arguing on the one hand that the lack of available engine thrust meant that such considerations were irrelevant for German development at the time, and on the other that Handley Page's chief designer for the Victor, Reginald Stafford, had seen Arado's wind tunnel models during the Farren mission and could not have failed to be inspired.

The latter points seem to ignore the German interest in very high-speed flight however, and the parentage of the crescent planform is not in doubt, given the existence of Kosin's patent. As would become painfully obvious later, it would indeed be the connection between thickness distribution and spanwise sweep distribution that would yield an effective transonic wing.

In the USA, NACA also undertook studies to ascertain

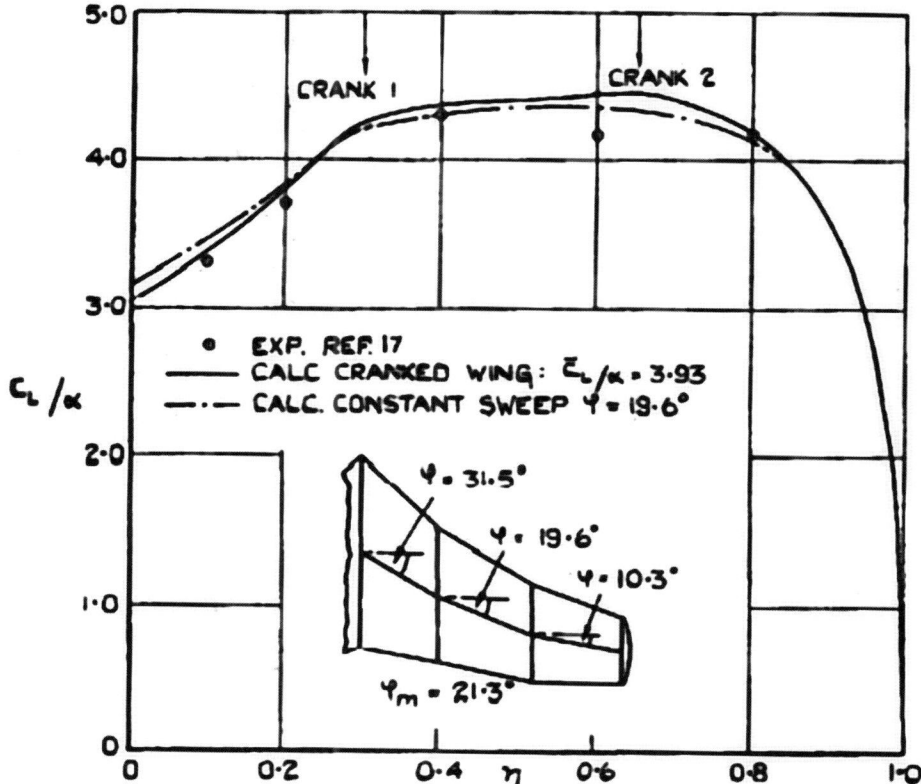

Brebner's calculation of spanwise loading compared with experimental data from Barnett and Lange.

the properties of 'cranked' wings. The work reported by Barnett and Lange[138] in 1950 is interesting for several reasons; in particular, they used a full-scale wind tunnel — NACA's enormous 40ft x 80ft section facility in which complete aircraft were tested. The striking size of the model that resulted allowed testing at a Reynolds number of almost four times the value of Ludwieg's tests in Göttingen.

However, comparison of the geometry between the German and American programmes shows a strong similarity in the range, if not precise values, of the sweep angles and thicknesses chosen. The American tests were concerned purely with the low speed, longitudinal stability characteristics of the wing, taking place at M = 0.07.

The wing was found to provide a broadly similar maximum lift to conventionally swept wings, of comparable thickness and aspect ratio, in the range 10° to 38°. Previous work had shown that for single sweep angle, the maximum sweep angle that could provide longitudinal stability was about 35°; the cranked wing was shown to be stable up to the stall, so showing that the reduced tip sweep enabled an increased root sweep. The report did not make the step, but the latter was in turn therefore an enabler of increased thickness inboard at high Mach numbers, while remaining stable.

When the wing did stall, it was from the leading edge on the outboard panel; this is probably not surprising in any case given the relatively sharp leading edge of the 9% thick symmetrical NACA 6-series aerofoil. It was shown that the situation in terms of maximum lift and L/D near the stall was considerably improved by a drooped leading-edge flap. Since it was the outboard panel that stalled first, perhaps there was effort to be saved by adding a leading-edge device on there. Both a flap and a slat were tried and, whereas the influence of the slot should have protected the leading edge to higher incidence, in the tests both options gave similar outcomes: a small increment of about +0.1 CL.

This was attributed to the effect of the slat's inboard edge, with either the vortex or losses from this interacting with the loaded crank region and prematurely stalling the wing. This interesting result helps to explain why the world leaders in slotted wing technology would adopt a plain flap for the leading edge of their bomber.

Sometime later, the American work would be used as an example by the RAE's G. G. Brebner, who was expanding the development of the organisation's theoretical methods for the calculation of aerodynamic forces on three-dimensional wings.[139] Having considered an earlier RAE test case of a single crank between a 45° swept inboard section

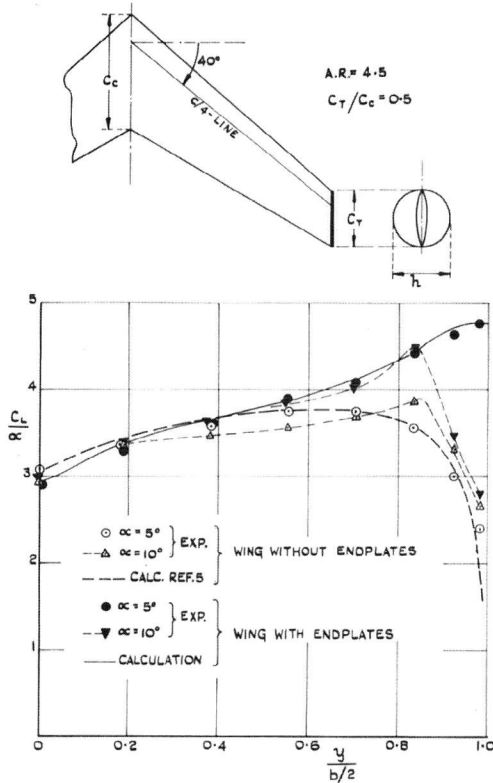

Küchemann's diagram of calculated and experimentally measured loading on a wing of Victor-like proportions and sweep, in the absence the crescent shape. The use of endplates as originally proposed enabled load to be carried further into the tip. With that region most vulnerable to stall, the detail design for aerodynamic robustness at high Mach number would have been challenging.

and straight outboard, he noted that the relatively gentle successive changes in sweep on the NACA wing were the likely reason that the spanwise distribution differed little from a conventional swept wing. He calculated spanwise loadings for equivalent wings with the same sweep as the mid panel and of the mean sweep (19.6° and 21.3° at mid chord respectively), finding them little changed.

However, his calculations showed that the chordwise loading at the cranks peaked further forward, such that these regions were always likely more vulnerable than their sectional geometry might otherwise imply. Along with the counterintuitive American result for the slat, both of these conclusions hinted at banana skins that lay around the detailed design of the crank geometry and its effect on high lift performance.

This work was reported somewhat later than the initial HP.80 designs, but certainly long before the aircraft flew and hence is a useful illustration of contemporary design capability. The published record suggests that Brebner appears to have worked closely with the German emigrees Dietrich Küchemann and Joanna Weber to advance this theoretical capability to design wings. However, if we retrace our steps back to 1947 and the start of the programme, it is also clear that situation was more primitive.

Küchemann and Weber's major contribution to the HP.80 was the use of theoretical methods for the calculation of the three-dimensional pressure distribution along the wingspan. This fundamentally was the enabling technology for the pursuit of the 'constant critical Mach number' concept, which was the important feature of the HP.80 as a high-altitude transonic aircraft and owed nothing to the Arado work.

By sweeping the outboard wing forward without reducing the thickness, Arado would have actively reduced the critical Mach number there compared to the inboard region. Küchemann and Weber's work enabled them to modify the wing sections and keep the isobars aligned with the geometry. As they were well aware, the really important point was what happened at the tip and root, where the wing aerodynamics would naturally 'unsweep' themselves and trigger premature shock formation. Küchemann's own text book made few references to specific aeroplanes, but perhaps as an indication of the importance of the work to him, the Victor received a lengthy description.

> "The next design aim was to keep the critical Mach number constant along the span at a cruise CL of about 0.3. Thus the shapes of the wing sections were specifically designed with suitable distributions of thickness as well as camber and twist to achieve straight isobars over the upper surface of the wing, with an angle of sweep of the peak suction line higher than the geometric angle; i.e. Cp min was located at about 0.3 c at the wing root and at about 0.6 c at the tip. The value of the t/c itself proved to be the most powerful design parameter and this led to the large thickness taper along the span, from t/c = 0.16 at the wing root to t/c = 0.04 at the tips. This, in turn, not only provided a large stowage volume for engines and undercarriage near the wing root, consistent with the basic crescent-wing concept."[140]

Küchemann and Weber reported in September 1947[141] on their work to calculate the velocity (and hence pressure) distributions on the proposed HP crescent wing. This used the methods described in two previous reports that Küchemann had written, both concerned broadly with ensuring that the benefits of sweepback were not compromised by the fuselage or tips. The mathematical technique was able

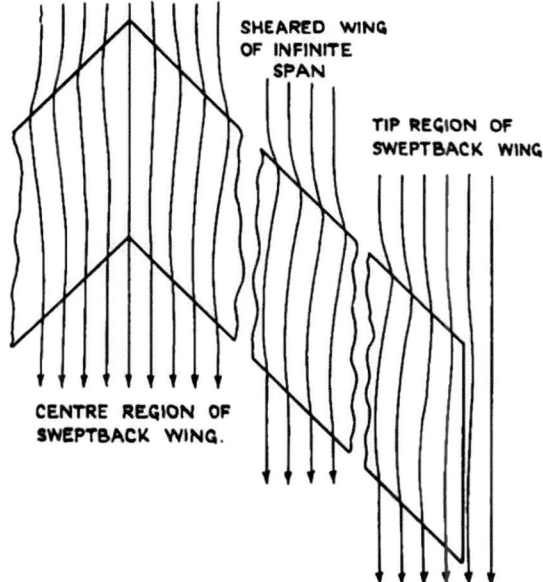

Küchemann and Weber's division of a three-dimensional swept wing into central, mid-span and outboard regions, with their characteristic surface streamlines.

to calculate the velocity distribution at the position of the 'kinks', where step changes in sweep occurred. In order to use this method at the root and tip, the local geometry was mirrored as if the wing were continuous, which of course had some physical reality at the root, but at the tip merely facilitated the assumption of an infinite span. In addition to these mathematical sleights of hand, the method considered zero lift overall (an approximation to a low lift coefficient) and assumed the effects of thickness to be dominant, neglecting camber.

In the parlance that would be adopted in future reports and will become familiar later in this document, the wing could be described in the format A/B/C/D, using the values for maximum thickness/chord ratio at the root (A), first kink (B), second kink (C) and tip (D) respectively. The geometry at the start of the work would therefore have been 14/12/08/08. Obviously, this nomenclature includes no planform information. The sweep angles for the initial wing were 45°, 35° and 25° on the inboard, mid and outboard panels respectively, at the 40% chord maximum thickness line.

The results of the calculation on the HP-supplied baseline geometry were reasonably positive. Particularly over the rear portion of the wing, where the flow was decelerated, the isobars were reassuringly straight. This held too for the region close to the leading edge, and while close to maximum thickness there were closed loops implying a degree of spanwise flow, the patterns were consistent and tidy.

More concerning was the dramatic unsweeping of the isobars at the root. In this region, the calculation suggested that the wing would lose the favourable high Mach number benefits of a swept wing; in a sense, this region then becomes the global critical point at which a shock wave might be seeded first and therefore reduce the drag divergence Mach number of the whole configuration, irrespective of how good the performance was elsewhere.

An initial modification was instigated to improve matters; the thickness of section A (root) was reduced to 12% and shifted forward, while the position of maximum thickness at section D was moved rearwards. The effect of these modifications ought to have been a local reduction in velocity increment at the root, and an increase in isobar sweep towards both root and tip. In fact, these objectives were successfully achieved, albeit with the inboard velocities still raised compared to the outboard.

Finally, the effect of increased thickness (and hence raised velocities) at sections C and D was investigated, resulting in thicker mid and outboard regions. This served to improve the velocity matching across the span, such that the closed isobars in the mid region are eliminated and the desired, fully swept pattern was achieved. Unfortunately though, it was at a cost that likely could not be accepted: the root was no longer deep enough to house the engines. Fine for the wind tunnel, but not for the real aircraft that Handley Page would have to build.

Küchemann and Weber proposed two possible solutions to this dilemma. The first involved a reversion to the 14% thick root and then, while maintaining absolute thickness, extending the chord and increasing the sweep to 50°. This was ineffective as control of the isobar sweep began to be lost, but their second proposal of modifying the local contour of the fuselage adjacent to the wing did work. With a similar strategy applied to the tip and its fins, the major aerodynamic concepts of the HP.80 wing were established. The analysis though relied upon the correlation between the zero-lift case that could be assessed, and the moderate lifting case that the aircraft would have to fly at. It was a starting point and a sound logical basis that gave confidence in the strategy, but there would be a long way to go.

Meanwhile, a model of the starting point wing used by Küchemann and Weber (the 14/12/08/08 wing) was tested in the RAE, as part of a complete 1/30 scale model of the HP.80.[142] On the face of it, the results were encouraging. Up to M = 0.9, that is beyond the design cruise speed, the lift curve slope was maintained at the same value as the low-speed condition, which implied the aerodynamics were well behaved. That said, the zero-lift

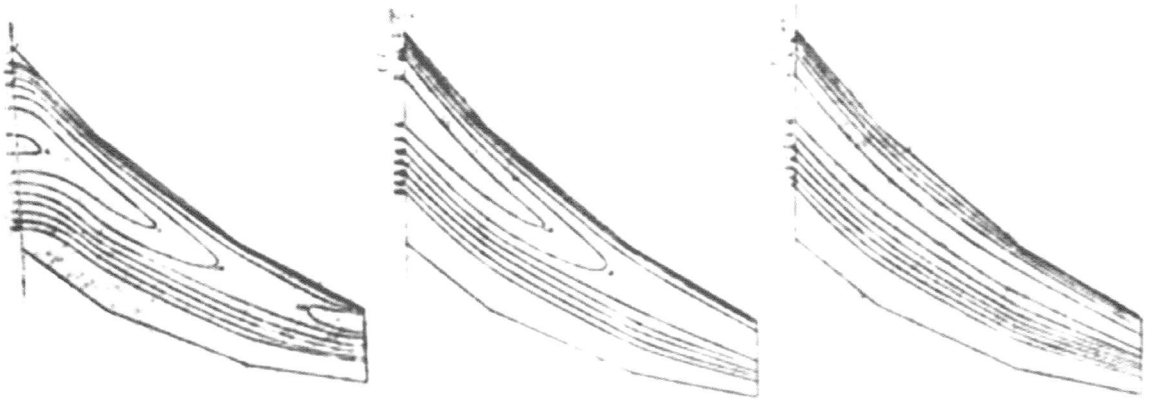

Configurations 1-3 (left to right) from Küchemann and Weber (1947). The isobars over the suction surface of three successive iterations of the wing are illustrated. This commences (left) with the configuration as supplied by HP. Noting that the target is for the isobars to continuously follow the geometric sweep, it can be seen instead that the calculated contours are closed both inboard and towards the tip. This implies that higher velocities (and lower pressures) occur in these regions then in the middle part of the wing. In turn, the pressure recovery demands are locally increased and these regions are both closer to stall and of lower critical Mach number.

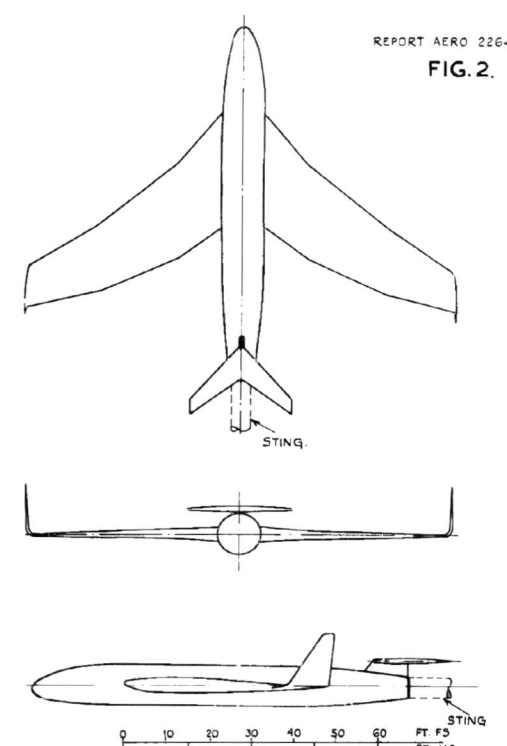

FIG. 2. G.A. OF HP 80 MODEL

GA drawing of the HP.80 model tested at the RAE High Speed Wind tunnel in October/November 1947. At this point, the wing was of the basic design provided to Küchemann and Weber for further development. The fuselage is approximately cylindrical in the region of the wing, the horizontal tailplane was mounted on a minimal stub fin, while yaw stability and control were provided by the wingtip fins. RAE AERO R.2264, 1948

pitching moment did increase dramatically, requiring in turn a large tail download to trim in the cruise. This was problematic from both a trim drag perspective, and also one of avoiding shock-induced separation on the heavily loaded tailplane.

In this configuration however, the tailplane was close to the plane of the wing, and subject to the idiosyncrasies of its changing downwash. Even without considering the engine position and the need to deconflict the exhaust flow, this would have been a driver to moving the location of the tailplane, either up or down. In retrospect, a low tailplane position might have avoided some of the pitfalls that came later, but there were good reasons overall to position it high, mounted on the fin.

ADVANCING DRAG DIVERGENCE

By February 1949, the HP.80 wing had evolved to a planform in which the familiar final form was recognisable. The thickness distribution at the four spanwise control stations (14/12/08/07) remained similar to the original HP wing of Küchemann and Weber's investigation (14/12/08/08), with the exception of a slightly thinner tip. The tip fins had been abandoned in early 1948 and an extended, curved geometry adopted from around 91% semi-span. Perhaps the most obvious fundamental contrast was the use of taper, which remained constant on each of the three panels but differed between. Testing of this geometry in the high-speed tunnel precipitated further analysis by Handley Page, identifying opportunities to promote the desired constant critical Mach number strategy.

Identified as Modification 1[143] and tested in

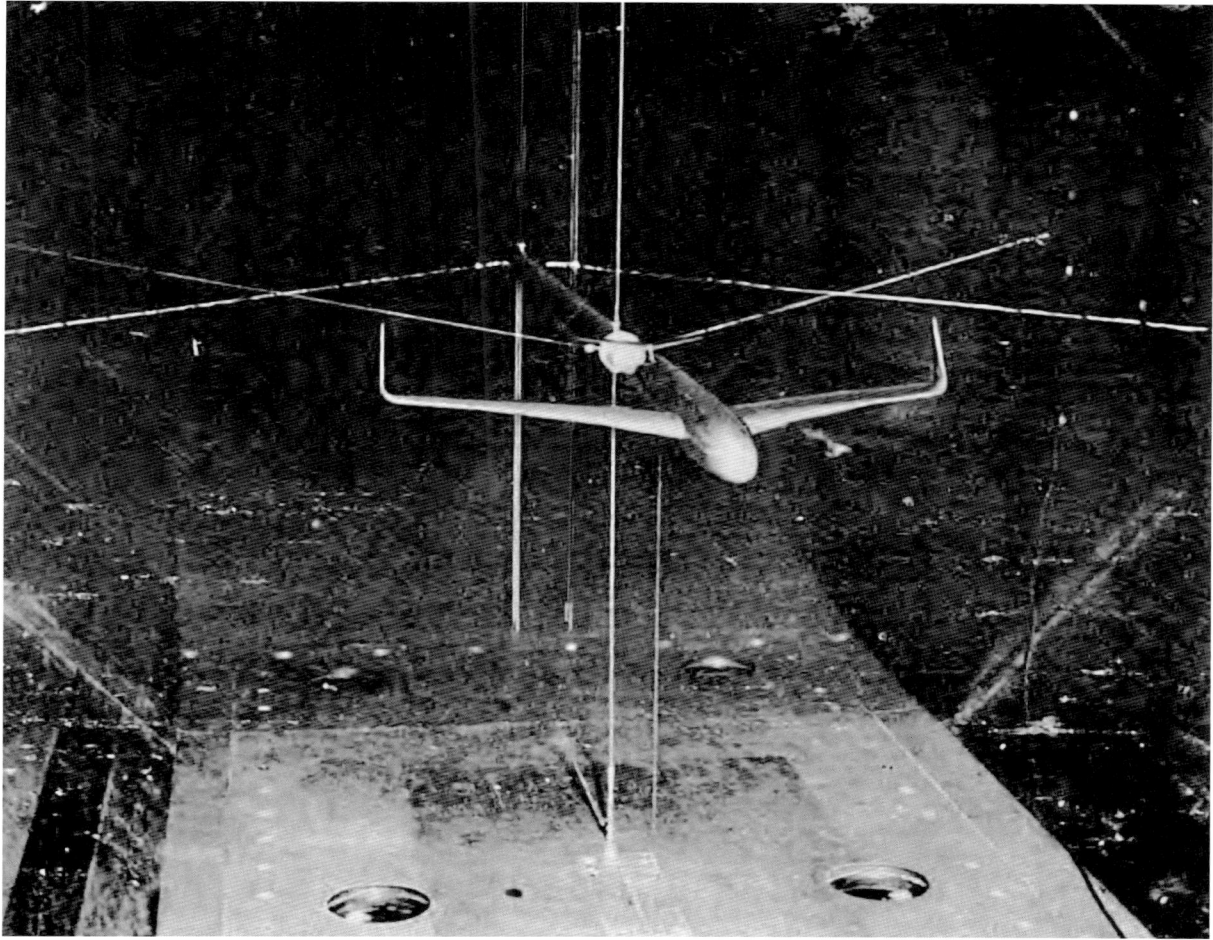

October 1947: The initial HP80 concept windtunnel model (1:30 scale) mounted on a sting, in the RAE HSWT. RAE AERO R.2264, 1948

A startling view of the HP.80 model under test in the Farnborough 8ft x 6ft WT, October 20, 1947. The annotation indications conditions of zero lift and Mach 0.6, while the port wing has been tufted in order to show the local flow features. VIA JIM SMITH

November-December 1949, this design incorporated,
* A small adjustment to the planform; the inboard kink shifted slightly outboard.
* Sweep on the outboard panel increased from 23° to 26°.
* Sections at root and IB kink reverted to symmetrical.
* Reduction by 2% of thickness on mid panel.

This modified version, now 14/10/06/07, did indeed demonstrate an increase in MD (defined as $\Delta Cd = 0.005$) of $\Delta M = 0.04$ in the range of lift coefficients that might be required for cruise, say from CL = 0 to CL = 0.3. How had this been achieved? The modifications to the geometry were observed to have resulted in:
* A reduction in local velocity on the mid panel, meaning that the airspeed at which supersonic speed was reached in that region was postponed.
* An increase in isobar sweep on the outboard panel, similarly increasing the critical Mach number locally.

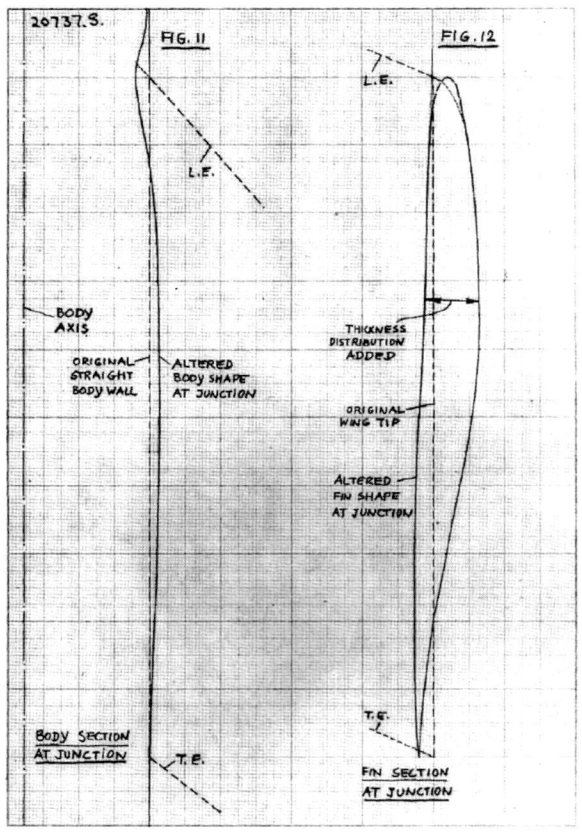

Küchemann and Weber's proposals for local modifications to the fuselage (left) and wing tip fin (right) planview shapes, in order to promote isobar sweep in these regions. Although the fins would be abandoned, the fuselage/wing junction would see much attention paid to this aspect. RAE TN Aero 0001, 1947

* A decrease in absolute velocity near the root, due to the increase in thickness taper (more rapid change in thickness from inboard to outboard).
* Complementary to this, an increase in isobar sweep inboard of first kink, also driven by the increased thickness taper.

With this step in predicted drag divergence in hand, it was time to address the inboard volume problem. First tested in January 1950, the new 16/10/06/05 design increased both thickness and chord on the inboard panel. The chord extension meant that while non-dimensional thickness was just 2% greater, the physical dimension of the maximum thickness was increased by around a quarter.

There was now sufficient volume to house the retracted undercarriage and projected Armstrong Siddeley Sapphire powerplants, but the geometric result was a small decrease in aspect ratio and corresponding small increases in sweep of the maximum thickness lines on the inboard and mid panels. Both of these incidental effects were compromises: the reduction in aspect ratio would be expected to be negative for induced drag, while the sweep would increase the torsional stiffness requirement and hence structural weight.

However, these would be expected to be balanced by the greater structural depth of the new wing root and the diminishing importance of induced drag at high speed. It was all a moot point of course, if there were no engines.

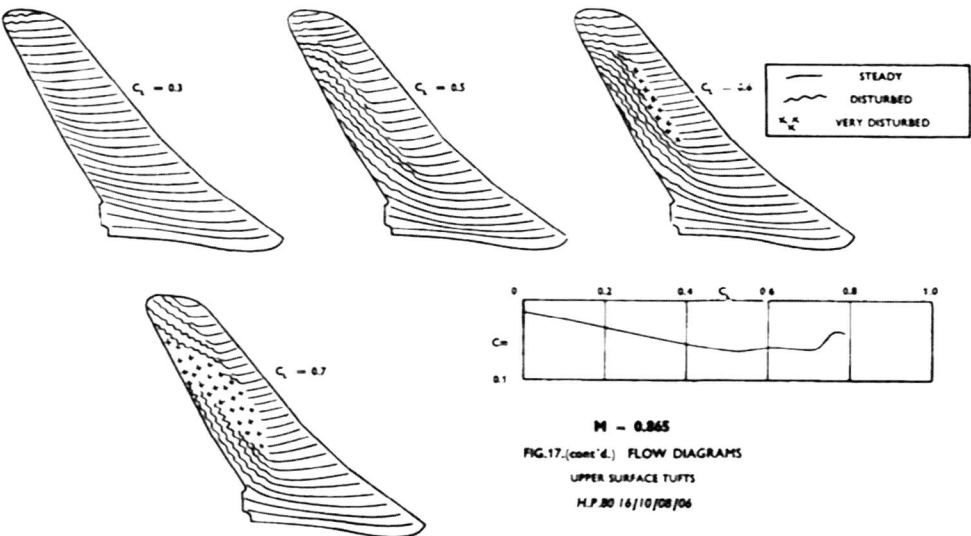

Flow patterns on the prototype standard Victor wing, as observed in the RAE High Speed Wind tunnel at $M = 0.865$. For cruise at 50,000ft, a lift coefficient (CL) of between 0.3 and 0.4 would have been required. It can be seen that the flow pattern was clear and attached at 0.3, while the curve for pitching moment coefficient is linear and increasingly negative in that range, which means the aircraft naturally wanted to pitch down as it approached the stall. The smooth flow patterns contrast with those seen for the B-52 in wind tunnel testing, which showed shock induced separation at similar M, Re and CL.

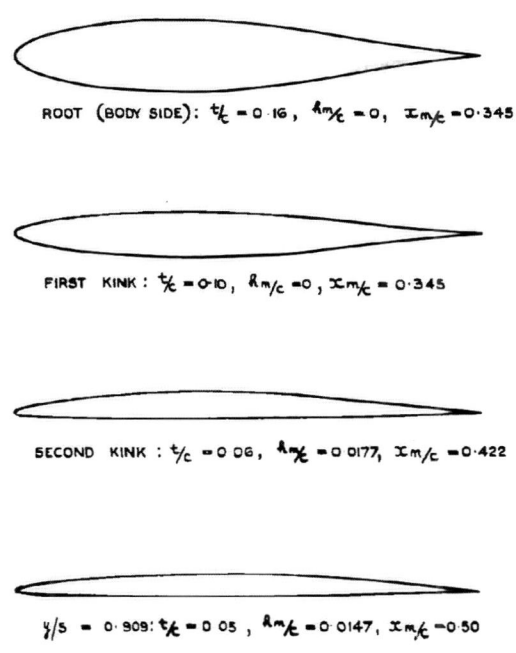

FIG. 4. TYPICAL PROFILES H.P. 80 16/10/06/05.

The wing profiles at the four control stations. Note that the sections are symmetric inboard, becoming much thinner and more cambered on the outboard section. Although custom designed, they had similarities to the NACA 6-series.

Nevertheless, this development had produced a design with drag divergence performance that was at least as good as the earlier, thinner root design at all relevant cruise lift coefficients.

The fact that these results were so impressive led to an unusual inclusion in Bob Collingbourne's Royal Aircraft Establishment report: a comparison with other wind tunnel data from simpler wing planforms.[144] Assuming that the inboard thickness of 16% was maintained across the full span, then a sweep angle of about 60° was estimated to be required for a similar drag divergence Mach number. Conversely, comprehensive results were available for 40° swept wings of similar aspect ratio to the HP.80.

This was close to the sweep of the mid panel, giving the conclusion that the bomber's wing would require a full span thickness of 10% to yield a similar performance. Collingbourne concluded, "Up to M = 0.89 the HP.80 is superior from the drag standpoint to any of the wings shown, more especially as none of these has an intake or nacelle represented." There was no doubt therefore that the complication was worth it.

By April 1951, a further development (Modification 3) finally gave the planform that we have become familiar with. The sweep and thickness of the mid and outboard panels increased, while the spanwise position of outer kink was moved inboard. The result was to reduce the movement of the aerodynamic centre position with Mach

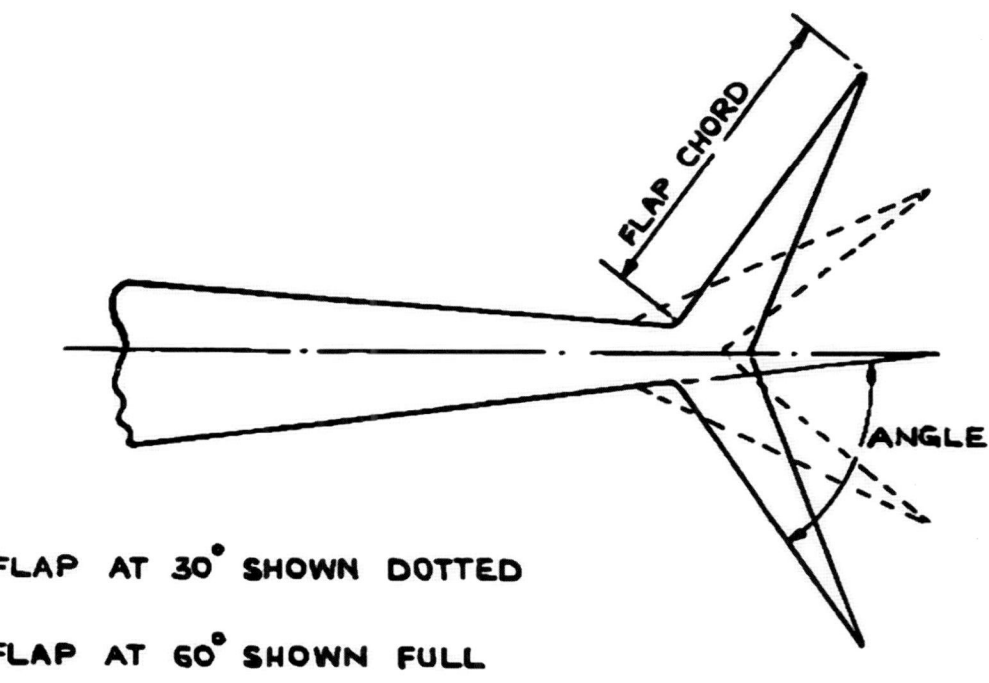

Briefly investigated during development, the contra-flap was a dual split-flap arrangement that would have occupied the space between the aileron and slotted flaps. The aim was to increase lift while minimising nose down pitching moment.

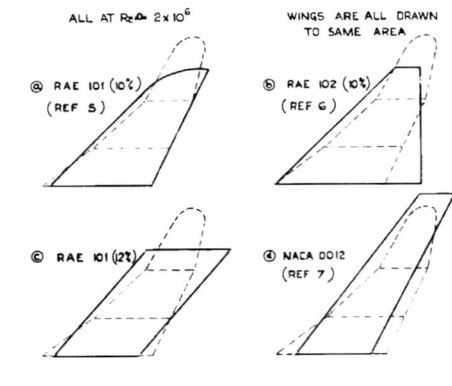

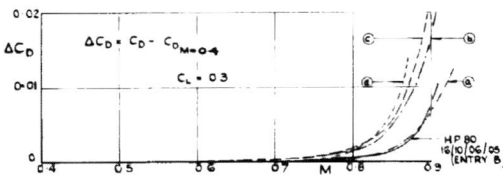

FIG.10. DRAG INCREMENT vs MACH No. AT $C_L = 0.3$ COMPARISON WITH VARIOUS OTHER WINGS. H.P. 80 HALF MODEL 16/10/06/05 WITH DUCT.

Collingbourne's comparison of the HP.80 wing to four other contemporary designs. The relatively lower aspect ratio and much thinner inboard swept wing (a) was the closest in drag-divergence performance to the crescent, although it had insufficient volume for engines. The delta (b) is of course of interest in a V-bomber comparison, being clearly inferior at M = 0.87. The Vulcan was designed around a larger area, which enabled it to operate at a lower lift coefficient and derive a drag-divergence advantage in a different manner.

number, aiding longitudinal stability and control.

The high-speed tunnel tests furthermore demonstrated improved stalling characteristics at low speed, attributed to increase in local chord at 60%-70% semi span where stall was initially triggered. It may have been a long time coming, with less than a year and a half to the first flight of the Vulcan, but what was surely one of the most remarkably integrated aerodynamic designs up to that point in time had achieved its final form.

LOW SPEED AND HIGH LIFT

"All three types of V-bombers are magnificent examples of British engineering. No one complains of the hard work that has gone into designing and making them, nor the skill with which they are flown, but when one looks at the lumbering big Vulcan one feels that there must be an awful lot of the aircraft that is not used for lifting the machine at all but is actually retarding its flight—certainly when at full altitude—and that that proportion is only there to get the machine off the ground".[145]

A major difference between the Avro and Handley Page philosophies was in the provision of high lift for the field performance requirements. The combination of large area and available lift coefficient removed the requirement for a high lift system on the Vulcan. This was clearly not the case for the Victor's crescent wing configuration, which would need variable geometry at both the leading and trailing edges.

Two observations might be made immediately. Firstly, by virtue of its technology development effort in the field of slotted wings stretching back for three decades, the theoretical basis for the work required should have been readily available to Handley Page. Secondly and as is familiar to this day, the engineering effort, time and cost associated with the development of a high lift system is likely very significant. While it would be very difficult to robustly quantify this, it can be said with certainty that the work used time and resource associated with the national wind tunnel facilities that was by definition in very short supply.

"Owing to their limited camber, and small leading-edge radius, these aerofoils have poor low-speed, lift and stall characteristics. Slats were considered to pose greater development problems than did leading-edge flaps, which increase camber and follow the switch shift in stagnation point at high [lift coefficient]... Had the Victor been designed a few years later, it would have had a drooped leading-edge, as has been fitted to the Comet and the Caravelle..."

Here again, as noted previously for the Vulcan, was a tacit admission of how close to the limit of aircraft design HP were working when engineering the Victor. Both leading edge (or nose, as referred at the time) flaps and slats would become commonplace over the subsequent 15 or so years, but at the time they certainly were not and few of the contemporary swept-wing aircraft designs used them, despite the potential high lift performance advantage.

The very thin and relatively sharp-nosed aerofoil used outboard was entirely appropriate for the high subsonic Mach numbers in the cruise, where incidence was low. At high incidence for low-speed flight however, the shape generated a high suction peak and rapid recovery, leaving it liable to stall. This was not helped of course by the natural tendency of the swept wing to impose an increased angle of attack on this outboard region. As the NACA full scale

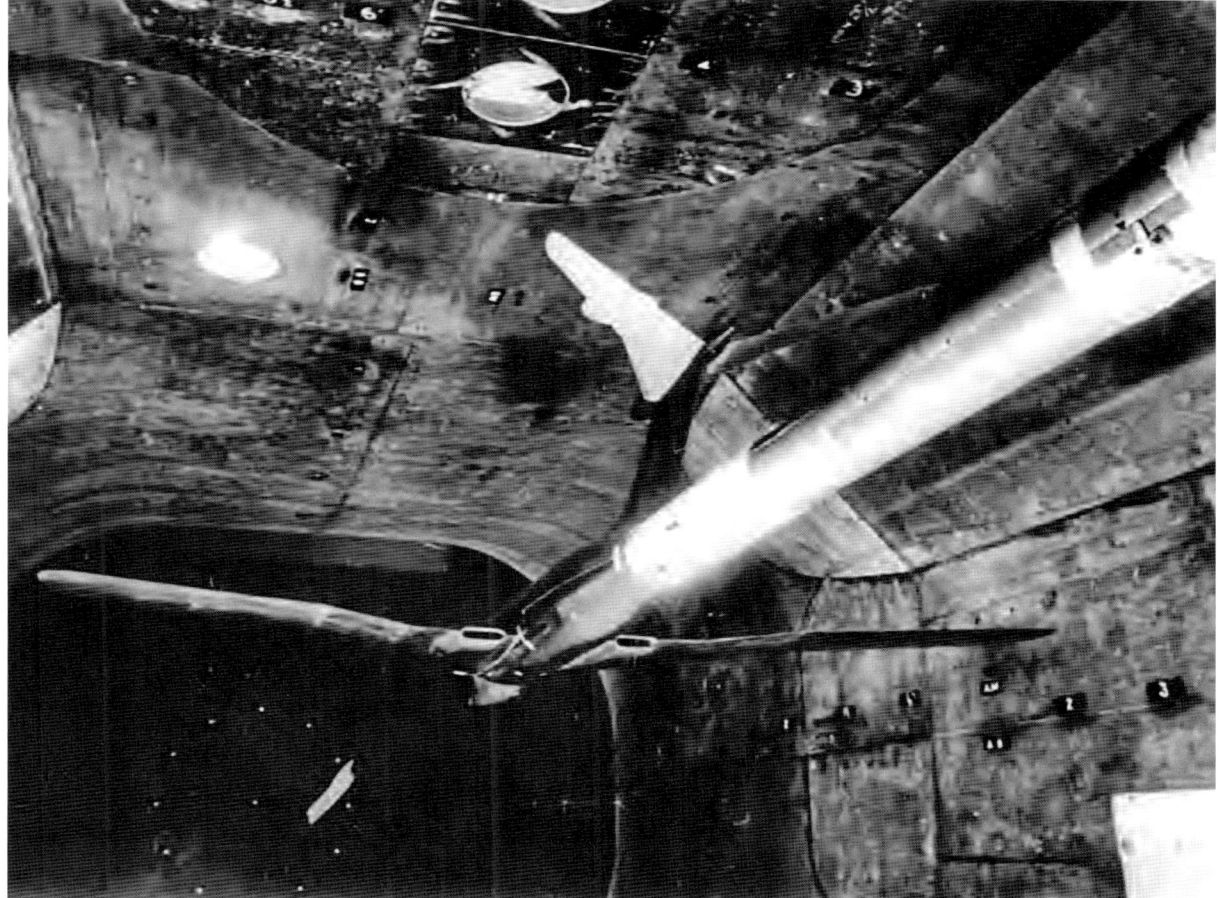

Testing of the prototype Victor configuration in the Farnborough high-speed wind tunnel, 1953. The model is sting mounted from the rear.

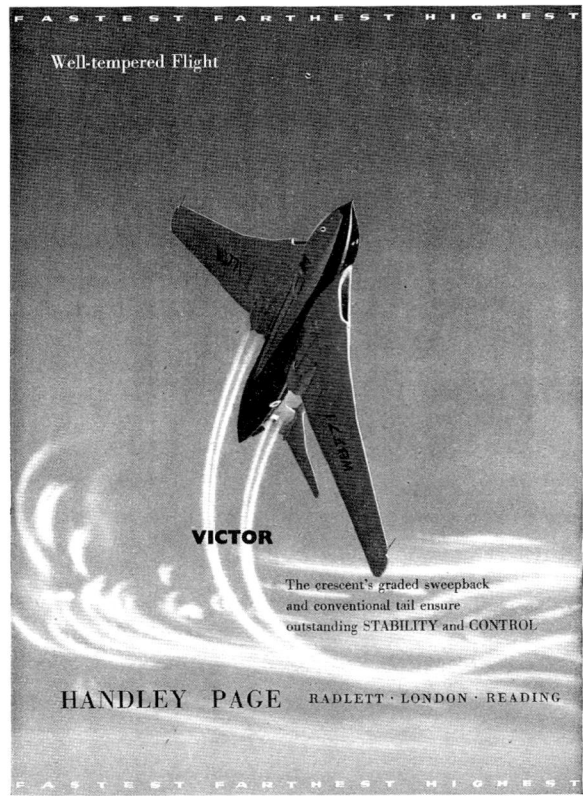

Fastest Farthest Highest — HP's claims for the performance of the Victor, as conferred by its novel wing design. Despite featuring the first prototype WB771, at the time this was published in Aeroplane *magazine (October 29, 1954), that aircraft had already crashed fatally.*

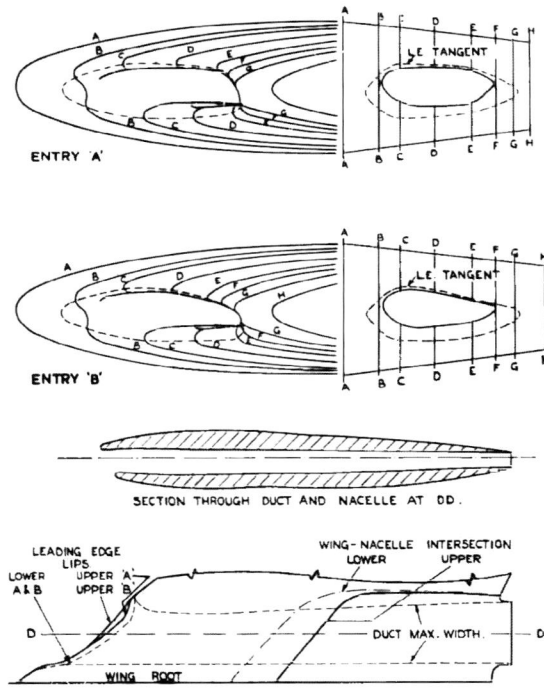

Development of the Victor inlet geometry at the RAE. Like the Vulcan, investment was made in ensuring that the external effect was tuned to the isobar-sweeping effect of the inboard wing.

cranked wing tests had shown, the solution at low speed would require the leading edge to be drooped and so better aligned with the oncoming flow. In turn, the technology of 1950 meant variable geometry of some kind, but even that was in its infancy.

Tests in early 1950 on the Modification 2 wing used a Kruger flap across the entire outboard and mid panels. Folding out from a hinge point close to the leading edge, this device was named for its German inventor, who had ironically carried out his early development on a wing section based on that of the P-51 Mustang. It gave the advantage of increasing the area of the wing as well as improving alignment, but having a hinge line positioned so close to the leading edge would require very high manufacturing tolerances, so as not to introduce surface steps in such a critical region when retracted. No production aircraft had used Kruger flaps at the time.

Within the year, the advent of the definitive Modification 3 configuration of the prototype, with the mid panel occupying a smaller proportion of the span, also precipitated a change in the leading edge device design. A plain hinged nose in two sections was used, eliminating the area extension, but also the challenge of the clean junction. The need to aerodynamically seal between two flaps with dissimilar sweep across the outer kink also went away, as the outboard panel span alone was sufficient, while the mid panel thickness and nose radius had both increased. A hint of a further potential advantage of the plain droop flap over the Kruger was given by an RAE report on high-speed tests:

> "In addition, tests were made to determine the effect on these quantities of drooping the wing leading-edge, since it is intended to use this high-lift device for manoeuvring at high Mach numbers."[146]

If the Kruger flap was a relative unknown in mechanical design terms, it also ran the risk of losing any advantage if it needed to also provide sensible leading edge shapes at cruise Mach numbers. It is likely that there was more confidence in retaining the extensively tested cruise leading edge shape, actuated internally at a hinge line of reasonable thickness, set back from the most sensitive region for surface quality.

Flight magazine's original coverage of the 1954 lecture encompassed the discussion afterwards, with a comment being made that the,

> "…nose flaps had proved necessary, but they were worth the complication."[147]

Aircraft design is very much a multi-disciplinary process and there is often a conflict between optimum aerodynamic shape and the resulting structure and implied weight required; the additional performance has to be worth the weight and cost that follow. The leading edge of a wing has to be strong enough to resist bird strike and will contain systems for de-icing, for example, before the additional weight, power and maintenance requirements of the actuation system are considered. Somewhere along the line, the complication of the nose flaps was viewed as an opportunity for development.

Test pilot Johnny Allam described how four stages of buffet had been identified, with nose flaps allowing safe operation all the way down to the maximum 'buffet phase 4'. Unfortunately a flap was lost during flight while, in service, accumulator failures occurred and left the nose flaps unable to retract, hence were a critical source of unreliability.[148]

The final five production Victors were built with a new, extended and cambered but fixed, leading edge and again, we can turn to a contemporary technical report to explain how HP was able to introduce this change. R.A. Wallis's report, published in 1965 but discussing work from nearly a decade earlier, starts with the statement that,

> "This project had its origin in a request from Handley Page Ltd of the probable effectiveness of discrete air jets as a method of boundary-layer control in the 'Victor' bomber aircraft. This was associated with a desire to replace the leading-edge flaps with a less complex device while maintaining an equal degree of effectiveness as regards maximum lift and aircraft stability".[149]

As the *Flight* article had said, the challenge was to adapt the high-speed outboard wing section to a different situation; with the angle of the wing to the incoming airflow increased from the cruise position to one that gave high lift, the sharp leading edge caused a rapid acceleration and pressure drop. The air was then faced with an uphill struggle to recover over the remaining chord of the wing back to atmospheric pressure at the training edge, while remaining attached and efficiently maintaining lift.

However, at some point as the angle of attack was increased, this pressure gradient would become too great for the boundary layer to negotiate. As this point was approached, the flow would be tempted further outboard as the swept isobars made it see a lower pressure region and hence path of least resistance there. This tended to load the tip of the wing and bring it closer to stall.

While Tip Stall was a major problem for early swept wing aircraft at high speed or when manoeuvring in the cruise, with the still lifting inboard and hence forward regions of the wing causing pitch up and exacerbating the situation, the problem for the Victor was one of widespread leading edge stall as the wing attempted to generate enough lift.

Energising the boundary layer by adding momentum from air jets would have seemed a viable solution. Compared to a fixed droop leading edge, it left the wing section in the optimum shape for cruise and adapted the aerodynamics of the low-speed condition to suit it. The report recounts that although the blowing worked to an extent, it would not have been enough to allow replacement of the flaps. The mass flow of air required was more than anticipated, but it was thought worthwhile to continue the tests to find out what would work, in conjunction with new leading edges designed to be drooped and hence meet the oncoming airflow at a reduced angle.

Wallis states that these leading edges alone produced such good performance that the idea of blowing was abandoned. A solution to the problem had been found, in a roundabout way. To avoid the problem with reliability, it was decided that in service, 'buffet phase 2' would be the limit for normal operation, while each production aircraft would be flight tested by the manufacturer at 'buffet phase 3'. The fixed droop leading edge proved capable of 'buffet phase 3' operation, which was by then sufficient. By removing the requirement for the final stage of low-speed performance, which was not used by definition in service, then the reliability problem could be solved.

The actual shape adopted for the Victor was the result of yet more design iterations with RAE and HP involvement, further building upon Wallis's work. Reattaching the flow after separation quickly was of vital importance to stability, as was ensuring that the stall occurred somewhere onboard first to avoid pitch up. The air jets had been found to reduce hysteresis and so might again have become part of the solution. However, the new 3% chord extension that would be fitted to the production aircraft solved both these problems by simply ending the extension abruptly at its inboard end.

The first production Victor was XA917, flown on February 1, 1956. Famously, it exceeded the sound barrier during a test flight in 1957, thought to have been the largest aircraft to have done so at the time.

Victor B.1 XA931 was the first such aircraft delivered to the training unit, 232 OCU at RAF Gaydon. Note the deployed outboard wing nose flaps in the view. ALISDAIR MACDONALD

The relatively conventional design of Vickers Valiant WZ391 contrasts with the exotic crescent wing and dihedral tail of Victor B.1 XA935.

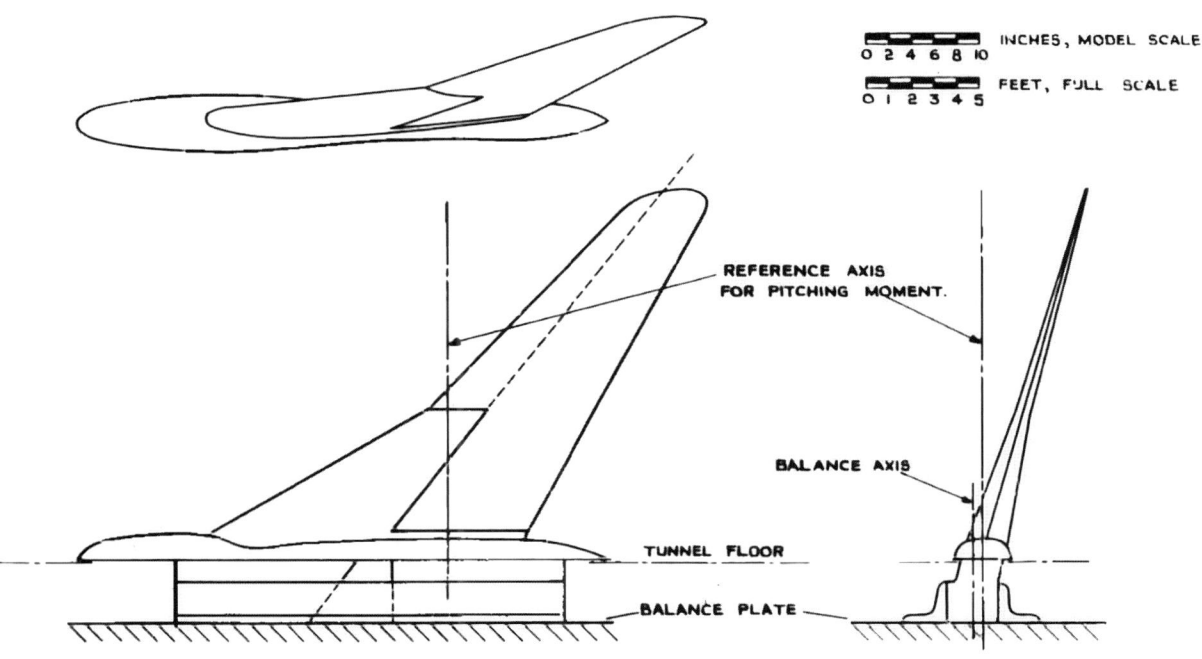

The Victor's impressive tailplane was tested as an individual assembly by the RAE.

This appeared to trigger stall locally, although the mechanism was not discussed. It might be expected that a local load increment from the edge vortex was generated, resulting in an initial separation, or perhaps the exaggerated discontinuity from the isobars of the inboard section, now that the peak suction outboard had moved forward, was the trigger. The Victor had acquired its very subtle dogtooth.

Either way, this work resulted in the required low speed lift being provided without the complexity of either leading edge flaps or air jets. Wallis's report makes several references to the ease with which the air jets might have been installed. Perhaps! They would have required bleed air from the engines to be ducted to them, which could possibly have been combined with the existing hot air de-icing. That said, the Victor B.Mk.1 in particular was not renowned for having spare engine mass flow capacity available on take-off. Provided that the cruise performance was not compromised (which was not discussed in the report), then the fixed leading edge would have been hard to beat by any powered system.

Late production Victor B.Mk.1 XH614, in anti-flash white. First flown in March 1959, it served initially with 57 Squadron.

THE TAIL

It is perhaps ironic that one of the defining and visually distinctive features of the final HP.80 configuration was the enormous dihedral tailplane, mounted in a then-novel position on top of the swept fin. Clearly, this was some distance from the visions of a tailless bomber espoused by Godfrey Lee and Sir Frederick Handley Page himself.

As noted previously, HP had been pushed in this direction by the relative reduction in longitudinal moment arm inherent in the crescent configuration, together with the realisation that both a powerful high lift system and provision of stability and control in the transonic regime would require the ability to accommodate significant changes in trim.

Providing the necessary stability and control authority was of itself not difficult. Nonetheless, the primary goal of HP's tailless studies had been to eliminate the parasitic mass and aerodynamic effects of the fin and tailplane as separate entities. For an aircraft chasing a specification at the edge of available technology, optimising the assembly to minimise its impact was vital.

With the wing root engine installation, the tailplane had to be positioned above or below the fuselage to avoid conflict. The initial HP configuration retained the tailless HP.75's wingtip fins on the new crescent wing but featured a swept tailplane without dihedral, mounted on a very short pylon above the rear fuselage. As noted previously, the redesign work undertaken by Küchemann and Weber included attention being paid to the wingtip fin profiles, but already doubts were present about them. If the fear of tip stall and subsequent loss of control was ever present and a key design driver, then it was surely of concern that primary geometry providing lateral stability and control was immediately adjacent to this aerodynamically vulnerable region of the aircraft.

Sir Frederick himself expressed concern over the possibility of asymmetric shock wave development, with a sudden and unstable change in drag occurring on one fin and generating an uncommanded yawing moment, inevitably coupled in both roll and pitch. The tip fins subsequently disappeared from the HP.80, replaced by a conventional vertical fin on the rear fuselage.

Intermediate development shows a tail configuration reminiscent of that eventually adopted for the Sud Aviation Caravelle, with a round tipped, leading edge filleted vertical fin and mid-mounted horizontal tail. Between this and a T-tail, it is likely that the lower mounting would offer a weight saving opportunity, with the horizontal tailplane loads carried closer to the fin root, hence reducing the bending moment. However, from an aerodynamic perspective, the endplating effect of mounting the tailplane at the fin 'tip' would allow a reduction in the height or area required to produce the same stability and control authority. A further aerodynamic advantage of the T-tail arrangement would have been the opportunity to minimise further the interference drag, with a junction between the fin and lower surface of the tailplane only.

The T-tail mounting may have allowed the fin to be

Victor K.1 XH621 landing at Luqa in 1973, making use of the impressive braking parachute. It wears the tail markings of Marham-based 214 Squadron, the final operator of the Mark 1. GODFREY MANGION/AVIATIONMT

Taking off from Luqa in November 1972, Victor XH667 shows its extended trailing edge flaps. By this time, the fleet had all been retrofitted where applicable with the fixed droop outboard leading edge, replacing the original flaps. 667 was the final Victor Mark 1, with the next serial number in series (XH668) being allocated to the ill-fated prototype of the Conway-powered B.2. GODFREY MANGION/AVIATIONMT

A superb study of XH667 about to touch down at RAF Luqa, in November 1972. The trailing edge flaps are fully extended and airbrakes open. This view particularly emphasises the size of the tailplane, while on the fin the badge of 214 Squadron is carried. This would be final unit to operate the Victor K.1, eventually disbanding in 1977. GODFREY MANGION/AVIATIONMT

more area—and potentially mass-efficient, but in turn it provided an opportunity for the horizontal tailplane itself to be more effective. By moving the whole assembly further out of the downwash field of the wing, it would become more powerful and would remain so to a greater aircraft angle of attack, as immersion in the low-energy wing wake was delayed. The rear fuselage mounting of the airbrakes reduced the impact of their deployment on the wing (and hence potential tail buffeting, as encountered on the Valiant), but also required sufficient physical distance between them and the tailplane to avoid negative interactions. Again, this pointed towards a system benefit of positioning the tailplane as high as possible.

Incorporating tailplane dihedral (sometimes referred to a Y-Tail) was not analogous to the roll stability in sideslip rationale of using this on the wing. From a static stability and steady state aerodynamic perspective, there were several potential advantages. The superimposition of local suction peaks in the junction would be reduced, in turn modulating a source of shock-induced separation. Importantly for maintaining control, the tailplane tips would be elevated further out of wing wake.

This would give increased control authority at high angle of attack, while in the event of loss of downwash from the wing, there was a measure of protection from the more progressive immersion of tailplane into wake.

Some clues as to the contribution of tailplane dihedral to the aeroelastic problems encountered by Victor may be gleaned from analogous work that was undertaken a little later in the United States.[150] In the early 1950s, the US Navy faced budgetary priority challenges due to the primacy of the USAF's role in strategic warfare. The intercontinental bomber was the only viable nuclear weapon delivery system, but fundamentally sat outside of the Navy's remit. Whether contrived or otherwise, an operational concept of a high-speed seaplane that could be used in either the strategic nuclear or minelaying roles was developed, eventually yielding a V-bomber sized jet-powered flying boat. Against competition from Convair, a development contract was awarded to Martin for what would become the P6M Seamaster.

It is interesting to observe that the particular requirements of its marine operation incidentally drove the configuration to be probably the most similar to the UK-built aircraft of any of the contemporary US-built types. Underwing podded engines were clearly not plausible; the four J71 turbojets were instead protected from spray, if not complete immersion, by being mounted above the inboard wing. Similarly, the horizontal tailplane had to be moved away from both seawater and jet blast. A T-tail was a natural solution, and perhaps an inevitable one with the involvement of Hans Multhopp as a consultant to Martin. Once again, the design evolved to one with 15° dihedral and NACA analysis showed very similar problems to those apparent in the UK:

* Flutter driven by fin bending and fin torsion.
* Dihedral likely reduced the impact of Mach number on fin structural modes.
* Dihedral significantly reduced the flutter boundary in terms of dynamic pressure.

* Fin torsional stiffness needed to be increased over the initial design, to give a viable structure.

NACA's overall conclusion was that tailplane dihedral and sweep (the latter having a similar rolling moment effect to dihedral), would definitely alter the relationship between fin bending and torsion compared to the 'flat' front view case. Fin torsion would create a stabilising yaw, producing in turn a rolling moment that would result in fin bending. Overall, the answer was the same as had become apparent on the other side of the Atlantic: if dihedral were not considered, then the dynamic loads encountered by the fin would not be predicted well enough to allow a safe structure to be designed.

The Victor's tail was clearly a step into the unknown, for good performance reasons. Was there a better alternative? For an aircraft with inboard wing-mounted engines, a low tail was probably ruled out, while a Valiant-style cruciform tail would not have provided the same endplating effect on the fin. Handley Page were—as Martin showed—not the only ones to be seduced, and it must always be remembered that the configuration was signed off by the scientific establishment of the UK government. Also, it looked amazing.

THE VICTOR CONCEPT IN RETROSPECT

So often it has been asserted that the Victor was a German concept, implemented by the Allies after hostilities using the information that they found. Yet the fundamental characteristic of the Victor's aerodynamic philosophy was a constant critical Mach number across the wingspan, which aimed to postpone the drag divergence Mach number to a point beyond the specified B35/46 cruise conditions (500kts at 50,000ft, equating to M = 0.873). In order to make this work, the reduction in sweep outboard was an enabling concept, and one that was convenient. There could have been other choices.

Handley Page had developed significant IP and detailed designs in pursuit of a large, swept wing (and tailless) aircraft, in advance of the availability of German data on high-speed aerodynamics at the end of the Second World War. The German development of crescent (or cranked) wing planforms did predate UK development, but as evident from both their approximately constant non-dimensional thickness and documented statements of intent by those involved, the designs were specifically aimed at improving longitudinal stability characteristics through alleviation of tip stall. Of themselves and in the forms discovered by the occupying nations in Germany, they did not provide the constant critical Mach number capability that was HP's target.

While designs existed in Germany that superficially resembled the Victor, it is clear from more detailed assessment that the ultimate solution was instead a far more advanced, transonic biased concept that was visually, but not aerodynamically, dominated by the crescent shape.

Victor K.1A XH621 refuelling Lightning F.2A, the latter in the low-level camouflage of RAF Germany.

The rivals — a Victor B.Mk.1 and Vulcan B.Mk.2 in New Zealand, in the early 1960s.

THE OTHER BROAD DELTA: THE GLOSTER JAVELIN

SOME MIGHT say that RAF Fighter Command remained set up to fight the Battle of Britain well into the 1950s. A vast multitude of day fighter squadrons, first Meteor and subsequently Hunter equipped, would play the part of the Spitfires and Hurricanes. Unlike Dowding's situation in 1940 however, the exploitation of the University of Birmingham's invention of the cavity magnetron, and its deployment in millimetric wavelength radar, would mean an interception in limited visibility was now practical. The Mosquito was gradually replaced in this role by the adapted Meteor from Armstrong Whitworth and it was hoped that eventually this too could be replaced by an all-new high-altitude aircraft: the Gloster Javelin.

It is interesting to speculate on how much Avro and Gloster, both HS group companies, discussed their delta developments and exchanged vital information. The integrated airframe and propulsion layout of the Vulcan and Javelin might at first glance appear to have much in common. It's fair to say that, at this separation in time from the events, the reputation of the latter does not compete well with that of the former.

Flight magazine of July 8, 1955, contained a telling snippet which revealed some of the pressure that Hawker Siddeley was under to make the RAF's new fighter, the UK's defence against nuclear attack in anything other than the daylight and good weather, work effectively. Like the Vulcan, it was one of two competing aircraft specified to fulfil the same specification. The other, the de Havilland DH.110, suffered a very public setback at the Farnborough Airshow in 1952, but was in no way a clearly better design prior to the disaster. RAE assessment showed that,

"There is no appreciable difference in performance between the two types so far as can be seen from existing evidence... With the same maximum lift coefficient the height at which it has a given turning radius will be lower by about 5,000ft than that of the Gloster design".[151]

In other words, the delta planform again provided the wing area that would mean more 'g' was available for manoeuvrability at altitude, obviously vital for the rear pursuit attack that either cannon or short-range infra-red-homing missiles would need.

By the mid-point of the decade, the new fighter was still not in service and questions were being asked. Under the title 'Javelin Rumpus', HS produced an eight-point rebuttal to allegations in a *Sunday Express* article, concluding with,

"There is no similarity between the state of the

High-speed wind tunnel model of the Gloster Javelin, to 1:72 scale for use in the RAE Bedford 3ft x 3ft tunnel, as part of the investigation into stability around M = 1. ARC-R&M-3403

Javelin and the [Vickers-Supermarine] Swift story, and any attempt to create it is misleading and irresponsible."[152]

How had HS ended up in a position where its delta wing fighter had to be defended from comparison with the unfolding disaster that was Supermarine's situation?

Flight testing of the Javelin was an occasionally unhappy experience, in common with many new high performance military aircraft of the time. One prototype and its pilot were lost, while another came close to killing Gloster chief test pilot Bill Waterton, when after courageously and skilfully landing his elevator-less example in order to retain the evidence for analysis, he was trapped inside. He received the George Medal for his considerable troubles.

Upon severing his ties with the firm, Waterton took his poacher's insight to the *Daily Express* as its air correspondent. He appeared to take seriously a remit to hold the British aeronautical establishment to account, and was scathing in his words. In particular, he bemoaned the inability of industry to deliver successfully matured manifestations of their promising conceptual designs. How could it be that the RAF aircrews were being asked to fly aircraft with fundamentally unsafe handling characteristics, that might reasonably have been found at the prototype, if not wind tunnel stage?

As an example, flight tests revealed in 1955 (when the aircraft was close to production) that the Javelin lost directional stability as it approached Mach 1. In practical terms, this meant that aircraft happily pointed into the wind until $M = 0.98$, from which point as the Mach number increased it would yaw towards 5 degrees side slip, triggering Dutch rolling motion. Passing through sonic speed, directional stability was restored. As this highly non-linear and undesirable behaviour was only just beyond the aircraft's design conditions, it needed to be understood and mitigated, if not eliminated.

Wind tunnel tests in the RAE Bedford 3ft x 3ft transonic tunnel revealed the cause. The rear shock on the inboard wing was strengthened by the recovery demand over the adjacent fin and rear fuselage; the local expanding channel created by these parts as they reduced in thickness rearwards, at sufficiently high Mach, resulted in the shock wave causing fin separation and loss of directional stiffness. Balsa wood and plasticene were deployed to create new geometry to fix the problem. A build up of the rear fuselage contour and the addition of vortex generators successfully did so. Clearly, it was still possible to miss some of these idiosyncrasies with the tools available at the time. Equally, when discovered they needed to be addressed as soon as practicable if the aircraft was to have any hope of being viable in its intended roles.[153]

There is another side to this of course. In the late 1940s, men and women living on rationed food left their homes in bomb damaged streets, travelling to war-weary factories on equally tired bicycles or steam-hauled trains. When

Advertisement from 1954, linking the Hawker Siddeley group's prowess in delta bomber design to that for future NATO fighters.

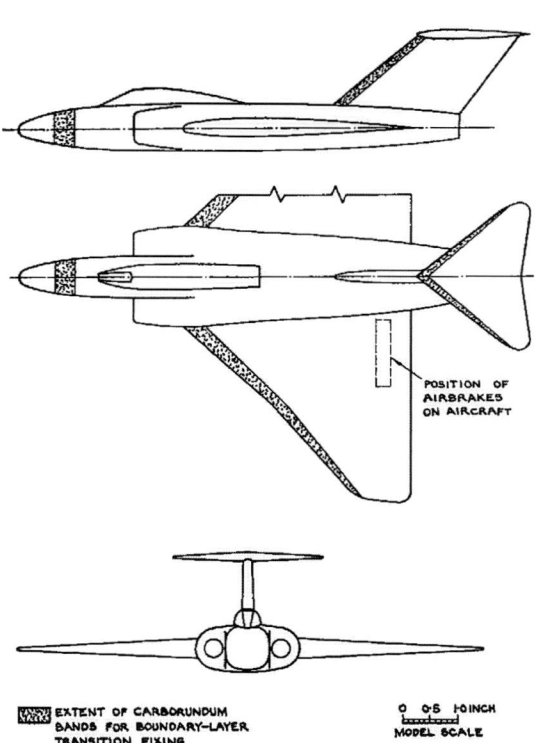

Layout of the Javelin high-speed model used at Bedford, showing in particular the roughness strips applied to the leading edges in order to trip the boundary layer to turbulent earlier on the chord, as would happen at the higher Reynolds number of full size. Otherwise, the interaction of the shocks with the boundary layer would have been representative. ARC-R&M-3403

they got there, they engaged intellect and slide rules to produce world-beating technology. They knew it had to be done; they knew what a war felt like and there was a new enemy. Even if peace prevailed, their economy was on its knees and exporting these incredible products was a way of balancing the books. Not all of these aircraft worked as intended, but such is life at the cutting edge.

The aerodynamic engineering effort discussed so far has focused on the requirement for high-speed wind tunnels to adequately characterise and allow development of these pioneering transonic aircraft. Wind tunnels allow a controlled environment to be created, in which repeatable tests can be performed. The models used can be made sufficiently stiff and strong, such that engineers can be confident that they are measuring aerodynamic loads from a myriad of internal sensors, rather than the effect of deflections of a flexible structure. They were expensive, complicated and of themselves approaching the limit of known technology; gathering reliable measurements at transonic speeds was fraught with difficulties associated with interactions between the shocks and the tunnel walls.

Wind tunnels were certainly not the only way to gather data however. One could choose alternatively to strap a large, instrumented model onto a big rocket and stand well back. This is what happened during development of the Javelin, using technology and techniques that had their origins in the aborted Miles M.52 supersonic project. The RAE explained the attraction of this technique as,

"*The rocket-powered model may be regarded as being complementary to the wind tunnel model. The advantages it offers are the absence of any modification to the model shape because of model support requirements, the absence of the wind tunnel constraint corrections and the ability to attain high Reynolds number. For models of typical size, chord Reynolds numbers in excess of 10 million can be obtained at supersonic and transonic speeds. The range of lift coefficients which can be obtained is, however, limited because of the low wing loading inevitable on models.*"[154]

The basic planform of the Javelin's delta wing was not dissimilar to that of the baseline Vulcan design (48°

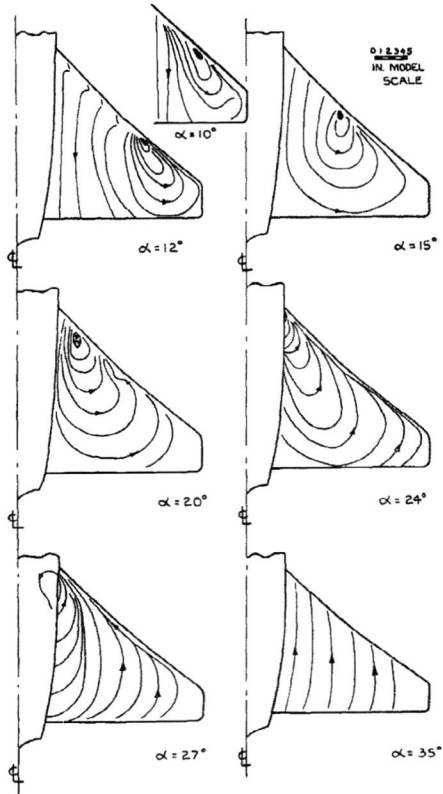

Analysis based on wind tunnel visualisations and pressure measurements, of the flow over the Javelin wing at low speed. A leading-edge separation began to appear outboard from about ten degrees angle of attack, initially as a closed bubble, but quickly forming a strongly out-washing region of completely separated flow. This was in marked contrast to the aerodynamics of a 60 degree leading edge sweep delta with a sharp leading edge, as is typically of highly-manoeuvrable fighter geometries of the intervening period. For those wings, stable separation occurs from the apex of the wing, forming a full-span vortex and a significant contribution to the wing suction and lift. Like the Vulcan, the concept pre-dated this understanding. ARC-R&M-3078

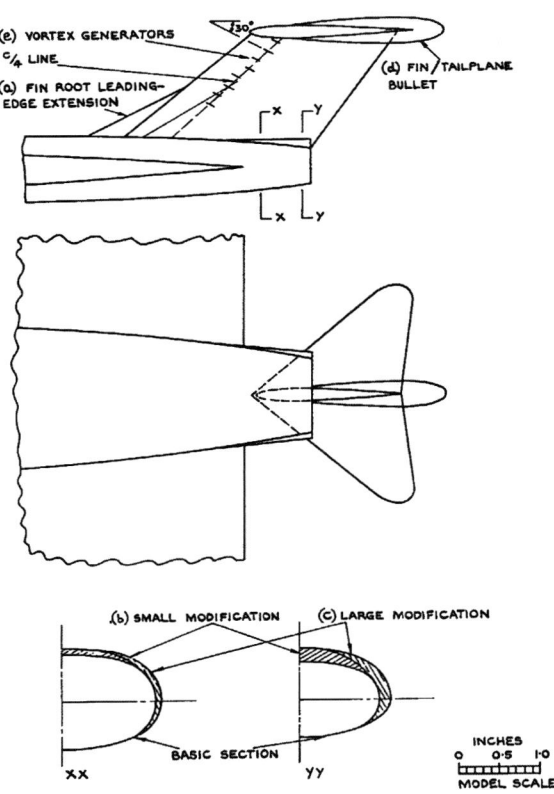

Measures taken during development of the Javelin to correct the lack of yaw stiffness around Mach 0.99, then tested in the RAE Bedford 3ft x 3ft wind tunnel. Building up the rear fuselage proved to be the fix, as it reduced the rate of pressure recovery needed and so suppressed shock formation. The Vulcan wing trailing edge was relatively rearward in alignment with the fin, when compared to the Javelin, and obviously was not designed for quite such high speeds. Both of these factors would have helped the bomber avoid these issues. The tailplane was essential to provide the increased pitching moment for 4G+ manoeuvres, together with longitudinal trim at high Mach number. ARC-R&M-3403

leading edge sweep compared to the bomber's nominal 50°, yielding similar aspect ratio) and it used the RAE 101 transonic aerofoil, a more modern approach than the relatively conservative modified NACA aerofoil of the Vulcan's wing. We know though that the bomber benefited hugely from the section and planform modifications proposed by the RAE engineers, with the swept root and tip of the maximum thickness line and the extended, dropped leading edge. The Javelin wing was not modified from its vanilla design to anything like the same extent. This technology level seems consistent with published performance data and anecdotal statements from those associated with operating it: the Javelin would have struggled to catch a high-speed, high-altitude contemporary bomber like the Vulcan. This led to a development project that became known as the 'Thin Wing Javelin'.

A modified, thinner wing was the major change considered, with the rest of the aircraft initially remaining similar. Unsurprisingly, the development became more ambitious and ultimately little of the original was left, however performance was still likely to have been, "just supersonic in level flight at altitude." Minister of Supply Reginald Maudling (responsible for the provision of military aircraft) became aware that HS Group was building a much more advanced and potent fighter aircraft, the Avro Canada CF-105 Arrow, in December 1955. The Thin

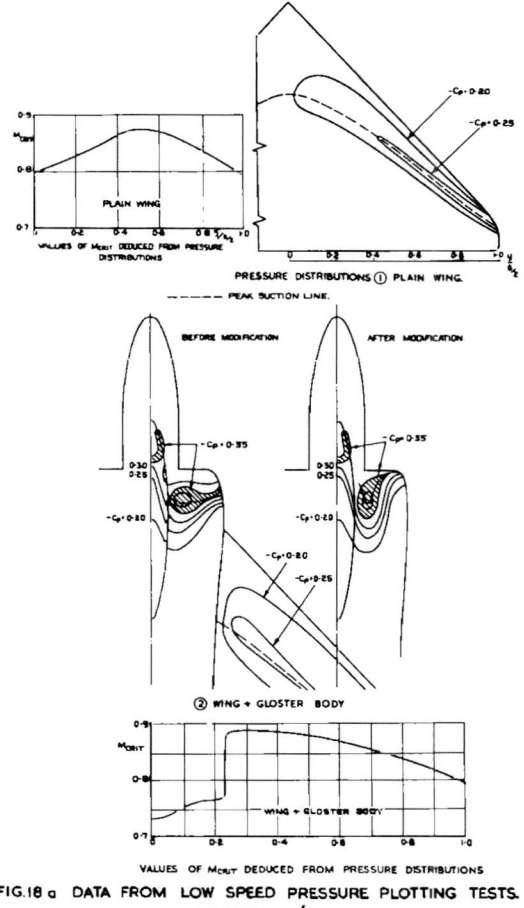

Pressure plotting tests from early Javelin testing in the RAE HST wind tunnel. The combination of fuselage and wing was seen to extend the isobar sweep further into the central region, advancing drag divergence compared to the wing alone. However, the critical region was the cabin. RAE AERO R.2359

Wing Javelin was obviously out of date and was therefore cancelled in June 1956; Hansard states an expenditure of £2.3m on the project. Was that a lot? Certainly not compared to the £22m which the same source records as having been spent on the Supermarine Swift fighter, for much the same outcome.[155]

The free flight model tests were divided into two strands. The first used a baseline Javelin fuselage and nacelle centre section and tested four sets of wings. Wing 1 was the baseline Javelin wing, referred to as the 'Mk.1 Javelin wing', although confusingly not actually the wing that would be fitted to the Javelin FAW.1 when it entered service. This was of constant RAE 101 10% thick section. Wing 2 was a baseline, constant section 'thin wing', similar but with a thickness of 7% and intended to allow postponement of drag rise to an increased Mach number. Wings 3 and 4 were much more interesting aerodynamically, but illustrated the scope of the challenge. They were designed to exploit the structural advantages of tapering thickness, which should have meant that they were lighter for a given stiffness, when scaled up and stressed for full size weights.[156]

Both were designed aerodynamically to the 'Haines' concepts which were implemented in the Vulcan redesign of 1949. A. B. (Barry) Haines demonstrated that by tailoring the chordwise position of the maximum thickness of the wing sections across the span, the aerodynamic sweep of the isobars could be displaced from the geometric sweep of the wing itself. This was important at the root and tip in particular, where the isobars would otherwise unsweep themselves as they passed through the fuselage or into the freestream flow respectively. In this case, the geometry lofted through appropriate sections, including the baseline RAE 101 with maximum thickness at 31% chord, to the RAE 104 which placed it at 41% chord. Towards the root, a section with the max thickness moved far forward to 12% chord completed the effect.

However, the thickness variation was not only of position, but of magnitude. Wing 4 was designed to be optimised aerodynamically while exploiting the structural benefit of tapering the thickness outboard. Wing 3 was based on this optimum design, but made more pragmatic allowance for actual internal stowage requirements. As such, while root thickness was 10-11%, both tapered down to 5% at the tip.

What is remarkable was just how much better all three of these developments were, in terms of delaying drag rise. The largely impractical constant 7% thick wing could essentially be discounted, but the implementable option of Wing 3 demonstrated low speed drag that was much lower, while the onset of drag rise was postponed from about M=0.88 to M=0.96.

An outcome of the test results related in the report was the statement,

"For all practical purposes the mean thickness (other things being equal) is an adequate guide to the transonic performance of the particular configurations tested."

This is of course of great relevance to an assessment of the high subsonic Mach number performance of the Vulcan, which had demonstrated effective high-speed performance in its B.1 form with a wing of relatively high mean thickness. While such a gain as seen in the Javelin wing tests would have been immensely valuable in pushing out the performance boundary of a firmly subsonic bomber like the Vulcan, it was also clear that the advantage to be conferred on a fighter aircraft which aimed to be

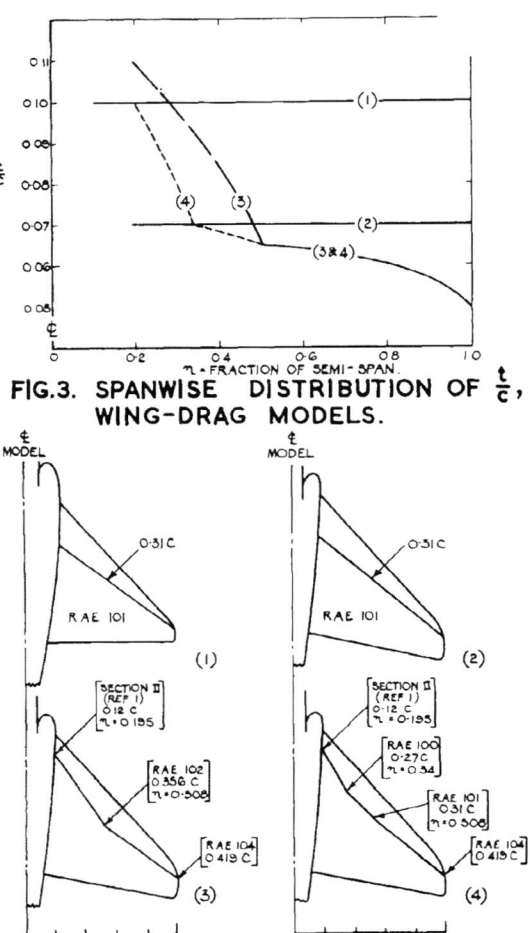

FIG. 3. SPANWISE DISTRIBUTION OF $\frac{t}{c}$, WING-DRAG MODELS.

FIG. 4. LINES OF MAXIMUM THICKNESS. WING-DRAG MODELS.

Investigation of advanced wing designs, using rocket powered models to collect data. The baseline Javelin (model 1) was advanced through predominantly thinner geometries, with a tapered thickness wing proving both effective and practical. The proposed designs were not unlike those proposed for advanced Vulcan development. ARC-CP-0678

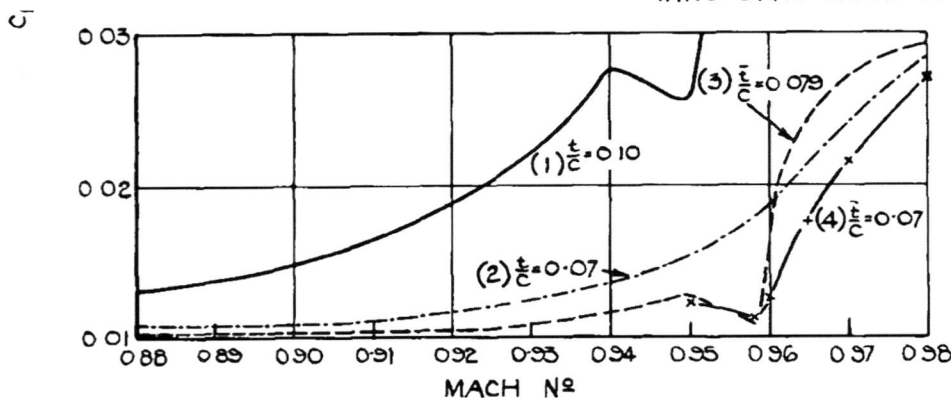

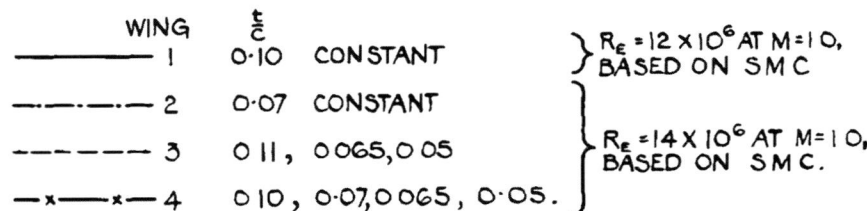

FIG. 8. ZERO-LIFT DRAG — WING-DRAG MODELS.

Applying fairings to the modified, high speed Javelin design to conform more closely to area ruling proved moderately successful supersonically, but was poor in terms of subsonic drag.

supersonic was not significant enough to justify such vast engineering effort.

There is a second reason why these geometries may not have been sufficiently advantageous. By the time the Javelin entered service, problems with tip stall identified very early in the flight test programme had led to the necessary adoption of a new outboard wing for stability and control reasons, rather than pure performance. From 57.5% semispan, the wing leading edge sweep was reduced to 41.5°, resulting in an increase in tip chord. At the tip, the thickness was reduced to 7%, the leading-edge radius significantly sharpened and the maximum thickness moved back to 51% chord. This region was then defined by smoothly lofting between the basic inboard section. Consequently, the production Javelin took a very modest step towards becoming 'thin wing' and sweeping back the tip isobars; its drag divergence Mach number was presumably already better than the baseline case considered for the free flight tests.

While the tests described above were conducted using wing models attached to a bespoke fuselage containing the rocket motor, a second strand of tests was undertaken in which a complete aircraft model was mounted on a separate booster. This looked at adopting a form of the area rule, which was being found (at roughly the time of the initial tests) to reduce transonic drag. This required the rate of change of cross-sectional area of the complete aircraft with respect to the direction of flight to be minimised; in other words there should be no abrupt discontinuities. The tests all used the favoured Wing 3, the best overall practical solution.

The development of the whole aircraft resulted in substantial changes. One example was the use of highly swept intakes to replace the baseline Javelin's straight pitot inlets. Clearly, the latter change removed a major discontinuity that may have been a significant player in this aspect of the actual aircraft's performance. Large fairings were added to the rear fuselage outboard of the jet pipes, aiming to compensate for the sudden end of the wing's contribution to cross sectional area. It was noted that a blended delta is inherently suited to the area rule, but that the wing trailing edge was one of the regions where action such as this would be required.

Ultimately, the results of the tests showed little effect subsonically, although all options indicated a considerably better trend supersonically. Two results of particular significance from the perspective of the V-bomber designs were variations in intake sweep. Both an intermediate and final step showed considerably better performance in terms of drag at both subsonic and supersonic Mach number. Although the Vulcan (and Victor, despite appearances) were far too early to consider the area rule, in effect Avro lucked in with their application of the swept, slotted inlet to sweep the inboard isobars as per Haines's theory, whereas Gloster's use of the pitot inlets had the opposite effect. Once again, the perils of being the pioneer, but in this case the effect was important, given that the fighter's maximum operating Mach number needed to be beyond M = 1, in order to give margin for diving above the specified maximum level speed.

A further point drawn by the report was that the subsonic drag increase (not related to wave drag) for the rear fuselage modifications was considerable, despite their positive effect supersonically. This was surmised to be due to the increase in base area from the additional volume; a properly designed solution could have addressed this to some extent.

This brief look at the Thin Wing Javelin through the prism of contemporary technical data, confirms that the changes required to produce any worthwhile performance delta would have amounted to a new aircraft. After initially testing a completely new wing, it became obvious that any improvement in transonic performance would require application of the area rule, and that in practice would have meant a new fuselage. Given that the manufacturer would need to start again and that the basic aircraft itself was proving difficult, the decision not to proceed with the Thin Wing Javelin can be seen to be entirely logical. In spite of any legend that may have grown to show this as one of the great lost opportunities of Britain's 1950's aerospace programme, real technical data fundamentally shows that was unlikely to have been the case.

On the other hand, it has given some more background on just how advanced the eventual application of some of these strategies during Vulcan development was. The Avro bomber clearly benefitted from the RAE's theories on thickness distribution with span, particularly in the inboard region. It may too have improved transonic wave drag delay due its swept intakes. What the work does clearly illustrate though, is where Avro had the potential to go with improving the Vulcan's wing design. A wing of, say, 7% constant thickness would require a new aircraft. A wing of 7% average thickness, retaining the hard-won benefits of deep inboard wing and engine installation, might just be the answer to higher altitude and improved manoeuvrability, the keys to the bomber's safety and deterrent potential.

10

SECOND GENERATION: THE VULCAN AND VICTOR MARK 2

Neither the Vulcan nor Victor in their B.Mk.1 forms delivered the payload-range-altitude performance that B35/46 had specified. At the same time, the continuous development of fighter aircraft, as well as guided missiles, meant that a similarly continuous improvement in bomber performance was sought. Compared to earlier aircraft, the development time of the V-bombers had been long and the production ramp-up slow. Nearly three years would separate the initial delivery of a Vulcan B.1 to the RAF, on July 20, 1956 when XA897 arrived to Waddington as 230 OCU's first jet, to the 45th and final B.1 production aircraft, XH532, on April 30, 1959. Several thousand Lancasters had been produced and supplied from the same factories in a similar timescale, a little over a decade earlier. But, in the same way that the capability of their piston-engined forebears had been boosted by the development of engine power and exploitation of their inherent strength, both the Vulcan and Victor were well placed to take advantage of the rising tide of the UK's gas turbine prowess.

With the bomber's defence relying on altitude (which might be increased) and speed (which was effectively constrained by the subsonic design), improvements to survivability would imply an increase in wing area and available cruise thrust from the engines. For the Vulcan, both of these were entirely within the realms of possibility, and were the subject of serious work even prior to the B Mk 1 entering service. The basis was further development of the remarkable Bristol Olympus, together with another conceptually advanced competitor from Rolls-Royce, the RB.80 Conway. As if to underline this nascent potential, in August 1955 Canberra WD952, equipped with Olympus 102 engines replacing the standard Avons, reached an altitude of more than 65,000ft (or more than 20km above the Earth), well out of the reach of any Soviet fighter.

THIN WING VULCAN – PHASE 3

In December 1954, Avro proposed a 'Thin Wing Vulcan', identified as the Phase 3 variant. From the 10% thickness section immediately outboard of the engine bays, the new wing linearly reduced to 6% at about mid span, then more gradually down to 4% at the tip. The brochure design has a relatively long chord tip, but without the curved planform that would become more familiar on the eventual Vulcan B.2. While the external aerodynamic changes were all in the direction of increased usable lift coefficient and postponed drag divergence, the changes were far from skin deep.

The wing would have been of multi-spar sandwich construction, more akin to the technology deployed by Handley Page on the Victor, but which was also being

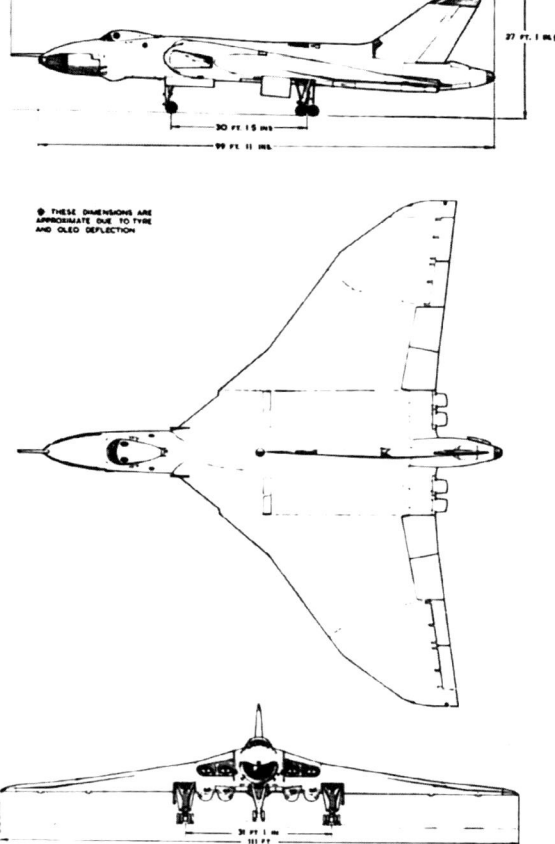

General arrangement drawing of the Vulcan B.Mk.2

investigated by Avro for its 720 rocket-propelled interceptor, being designed concurrently. The key advantages were a reduction in structural weight for a given strength, well supported and accurate external contour, and not least an increase in internal volume for integral fuel tanks. The aircraft was proposed at double the original B35/46 weight, now at 200,000lb, and to support this a revised main undercarriage with four (rather than eight smaller) wheel bogies would have been incorporated.

Compared to the Phase 2 wing (Newby's extended leading variant adopted for the production B. Mk. 1), area increased from 3,446sq ft to 4,060sq ft. With all other things equal, this would have given a proportional reduction in the lift coefficient required to support a given weight, in turn moving the wing further away from the point of shock induced separation. That could be cashed in a number of ways, with height over target nominally the most useful, but clearly improved manoeuvrability for evasive action was also of great value. In the latter case, when lift coefficient was by definition increased from the cruise value in order to provide 'g', the 14ft increase in span and change in aspect ratio from 2.84 to 3.14 implied relative reductions in lift-dependent drag too. Remarkably, the total wing weight was calculated to increase by only 1,330lb, despite being stressed to support a 38% greater fuel capacity in the integral tanks. With Olympus 6 power, carrying a Blue Steel equivalent stand-off weapon and fuel for 5,000nm, the aircraft would have achieved a height over a half-range target in excess of 56,000ft.

The performance that would actually have been realised by a production Phase 3 Vulcan is of course, something that it will never need to prove. The published drawings show a relatively simple planform, as might be expected of a design that was proposed prior to the Vulcan B.1's leading edge modifications. That said and as the Thin Wing Javelin data has illustrated, the reduced average thickness across the span in general, with an extended tip chord to exaggerate the achievable minimum thickness to chord ratio in the susceptible region, both combined to reduce the likelihood of the high peak suctions that had been experienced on the B.1. Of itself then, the new wing design was a logical method rooted in the assumed sound aerodynamic engineering knowledge base of the time, that would have been expected to push back the buffet boundary.

It does seem most likely though that the emerging learning on the supersonic expansion, isentropic compression 'peaky' phenomenon would have been applied to the Phase 3 design as it developed. Whether this would have taken the form of a leading edge extension, or a progressive, conical camber-like droop within the existing planform boundary is open to conjecture.

When the Phase 3 aircraft was proposed in 1954, there were still precisely no V-Bombers in service and the definitive machines were still some years from being so. Perhaps the diversion of available resource and effort into such a radical redesign was a source of concern. Relying partially on the innovative structural design of the new wing for its performance advantage, the realisation would in turn have needed Avro to solve the manufacturing and production problems that were already a headache at Handley Page. Undoubtably, it would have taken much longer to get into service than the eventual second-generation machine that did emerge, while the clock continued to tick down on the supposed window of viability for the high-altitude subsonic bomber. A simpler concept that leveraged more of the investment already made, in time as well as treasure, was obviously an easier sell.

PHASE 2C

Stuart Davies listed the changes introduced in the Vulcan B Mk 2 in his 1969 lecture on the aircraft as:

SECOND GENERATION: THE VULCAN AND VICTOR MARK 2

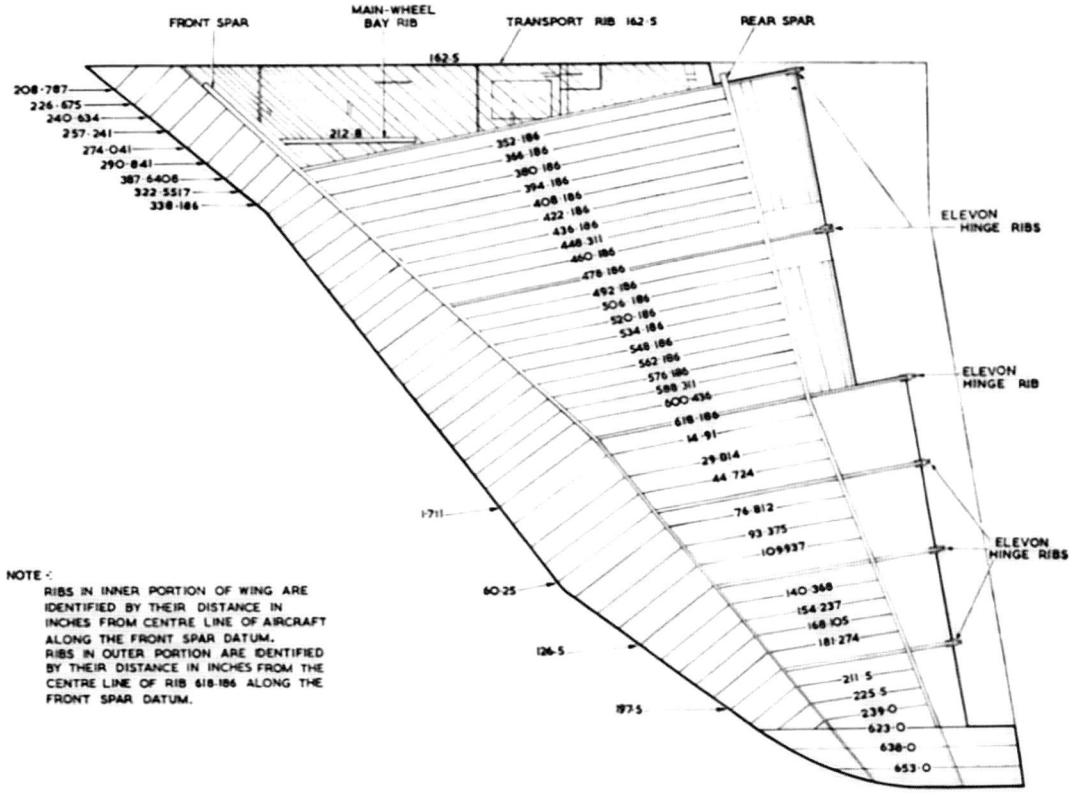

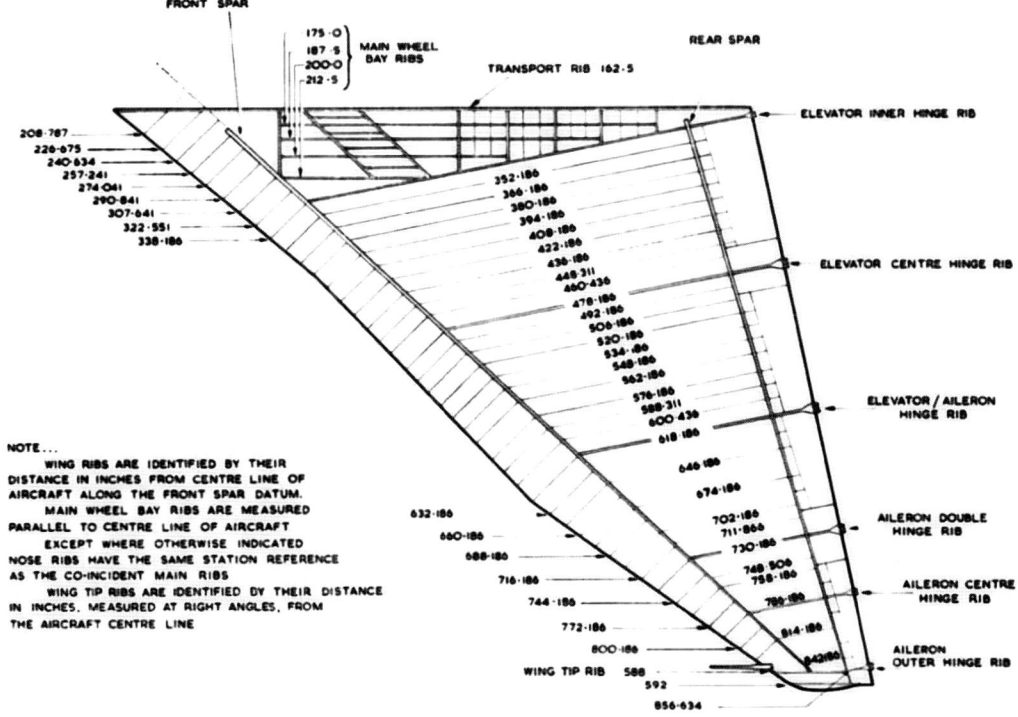

Detail of the Vulcan Phase 2 (lower, for Vulcan B.1A) and Phase 2C (upper, for Vulcan B.2) wings. These diagrams were included in the service manuals for the aircraft, as provided to RAF engineering personnel.

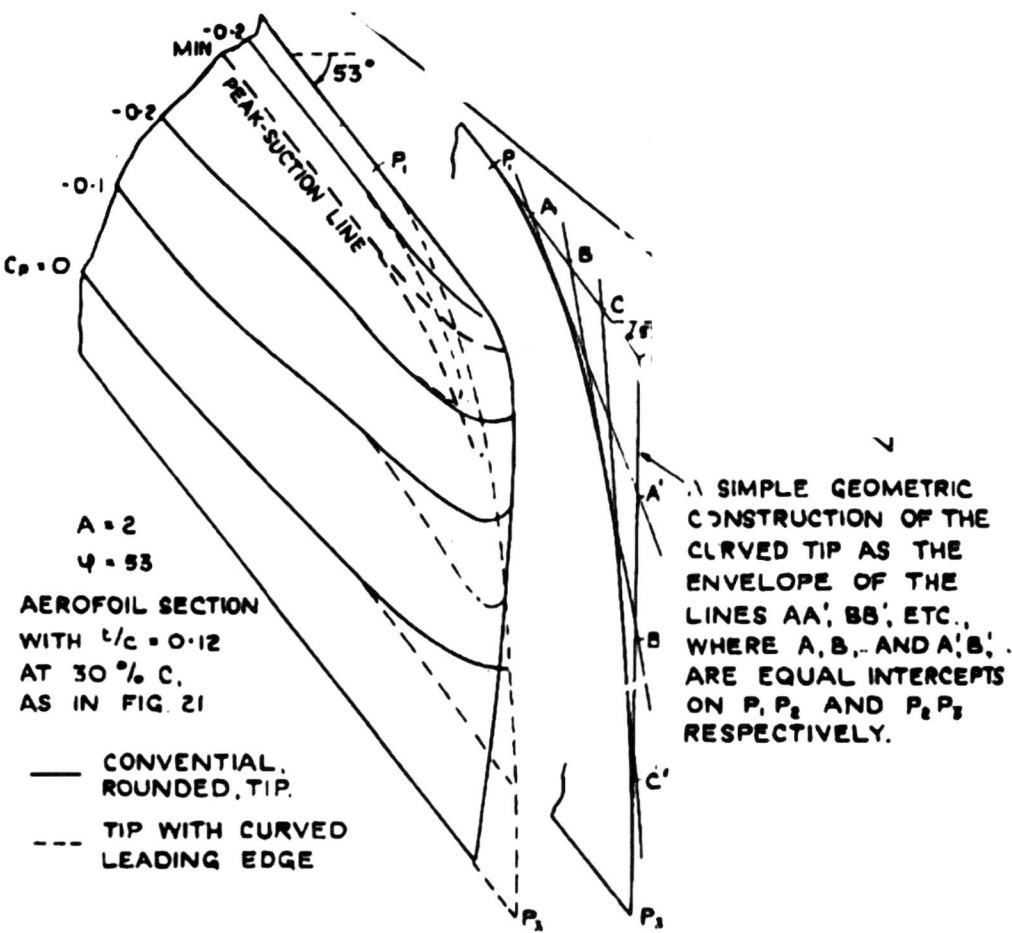

Description from Küchemann of the principle of his eponymous wingtip. The isobars of a basic design are illustrated as solid lines, which 'unsweep' as they approach the tip, having nothing beyond to retain the desired pattern. Whereas, by moving the region of peak suction gradually rearwards, the Küchemann tip exaggerated the isobar sweep and mitigated this effect. The overall effect was to avoid the early triggering of a shock wave that would propagate inboard, undoing the good work of the rest of the planform.

(a) More powerful engines: Olympus B016.
(b) New outer wing and leading edge.
(c) Elevators and ailerons of the Mk 1 changed to elevons.
(d) New and enlarged rear fuselage to accommodate additional equipment.
(e) AC electrical system.
(J) Rover auxiliary power unit.
(g) Strengthened main and nose undercarriages.
(h) Larger air intake for future Olympus, e.g. B0121 approximately 20,000lb sea level static thrust.

Further changes would include shortening of the nosewheel leg, which increased clearance under the rear fuselage. In turn, this connected with a very significant development in the aircraft's envisaged role, as free-fall weapons gave way to stand off missiles. A brochure of November 1956 introduced the proposed Vulcan Phase 2C and Blue Steel missile combination, nominally a fearsome advance in Bomber Command's deterrent capability.

Commentaries on the history of the Vulcan's development almost invariably include a statement to the effect that more powerful engines needed a larger wing to realise their advantage. Is it immediately obvious why this should be so? Firstly, the V-bombers were characterised by their need to cruise at high speed and altitude, so the key requirement was one of level flight. In that case, to satisfy Newton's second law of motion, we know that the aircraft must have an equilibrium of forces about its centre of gravity. It is not changing its velocity with time, that is accelerating and hence requiring a force to be applied. Normal to the flight path, the lift force generated must be equal and opposite to the weight for this balance to be true, while along the path thrust must be equal to drag.

Both lift and drag are forces generated by the aircraft's

Vulcan B.2 XH558 in flight over Old Warden in October 2015, clearly showing the curved Küchemann-style wingtips adopted as part of the Phase 2C design.

aerodynamics, and are proportional to air density and the square of the air velocity. The other factors are the lift or drag coefficient, a non-dimensional quantity describing the effect of the aircraft's shape, together with wing area, which effectively is the scaling factor that gives the coefficient dimension. We know that lift is equal to weight and for a point in the flight, the latter is fixed.

If the aircraft were to cruise at a higher altitude, then it would find itself in a condition of reduced air density. Lift would have to be increased, but due to the Mach number limit this cannot be achieved through an increase in speed. The incidence of the wing could be increased, to change the lift coefficient to a higher value. This unfortunately has a knock-on effect of generating more lift-dependent (induced) drag, reducing efficiency and shifting operation away from the optimum lift-to-drag ratio. In the transonic world, it also at best brings the buffet boundary closer but in all likelihood would generate more wave drag, as if the Mach number had been increased. Fundamentally then, the V-bombers were stuck in a corner, limited by their fledgling transonic technology and by definition close to their operating limits. In time, improved aerofoils and three-dimensional wing design capabilities would allow efficient operation at high lift coefficients, but in the near term, an increase in wing area was the most expedient method of increasing altitude over the target.

Avro's solution to a minimum change, maximum reuse thin wing was to retain much of the existing inboard region. The outboard wing was extended in span, from 99ft to 111ft, and in order to retain sensible sweep the planform increased the trailing edge angle outboard of the second control surface line. Like the Phase 3 shape, the wingtip was of relatively long chord, so for a constant physical thickness, the thickness to chord ratio was reduced. The tip was just 5% thick as opposed to the original 8%, while the leading edge kink (a necessity in order to incorporate the drooped and extended leading edge shape in the adapted design) was just 4.25% thick. The section through this region was reminiscent of the cut-and-shut geometry developed by Newby for the Vulcan B.1, but was designed in from the start using the lessons of that programme. The tip used the RAE 104 section, which

A long way from home. This unidentified Vulcan B.2 was photographed at Elmendorf Air Force Base in Anchorage, Alaska, probably in December 1961. It lacks a refuelling probe, but has the definitive inlets, of which XH557 was the earliest example so configured. JACK KELL

Almost certainly not the background that the dark green and medium sea grey camouflage scheme was designed to be effective against, nonetheless Vulcan B.2 XL445 was doing a good job of hiding among the colours of the Maltese landscape. GODFREY MANGION/AVIATIONMT

A Vulcan B.Mk.2 about to touch down on runway 24 at RAF Luqa, Malta, in October 1967. GODFREY MANGION/AVIATIONMT

moved the peak thickness rearwards compared to the RAE 101 section that had been used previously. Because it was thinner, the reduction in leading edge sweep was acceptable aerodynamically; one parameter was traded for the other. A second important effect was the vortex generators on the upper wing surface; a band-aid to postpone shock induced separation, they were no longer required.

Towards the tip the blend to the RAE 104 aerofoil, with its rearwards displaced maximum thickness, served to further sweep the isobars. This was still more refined by the curved tip leading edge, providing locally increased sweep but still retaining a finite chord and hence sufficient taper ratio to avoid tip stall. It is also interesting to note that this tapered thickness approach was part of the basis of the crescent wing used on the Victor, although the implementation was of course very different.

A consequence of the thinner wing was that the PFCU systems for the outboard control surfaces would no longer fit with it. Instead, they were mounted within fairings below, each drawing cooling air from a NACA inlet in its forward part. All Vulcan B.2s sported a curious feature associated with this: the fairing's mounting flange extended to the trailing edge of the wing, whereas the fairing itself was cut back to provide a cooling exit. As might be guessed, this had literally been the case on the first aircraft; in operation, the exit provision had been found to be insufficient, leading to the evidence of design by hacksaw on a nuclear bomber.

In a major revision to the control strategy, all eight of the wing trailing edge control surfaces were dual function elevons, as opposed to the specific inboard elevators and outboard ailerons of the B.1. Elevon control of the Vulcan was discussed at the embryonic stages of the 698 programme, and analysis of its use was written by Newby following RAE wind tunnel testing. Like much of the early testing, the results were contaminated by the effects of the low Reynolds number. This was particularly serious in the case of the ailerons, due to the early onset of tip stall.

One effect observed was flow separation on the lower aileron surface, seen at high up deflection. Reducing the angle required for a given rolling moment could be achieved by using the full span of the control surfaces, albeit the inboard surfaces (the elevators on the Vulcan B.1) had a much shorter moment arm. The RAE tests were conducted on a pre-1949 redesign configuration, showing that the elevon control was twice as effective as the aileron alone. The total area of all the surfaces was two and half times that of the ailerons alone, demonstrating why the lateral control surfaces were typically outboard in the first place.

If using the elevons in a lateral mode was useful if not groundbreaking, in a longitudinal mode, where the full

A Vulcan B.1A in low-level camouflage. The introduction of the ECM tailcone on this variant markedly reduced a pitching tendency with open bomb bay found on the original B.1 configuration. ADRIAN PINGSTONE

span was at a similar distance from the centre of gravity, there was much to play for. At low deflections, elevon control in terms of pitch was about twice as effective as the elevators alone, whereas the area was increased by only two thirds. The significance came from the peculiarities of trimming a tailless aircraft. While the Vulcan obviously incorporated sweep, it had no twist in its early form and some download at the trailing edge was always likely to be necessary at cruise lift. Indeed, the increase in nose-down pitching moment that was associated with an increase in lift coefficient was a major problem for a statically-stable moderate aspect ratio delta, in terms of its impact on maximum lift. At a similar time to the 707 programme, flight tests were conducted in the UK on another delta research aircraft, the Boulton Paul P.111. This has a 45° leading edge sweep configuration and tips that could be removed in two stages, allowing three different combinations of taper- and aspect-ratio to be investigated. A report on early flight tests recorded,

"CL_{max} was 0.86 for all taper ratios, but was reduced to a trimmed value of 0.65 with a static margin of 0.10c, due to the large loss of lift caused by the elevons."[157]

This was the basic compromise of a tailless configuration compared to a conventional layout, in which the tail surfaces provided the download independently. The production Vulcan had the control surface datum at 2° up, which meant that any significant further deflection upwards started to move into a non-linear region where, according to the wind tunnel tests, effectiveness would reduce. As with the ailerons, by sharing the download across the span, the deflection angle could be reduced relative to that from using the inboard pair alone. The importance of this was again emphasised by the RAE's wind tunnel test programme on the original Vulcan prototype, also documented by Newby.

"Apart from introducing the instability discussed above, the backward movement of the aerodynamic centre at high Mach numbers has a marked effect on the trimmed lift coefficient of tailless aircraft, because of the large loss of lift over the inboard part of the span associated with the larger elevator deflection required to trim.

Moving the C.G. back by 2.4% considerably improves the trimmed-lift curve slope because of the smaller elevator deflections required... Thus it is

The ECM tail fairing of Vulcan XH558. Unusually, this aircraft is not fitted with the cooling system (VCCP) for the electronics, which was evidenced by an external intake. Because this Vulcan was operated briefly as a K.2 tanker version, the ECM system was removed to allow the fitting the HDU in the now empty bay.

desirable to keep the static stability of the aircraft as small as possible in order to reduce the loss of lift associated with the elevator deflection required to trim; this will improve the CL at which tip stalling occurs, and also the drag divergence Mach number (because of the smaller wing incidence for a given value)."[158]

Longitudinal trim and control was achieved more efficiently than had been the case on the B.Mk.1, but the need for automation to make the aircraft viable to fly for service crews was an inherent part of its shape. Air Commodore Ed Jarron was a Vulcan B.Mk.2 captain on 35 Squadron in the 1970s.

"Having over 2,000 hours on the Mk 2 Vulcan I can personally vouch for the aircraft's stable and predictable handling at high subsonic Mach numbers, certainly up to its limiting 0.93 Mach; however, it could be quite entertaining for the few heroes who (usually inadvertently) strayed above that!

Above 0.93 Mach the aircraft would enter a 'divergent phugoid', which would feed in increasingly nose down trim as the speed increased. This was countered by 'auto Mach trim', a mechanism in the controls that fed in increasingly nose-up trim to counter the effect. Clearly if this occurred the obvious thing to do was to cut power and deploy the airbrakes to reduce speed as a matter of urgency. Unfortunately, the airbrakes also introduced a nose-down change of trim!"[159]

A renewed emphasis on the Vulcan's ability to defend itself through electronic warfare, or radio countermeasures (RCM) as it was referred to, drove another obvious change in the form of the inflated and extended rear fuselage. From an aerodynamic perspective, this at the very least introduced more surface area and skin friction drag, but would have required careful shaping to avoid triggering the interaction between the wing shock system and the tail fin, as experienced by the Javelin. On the other hand, it represented an opportunity, as the addition of cross-sectional area aft of the wing was a source of concurrent research.

Küchemann's ideas about wing-fuselage junction design

Wearing disruptive camouflage over their anti-flash white lower surfaces, Vulcan B.2s XJ781 and XH562 await their next tasking at RAF Luqa, Malta, in November 1971. Both aircraft are equipped with the Terrain Following Radar (TFR) in a nose pod, ahead of the H2S scanner. GODFREY MANGION/AVIATIONMT

to promote isobar sweep, incorporated by Handley Page at the design stage in the HP.80, were investigated as potential additions to existing aircraft. Hunter F.1 WT571 was modified with a bulged fairing around the rear fuselage, created using the RAE's design rules and demonstrating a small but worthwhile improvement in drag-divergence Mach number. This benefit was strongest at low wing load, where it would be needed most for high-speed flight, but crucially was almost invisible in the drag trace below transonic speeds. This provided part of the clue that the fairing, undoubtedly adding a small amount of drag itself, was acting on the pressure distribution of another part of the aircraft—the wing—and improving its efficiency.

The Vulcan's new tailcone therefore had the potential to be all but drag neutral if carefully designed. This was not an opportunity open to the Javelin, which had its own 'area ruled' rear fuselage modification compromised by the position of the engine nozzles, and the consequent base drag. To be clear, Küchemann's theories as applied to the Hunter were not the area rule as advanced by Whitcomb in the United States; rather, this was a strategy to favourably control a specific region of the aircraft's surface pressure in subsonic flow, as opposed to the global flow field at supersonic speed.

The new RCM system was to be retrofitted to the B.Mk.1 fleet, which subsequently be known as the B.Mk.1A. The RAF's earliest Vulcan, XA895, was converted to the new specification and test flown by the A&AEE, leading to the new variant's Release to Service in February 1960. In fact, XA895 retained its original Olympus 101 engines of nominal 11,000lb static thrust, whereas the service fleet standardised on the 13,400lb thrust Olympus 104. The heavy system in the extreme rear fuselage meant that the usual centre of gravity position was correspondingly shifted rearwards. To compensate, the aileron control circuit was altered to bias the datum pitching load further to the elevators, giving them a larger required deflection for a given 'g'. This had the effect of desensitising them, important as reduced distance to the between the aft CG and the aircraft's aerodynamic neutral and manoeuvre points could otherwise have resulted in overcontrol.

The final B.Mk.1A specification also increased the authority of the Mach trimmer, now able to use 7.4° of up-elevator rather than the 6° of the baseline jet, and the pitch damping system. Both of these can be seen as consistent with both the increased available thrust and more aft CG location. Slightly more elevator force was indeed found to be required at equivalent load, which was deemed satisfactory. One unusual effect noted was a reduction in trim change with opening of the bomb doors.

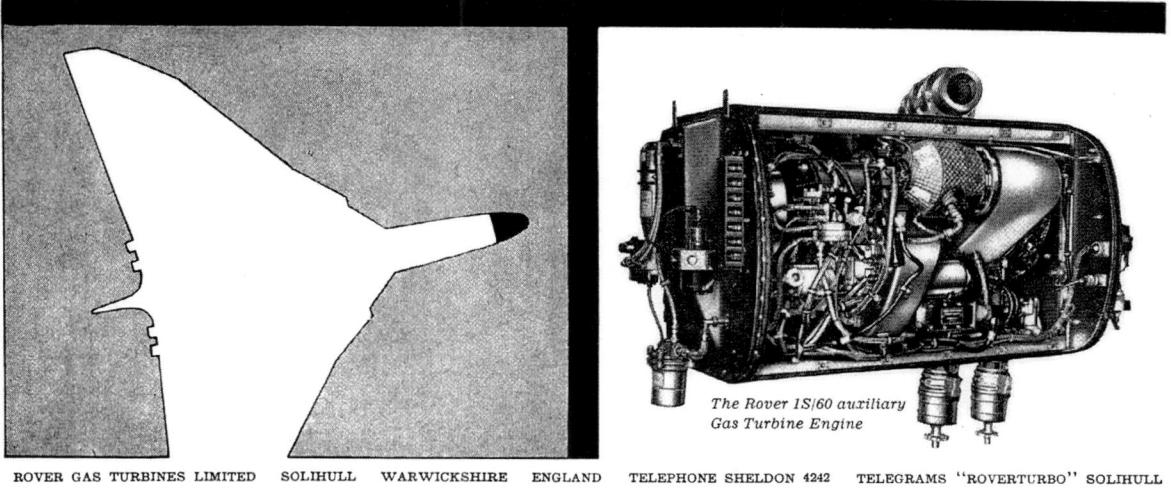

Rover were the suppliers of the Vulcan B.2's Airborne Auxiliary Power Unit (AAPP). In an emergency situation, this was a major advance over the back-up battery power that was relied upon by the B.1. DAVE ROBINSON — AVIATION ANCESTRY

On the B.Mk.1, this effect was pronounced enough that a link to elevator actuation was added, to make it invisible to the pilot. For the B.Mk.1A, the A&AEE advised that it might be removed.

Outwardly more worrying seems to be the finding that the aircraft was longitudinal unstable above an IMN of 0.85, however the report described this as not serious within the proposed speed limits and noted that the service Mach trimmer spec was not fitted. It was found more difficult to trim when using autopilot without height lock (as might be the case for a cruise-climb, with the aircraft steadily gaining height as fuel was burned off), but recommended instead using the very effective height lock capability and operating a step-climb procedure. This would have been likely to give a small range penalty. It is interesting that testing was conducted up to at least 43,000ft, even on this early aircraft comparatively weighted down by its RCM fit, and handicapped by low thrust engines. The early fears of poor altitude performance had not materialised, and there was more to come.

One difficulty with integrating the RCM system into the B.Mk.1A came from the aircraft's DC electrical system. To interface with the new load's AC electrics, a new alternator powered by bleed air from the starboard was fitted, just aft of the main undercarriage bay.

The V-bombers were predominantly 'electric' aircraft in terms of their systems, with limited use of hydraulics compared to the architectures that would become mainstream for large aircraft over the subsequent decades. Powerful, reliable electrical generation was therefore a vital aspect of the designs, but the aircraft's operating regime introduced a significant problem. The use of Alternating Current (AC) electrical systems was in its infancy, which meant that Direct Current (DC) systems using brushed generators were the norm. To deliver the power required at acceptable weight, voltage needed to be as high as practicable, which in the late 1940s meant a 112v DC system.

As some of the highest flying (and therefore operating in the lowest ambient pressure) aircraft conceived at the time, the margin to two major failure modes inherent in this generation technology was substantially reduced, when compared to previous types. Firstly, the wear and creation of dust from the commutator brushes was increased by low pressure and humidity. This had the dual effects of decreasing brush life (or time to failure) and promoting the conditions for electrical arcing with subsequent generator flashover. The second, related problem was that the voltage required for sustained arcing was reduced with air

Vulcan B.Mk.2 XM653, an aircraft from right at the end of the production run, was delivered from new in camouflage. GODFREY MANGION/AVIATIONMT

Vulcan B.Mk.2 XL317 landing in Malta in 1973. Delivered to 617 Squadron as a Blue Steel carrier, it was still operational with the Dambusters at the time and carrying their fin badge. GODFREY MANGION/AVIATIONMT

SECOND GENERATION: THE VULCAN AND VICTOR MARK 2

Vulcan B.Mk.2 XJ784 at RAF Luqa, Malta, in October 1973. The gradual toning down of the aircraft has started to take place, with red/blue national markings only, and the TFR is carried for low level operations. GODFREY MANGION/AVIATIONMT

density. Major physical damage to the system could therefore result, which at best might take a generator offline but could easily result in fire.

As a military aircraft subject to combat damage, the potential for gaps to suddenly appear in high voltage cabling represented a further risk for starting fires, again exacerbated by the environment. To illustrate this, much later, in the early 1970s, work reported by the RAE on future aircraft electrical systems would return to high voltage DC systems. As part of their assessment, arcing with simulated altitude conditions was investigated. The tests showed that at 50,000ft, a 200v DC arc could be sustained over a gap of up to six inches. However,

> *"With a 200V, 400Hz, AC supply it was not found possible to sustain an arc with any gap down to 1/16in even at altitudes of up to 80,000ft".*[160]

This pointed towards the eventual solution. The power density of AC alternators was significantly better, often doubled at 1950s technology level, compared to DC generators.[161] HP from the start used AC alternators in the Victor, albeit variable frequency machines that were geared to the engine drive and therefore subject to its speed of rotation. As the aircraft's main Medium Voltage (MV), systems were DC (with the exception of anti-icing), rectification was required, which introduced another mass, process stage and potential failure mechanism.

Generating at a particular constant frequency needed a method of compensating for the varying speed of the engines that drove them. This problem was solved in 1946 by Sunstrand, who introduced the Constant Speed Drive (CSD), a hydro-mechanical system that converted the varying accessory drive shaft speed to one that provided a constant input to the alternator, typically outputting 115VAC at 400Hz.[162]

The CSD, as a hydraulic coupling, absorbed transmission bearing thrust loads via the fluid rather than metallic, mechanical connection. This substantially reduced wear and also enabled very high peak loads to be safely absorbed for short times. This improvement in reliability compared to the mechanically driven DC or variable frequency AC machines was one of the keys to the system improvement offered by the new AC electrics, which could also be optimised around a known operating condition.

Transmission of electrical power at high voltage and hence lower peak current had the potential to reduce the mass of cabling, although due to the peak loads that needed to be accounted for, this was not as clear cut as is often portrayed. To get the best weight advantage from the complete system, elimination of DC engine starter motors and associated batteries was necessary; this could be achieved by incorporation of an Auxiliary Power Unit (APU), which itself represented a much smaller electrical starting load requirement. An APU conferred various operational advantages for an aircraft that deployed

The antepenultimate V-bomber to be built, Vulcan B.Mk.2 XM655 was delivered to the RAF in 1964 and is seen here RAF Luqa, 1971. The aircraft survives operationally (if not airworthy) to this day, in the care of 655MAPS at Wellesbourne Mountford. GODFREY MANGION/AVIATIONMT

away from main base facilities, as the V-bombers did, and the potentially improved starting capability at low temperatures compared with batteries furthered this aspect of self-sufficiency. In the language used at the time, the APU was known as the Airborne Auxiliary Power Plant, or AAPP.

The next problem in turn to solve was the provision of adequate emergency electrical power for essential services, now that battery capacity had been reduced to a minimum. The answer was the provision of a Ram Air Turbine (RAT), coupled to its own alternator. Rather than the 20 minutes or so expected of the batteries with the original DC systems, this device could generate the required AC power so long as it remained deployed and working. Avro installed the RAT behind a deployable panel under the port wing root; it was a one-shot system and could not be retracted in flight. On the other hand, HP would mount two RATs in the rear fuselage of the Victor, fed from two retractable inlets on the upper rear fuselage. These inlet doors opened when engine speed on the respective side of the aircraft fell below 52% rpm. The second-generation V-bombers therefore had the assurance of electric power generation from four engine-driven, one AAPP-driven and one (or two in the case of the Victor) RAT-driven alternators.

VICTOR B.2

Like the Vulcan, the possibility of increased engine thrust implied more weight could be carried from a given runway for an improved Victor. Handley Page was aware of Armstrong Siddeley's development of the Sapphire, from the 11,000lb thrust AS.SR.7 (Sapphire 7) into the Sapphire 9 and 10 with perhaps 14,000lb thrust, aimed at advanced Javelin variants. These engines used turbine cooling, allowing an increase in turbine inlet temperature from 1,125K to 1,250K, which in the real world of irreversible and lossy flows meant an increase in fuel efficiency for a given thrust, and was combined with an increase in engine mass flow greater than 10% to deliver the step in maximum thrust. For a turbojet, the fall in thrust with altitude was approximately proportional to the ratio of air density. A sea level static thrust increase was therefore valuable in terms of providing the required thrust at a higher altitude — the key to the V-bomber's survivability.

HP's initial proposals were of similar form to what would actually happen for the Vulcan, a 'Victor II' with basically the same airframe planned as the Victor B.Mk.1, but taking advantage of the Sapphire's growth. However, also as Avro had done, substantial airframe modifications began to be proposed, settling on a small increase in wingspan of 5ft in the first instance. Further work delivered a pair of fuselage plugs at either end of the bomb bay and an extended wing centre section for a span of 126ft and length of 137ft. The latter was described in a brochure of July 1954 as the Victor III, and by now had abandoned the Sapphire in favour of the promising Rolls-Royce Conway, projected as a 15,000lb thrust engine. Normal operating weight would have been 200,000lb, with a maximum take-off mass of 225,000lb if necessary, numbers that are again familiar from those projected for advanced Vulcan variants in the period leading up to the entry into service of the B.Mk.1 aircraft.

As with the Vulcan, however, the ambition was reined in and major structural changes were limited to those essential in order to take advantage of the increase in thrust. The outcome consisted of September 1955 proposals for Phase 2A and 2B aircraft, basically Conway and Olympus powered versions of the Victor II, with the wing extended to 120ft span (from 110ft), which in turn increased area to 2,597sq ft and aspect ratio to 5.545, from 5.03. The Conway powered 2A was calculated to provide a height over target of 59,000ft with the 10,000lb special bomb, at a B35/46-consistent range of 3,400nm: exceptional performance indeed.

The major modifications were anticipated to be nonetheless simple to embody. The engines were of larger diameter and required a greater mass flow, which would mean redesign of the three spar forgings and inlet geometry. The nacelle fairing above the trailing edge was also appreciably larger, while to provide the increment in area inboard, an 18in extension offset each wing outboard, while 3ft 6in was added to each tip. Local modifications were required for the trailing edge flaps as they encountered the jet pipes, while ultimately the extensive systems re-engineering embodied on the Vulcan B.Mk.2 was also a feature of the developed Victor.

As had happened with the Vulcan, production of the Victor B.Mk.2 began as soon as was practical, with existing orders for the B.Mk.1 being switched to the new variant as it became opportune. The final B.Mk.1, XH667 was therefore followed by the first B.Mk.2, XH668, clearly midway through an existing contract and serial block. As the nominal prototype, XH668 would sport some minor difference to the subsequent machines, perhaps the most obvious being the lack of the extended fin root, incorporating a cooling inlet. Above the visual bombing windows in the extreme nose, a NACA inlet on each side featured a covering duct, with the aim of reducing noise in the cabin. Like the other early aircraft, it lacked the oil cooler inlets below the main engine inlets which would be associated with an uprated alternator specification later retrofitted, and possibly had only the starboard RAT inlet above the rear fuselage. Its appearance then, with straight fin leading edge and enormous intakes, is unique but sadly it was not one that many would have the opportunity to see.

XH668 first flew in February 1959, within six months of the corresponding event for the first production Vulcan B.Mk.2, XH533, although Avro had the advantage of flying the new version's wing on the original second prototype, VX777. Handley Page's uprated and long-spanned new Victor cut a more aggressive stance than its predecessor, largely due to the physical size and angular lower lips of the Conway's intakes, which as the RAE noted lay behind a 53° swept leading edge.

It would soon become apparent though that all was not well with the installation. Testing on the ground and at altitude revealed problems with surging of the revolutionary bypass turbojets; the latter of course being particularly worrying for the bomber in its intended role. Unsurprisingly, the search for a fix would involve the engine manufacturer as well as HP and the RAE, and triggered a wind tunnel test programme initially with Rolls-Royce aimed at defining a quick fix; it would subsequently move to the RAE at Farnborough to establish a more fundamental solution.

The Derby firm rapidly came to the conclusion that flow separation from the inboard wall of the inlets was the root cause at high speed, while on the ground it was the tortuous path around the outer wall that was to blame. An array of four vanes within the inlet was produced as a viable fix, although the initial solution was for two per side, which was implemented and presumably proved adequate.

The RAE programme worked over a slightly more relaxed timescale, enabling it to explore the flow physics in greater depth. The Establishment team eventually came to understand the mechanism to be slightly different from the initial RR assessment. Because the inboard inlet was the first to become fully enclosed when moving rearwards, the off-body compression of the external flow occurred further forward than was the case with the outer duct. This forced inboard flow outboard, which placed the dividing wall between the two ducts at a high incidence. The problem then was the required pressure recovery from the developed suction peak on the outboard side of the dividing wall, which was immediately faced with the dramatic pressure rise behind it from the outboard compression within the duct.

This resulting separation was observed to be the first event in the sequence, with the inboard wall separation following at greater velocity ratio, which the RAE attributed to the proximity of the bypass duct and fence, in

Victor B.Mk.2 wind tunnel model in the ARA facility, mounted on a twin sting rig. The majority of tests used a single sting which cut away part of the rear fuselage. This test cross-checked the interference. ARA IMAGE

turn contributing to a healthier and more resilient boundary layer. However this was caused, the outcome was the same: a variable velocity flowfield of reduced total pressure. A characteristic exposed by the detailed RAE tests was the relationship with angle of attack. The duct losses became worse outside of the narrow band of 0° to 3° incidence, which implied that sustained manoeuvring flight at altitude would be compromised by engine surging too. For all the effort of providing the larger wing and vastly higher-thrust engines, the cruising altitude would potentially be limited—not by available 'g' or induced drag but by engine handling.

The modifications proposed by the RAE included cambering the central divider such that the leading edge angled inwards. Like the drooped outboard wings of the bombers, this improved the alignment of the geometry to the outwashing onset flow. It had very little effect in practice. A more drastic alternative was to cut back the central divider, which meant that the deceleration through compression happened to a greater extent upstream, and was more equalised between the inner and outer ducts. This gave very promising results, both in an initial form where a hole was cut between the first and second spars, then finally when the remnant leading edge and fairing onto the first spar was discarded completely.

With the practicality of this modification open to question, an alternative of extending the outboard wall forward was investigated. This also proved effective although not to the same standard, while a final concept was the use of bypass slot to spill excess flow. All of these options worked, with the short divider being the most effective, in its complete form linearising the pressure recovery in both inlets and offering 95% pressure recovery at a velocity ratio of two. Curiously though, the tests were not compared to the simple (and readily implemented) modifications proscribed by Rolls-Royce, while no quantification of the effect on the external flow was made either.

There is a hint that these results, gained in the autumn of 1959, were probably more applicable to a future version of the aircraft. The work highlighted once more the complication caused by the buried engine configuration. Not only had the inboard structure, including the three spars ahead of the engines, been extensively redesigned, but their very presence constrained the possibility for simple modifications in the event of subsequent problems, in a world where simulation tools were not yet up to the task.

Certainly by August 1959, well in advance of the RAE tests, XH668 had the first fix of a single vertical vane ahead of each engine duct. On the 20th of the month, it took off from Boscombe Down under the command of a mixed A&AEE and Handley Page crew, en route to conduct high altitude manoeuvrability tests over the Irish Sea. This would have involved flying close to the stall, into a region of very heavy buffet. It would never return; from its test altitude of 54,000ft, it dived into the sea off the coast of Wales. None of the test crew managed to escape.

The disappearance of the first example of Britain's highly capable new nuclear bomber was a major shock and the

SECOND GENERATION: THE VULCAN AND VICTOR MARK 2

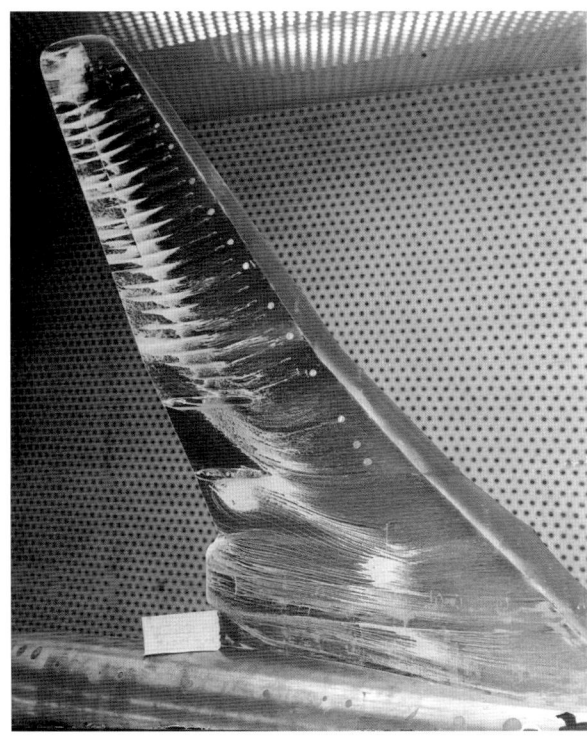

Victor wing flow visualisation in the ARA tunnel, at M = 0.88 and 2° angle of attack. The effect of the vortex generator array in suppressing shock-induced separation is clear, while the flap track fairings can also be seen to have a strong influence on the flow topology. ARA IMAGE

A similar image at M = 0.84, which was a more representative cruise condition. The pronounced outflow inboard of the aileron, and the effect of the upper surface flap track fairings in altering the shock position is again clear. The potential benefit of the Küchemann Carrot fairing in mitigating this, particularly at more adverse, high weight conditions, is implied. ARA IMAGE

An unidentified Victor B.2 in pre-retrofit form. ALISDAIR MACDONALD

Victor B.Mk.2 XL164 lands after its display at the SBAC Farnborough Airshow in 1961. The drooped nose flaps can be seen.

Shiny and new: Victor B.Mk.2 XL164 parked at the Farnborough Airshow in 1961, with an intrigued onlooker. The original form of the fin inlet duct and absence of the oil coolers under the engine inlets are of interest, together with the deployed RAT ducts on top of the rear fuselage.

event kicked off an enormous effort to find and recover the wreckage. More than 40 ships were involved over a 14-month period, eventually finding about 70% of the aircraft, although in more than half a million fragments. It was towards the end of this search that the critical evidence was found: the starboard wing tip was recovered, with the pitot head evidently detached through fatigue failure. This led to a plausible scenario for the aircraft's loss: the starboard pitot probe was connected to the Mach trim and stall warning systems, which would have interpreted the signal as a reduction in airspeed and hence cancelled the up elevator, followed by automatic deployment of the leading-edge flaps. The latter in particular would cause a nose down trim that might have been impossible to recover at high Mach, with limited elevator authority. At some point, the aircraft would have been overstressed and outcome inevitable.

Avro test pilot Tony Blackman has recounted that his opposite number at Handley Page, Johnny Allam, was sceptical of this reconstruction of events, and that the view was shared by others at the company. Rolls-Royce had found that the engines were at maximum continuous

Awaiting its turn at Farnborough, XL164 shows to advantage the much-enlarged engine inlets of the second-generation version.

cruising power at the time of the impact, which the crew would surely not have commanded themselves. The Mach trimmer retraction should not of itself have been a difficult control problem; although speed restriction was applied, there was a documented procedure in the pilot's notes for operating the aircraft despite such a failure. This had been carried over from several years of experience with the prototypes and B.Mk.1.

They postulated that a different control power problem had been at play, one that was not completely aerodynamic in nature, but connected to the fact that the hydraulic actuators for the ailerons ran out of authority by about M = 0.98. If XH668 had entered a spiral dive, perhaps following stall or high speed or simply through distraction of the crew, then the normal recovery using aileron would have been impossible. At the same time, directional control was also problematic, as the effect on the transonic flow over the fin of rudder deployment caused the sideforce to be generated in the opposite direction to that at low speed. If not understood and caught quickly, the attempt to recover using the rudder in extremis would quite plausibly have put the aircraft into an irrecoverable dive.

Perhaps the pilots thought that time—and reducing Mach number with increased air density—was on their side, and certainly they would have wanted to give the rear crew every chance of escape. Whichever scenario is chosen, the fix to the aircraft in terms of a more robust pitot head was relatively simple, and as recounted earlier was a driver towards the elimination of the nose flaps and their outboard edge buffet.

If the situation had any positives, it was that the second aircraft, XH669, had been flying for two weeks. Unlike the case with original Victor prototypes, further B.Mk.2 aircraft were already following down the production line at Radlett too. XH670 flew in November and would spend its life on trials work, while the slow but steady progress meant XH672 flew on April 6, 1960. This latter aircraft was also initially used for trials, being both loaned back to HP and used by the A&AEE. By August, it had flown at least five sorties in support of the continuing engine surging investigation, while its day job of autopilot and autoland system development had to wait.

The seriousness of the problem is illustrated by the fact that a further six flights were conducted in March 1961, more than a year and a half after the Conway-Victor combination had shown the first signs of trouble. The signs were better though, as in April 1961 it took part in release to service trials at Boscombe Down. In November, early aircraft from the second batch, the first to actually be laid down as B.Mk.2s, arrived at RAF Cottesmore for 12 weeks of intensive flying trials. By now it was 18 months since the first Vulcan B.Mk.2 had been delivered

Seen in Aden in 1964, this retrofitted Victor B.Mk.2 was engaged in low-level trials at the time. The upgraded machine features the additional inlet vanes and oil-cooler inlets below, together with provision for in-flight refuelling. CTHORNBOROUGH (CCA 4.0)

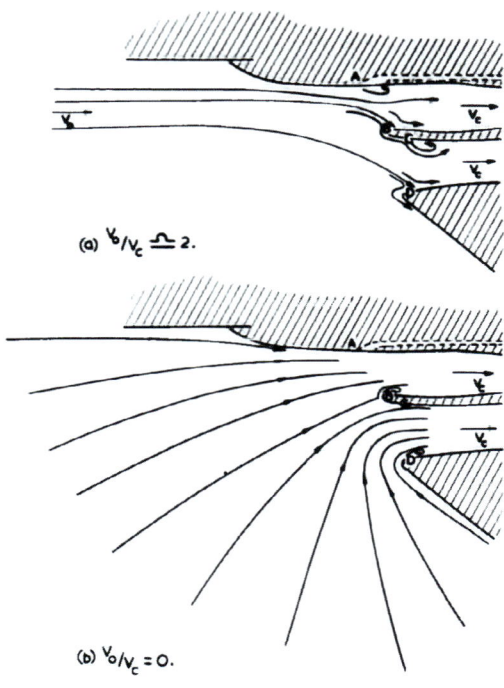

Schematic of the Victor B.Mk.2's inlets, showing the RAE's hypothesis of the cause of the aircraft's engine surging problems. At cruise (upper), the outfow caused separation on the central divider and outboard lip, while under static or low speed conditions, the flow was directed inboard and encountered the opposite alignment problem. RAE TN AERO 2979

to Waddington, and Victor's rival was firmly in squadron service.

XH669 provided the flight test data in support of the basic aerodynamic sign off of the new Victor, validating the calculations and wind tunnel measurements for lift, drag and longitudinal stability. Reported in 1961 just as 232 OCU began to get to grips with the powerful new machines, these data would be used later when the new Aircraft Research Association (ARA) transonic wind tunnel at Bedford was commissioned, as part of a series of tests to demonstrate the state of the art in drag prediction at high Mach number.

The tunnel operated at a higher Reynolds number than had been available when the V-Bombers were designed, and could comfortably allow a 1:25 scale model to operate up to $M = 0.9$, well into the transonic drag rise. The flight data was more sparse than might have been liked, but did include sweeps at $M = 0.82$, a representative cruise Mach number, and a lift coefficient of 0.35, which was close to the design case. The tests, with appropriate corrections applied for the difference between tunnel and full-scale Reynolds number, proved remarkably accurate in terms of absolute drag estimation, although the elevator angle to trim was found to match less well. This was assumed to be because of the aeroelastic bending of the fuselage in flight, for which the assumed corrections had proved less effective. What the process showed in retrospect was

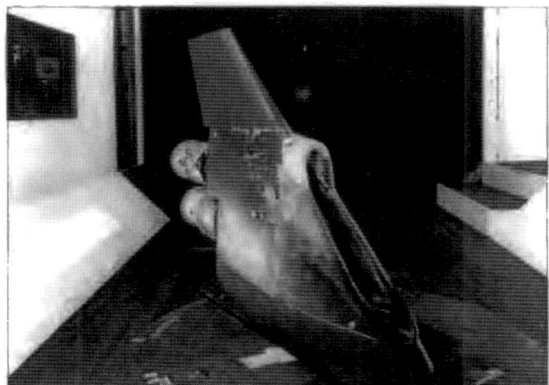

FIG. 4. MODEL WITH INTAKE OUTER LIPS EXTENDED FORWARD

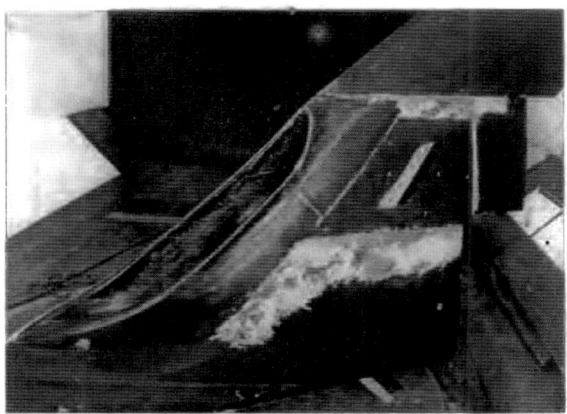

FIG 5 MODEL WITH SPILLAGE DUCT CUT IN LOWER WALL OF OUTER INTAKE

Options for a fix tested by the RAE. An extended outboard lip corrected some of the outflow issues, while a lower surface spillage duct allowed the inlet contours to remain unchanged.

what an excellent job HP and the RAE had done, in terms of using the less advanced experimental facilities actually available to them more than a decade earlier, in the initial design of the Victor.

As a test intended to correlate with a known full-size configuration, the model was detailed with excrescences to a greater extent than might have been usual, including items such as the nose NACA ducts and Q-feel probe, but not the inlet splitter plates or boundary layer bleeds, presumably because of the compromised modelling of the duct flow. An important conclusion was that the overall drag and value of the beginning of the steep drag rise was very sensitive to the elevator angle to trim, the parameter that had proven difficult. What the flight tests certainly did show was that the drag rise was well underway by M = 0.82, for CL = 0.35. This was for the cleanest form of the Victor B.Mk.2, without weight growth or additional external systems that might be anticipated through the aircraft's life. It implies that while performance was excellent, efficiency might be beginning to suffer compared to the B.Mk.1.

The relatively protracted timescale for the Victor B.Mk.2's entry into service had the effect of backing it into the period when both heavier, draggier weapons and still more powerful engines would become available. Certainly in the case of the Victor, this would be in the form of the Avro Blue Steel stand-off bomb. This was far too long to fit even in the internal space fit for 35,000lb of conventional bombs, and consequently had to be carried semi-recessed under the fuselage via bespoke bomb bay doors and fairings.

Blue Steel weighed in at more than 15,000lb, around 50% more than the freefall weapons it was intended to replace, and allowed some of the otherwise unused bomb-bay volume above to be occupied by shaped auxiliary fuel tanks. In addition, both the Vulcan and Victor had been designed with provision for external wing-mounted drop tanks, although it would only be the Handley Page bomber for which these would actually be specified. The tanks were trialled on HP's loaned test jet, B.Mk.1 XA930, but in service were only used on the mark two. The use of the tanks along with the heavy nuclear store could not have been realised without an increase in the maximum take-off weight available, in the release to service. However, the aircraft's ability to generate lift was still based upon the same wing area, so an increase in weight had to mean a corresponding increase in lift coefficient to compensate if altitude was to be maintained.

In the absence of compressibility, this would mean an increase in lift-dependent drag, complicated on a swept wing by the spanwise pressure gradient exacerbating the boundary layer thickening outboard from the aerodynamic perspective, and the torsional stiffness requirements when viewed from the structural design office. In the real, transonic world, matters were considerably worsened as loading increased by the earlier onset of wave drag and, perhaps, shock-induced separation. The pilot's notes for the Victor B.Mk.2 advised that buffet would be encountered above M = 0.88 with the tanks fitted, nonetheless meeting (just) the requirements envisaged back in 1947, but without speed margin.

One requirement driven by the prospect of stand-off weapons was a new location for the chaff dispensers, still referred to in Britain by its wartime codename of Window. These needed to be well clear downstream of any inlets associated with services for the new stores, in the region of the bomb bay. The wings would have been an ideal location, but no volume was available. Nonetheless, an intriguing and potentially high-Mach drag-neutral solution would present itself, drawing on work in the United States that was familiar to Küchemann and the RAE,

The most effective solution was to cut back the central divider, but this was restricted by the local wing structure. This image shows the model used to test the installation.

and ultimately had its roots, like the crescent wing, in Germany.

In the mid-1950s, NACA's Richard Whitcomb had expanded his transonic drag reduction work around the 'area rule', to consider more local effects. In 1958, a report was published on the application of 'special bodies' to moderately high aspect ratio swept wings, in order to delay drag divergence.[163] Within, he described two functions of the special bodies:

> *"The forward portions of these bodies decelerate the supersonic flow ahead of the shock wave above the wing with a resulting decrease of the strength of the shock and the associated separation. Furthermore, the local pressure fields produced by the bodies greatly reduce the adverse outward flow of separated boundary layer on sweptback wings".*

In other words, the bodies acted laterally on the boundary layer to limit spanwise flow, as a fence was designed to achieve, but also acted in a streamwise sense to provide a local compression in an efficient manner, analogous to Pearcey's isentropic recompression effort. In his most notable reported work on the topic, a 35° wing of taper ratio 0.38 and 9% thick, NACA 65-series section was the underlying subject, coincidently close to the Victor 'average', albeit at an aspect ratio of a little more than seven. Whitcomb noted that at moderate cruise lift coefficients, it was the pressure field above the wing that dominated drag rise; therefore, the geometry itself was added only above the plane of the local wing chord and this was the only area considered in the aerodynamic design.

Five trailing edge bodies were added, designed with a conical nose to a maximum thickness of 8% chord, close in dimension to the local maximum thickness of the wing itself. They were smoothly faired back into the upper surface, with fillets of similar magnitude radii. While the forward portion may have been delivering the desired effect, it was obviously necessary to have a pressure recovery tail surface, leading to the long extension of between 35% and 45% chord aft of the trailing edge. The large number of such bodies compared to future applications stemmed from previous results that showed the spanwise extent of the aerodynamic effect to be limited in the region of the drag divergence Mach number. The longitudinal positioning was explained by a further statement,

> *"A correlation of drag rise with airfoil shape for most usual airfoils indicates that the primary factor leading to a delay and reduction of the initial drag*

SECOND GENERATION: THE VULCAN AND VICTOR MARK 2

Inlet for the port-side pair of Rolls-Royce Avon engines on the production Vickers Valiant. These featured a deepened lower lip compared to the simple slots on the prototype. Visible on the central divider is an additional outboard vane, which helped to compensate for the effect of the sweep angle causing increased pressurisation inboard.

rise at moderate lift coefficients is a reduction of curvature of the upper surface near the maximum ordinate of that surface."

Whitcomb aimed to compensate for the curvature of the upper surface around maximum thickness by offsetting that of the added body to a different chordwise location. Taken together, the overall area from front to rear followed a smooth curve, related to the Sears-Haack distribution that theoretically minimised wave drag. The same clearly could not be true of the specific shape of an aerofoil. As a point of interest, Whitcomb also cited the by now well-known (in the right circles) work of Haines and his promotion of isobar sweep.

> "Results obtained by A. B. Haines of the Aircraft Research Association, Ltd. somewhat by providing spanwise differences in the shapes of the sections so that the sweep of the line of maximum induced velocities is increased. The differences in the total area developments… are the result of an attempt to provide a similar effect for the configuration of the present investigation".

This philosophy was already part of the DNA of both Victor and Vulcan, but not the long span swept wing that Whitcomb was working with. By gradually moving the maximum thickness of the five bodies rearwards, at their respective spanwise locations from root to tip, they could be used to promote this effect. This added complication of design was less helpful, and the need for a large number of bodies unlikely, particularly as the outboard thickness of the baseline wing was so low; well under 6% in the case of both types.

However, as had been seen through the development of the crescent wing, there was a particular overloading problem outboard of the second crank as the isobars rearranged themselves for the thinner, leads swept outer panel, but momentarily smeared the characteristics of the two flow patterns. If overloading at high Mach number was likely to have any local effect it was there, while conveniently that offered the only fixed region of trailing edge available on the wing—in between the flaps and aileron, where once the contra-flap device had been considered.

All of this clicked together when Handley Page aerodynamicist Peter White attended a symposium in Brussels and heard from Whitcomb himself on the concept. The special bodies could provide the volume needed for window, in the right place. While they would certainly contribute skin friction drag, this might be negated by an improvement in wave drag in the cruise. It appeared the best answer available; in a marginal gain scenario, the killing of two birds with one stone has to be seriously considered. The key was that the weight of the 'Window'

The starboard root region of Victor K.2 XH672, illustrating the adaptations that proved necessary for effective pressure recovery ahead of the Conway 201 engines. The central splitter was helped to remain attached in this post-retrofit form by four vertical vanes. Inboard of the inlet a further external fence acted to protect the inlet from ingesting the fuselage boundary layer.

Victor outboard wing model, as used by the RAE for development of leading-edge high-lift options to replace the nose flaps. In this flow visualisation, the separation and subsequent vortex formation are apparent. ARC-R&M-3455

was low and that the loading on the pods themselves would be relatively small. It consequently ought to be possible to conceive of a retrofit scheme that would not significantly alter the loading of the wing and require additional structure. It all came together and the distinctive fairings were designed, often known in British English after Küchemann as his 'carrots', for obvious visual reasons, but appearing in the technical literature as (surely another tongue in cheek pun) Window Boxes.

XL164, which was new when displayed at Farnborough in September 1961, returned to HP to become the prototype for the retrofit conversion. The shock bodies would be accompanied by uprated Rolls-Royce Conway R.Co.17 engines, now with a static thrust of 20,600lb and intakes protected by four vanes per side, further upgrades to the RCM system, increased capacity electrical generation system and provision for increased range through air-to-air refuelling. Although detachable, the underwing drop tanks would become an almost universal fit, to assist with the retrofitted aircraft's primary weapon. No longer in theory would the Victor need to overfly its heavily defended target, perhaps allowing some relaxation in the maximum speed it would need. Instead, it would launch the fearsome Blue Steel to complete the final 100 nautical miles at up to three times the speed of sound.

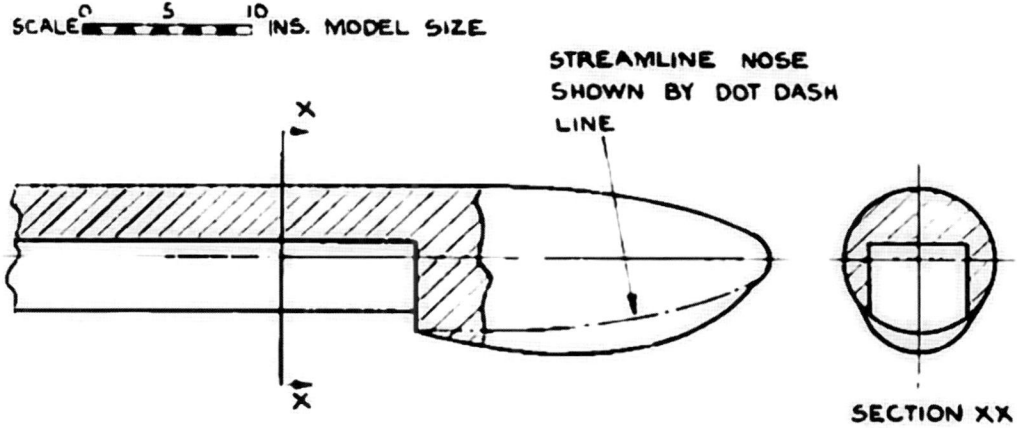

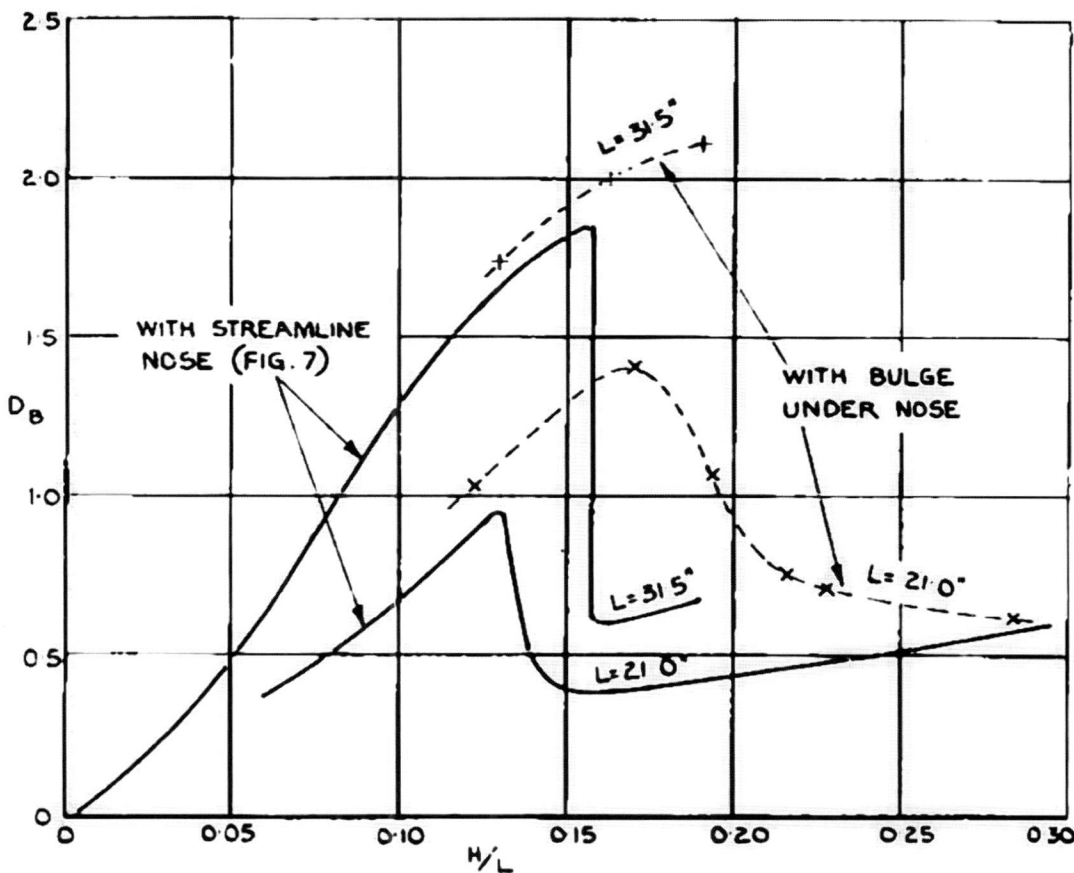

FIG. 27. EFFECT OF BULGE UNDER NOSE ON BOMB BAY DRAG (NO DOORS) BOMB BAYS IN 9INS. DIA. BODY (W=6INS.)

An indication from an RAE study on bomb bay aerodynamics as to why the Victor may have needed the forward vanes. A shallow bay (low height to length (H/L) ratio showed an increasing drag up to a point where it dropped abruptly. This was where the bay became deep enough to trap an internal vortex, with the outer flow passing over and reattaching behind. This effect was lessened by the presence of the bulge in front, as the Victor had, but improved by a forward vane array. RAE AERO REPORT 2511

Detail view of the inboard end of the fixed droop leading edge, on the outboard panel of Victor K.2 XH672's port wing. Originally built with a nose flap, the extended end and free lateral edge gave a subtle dog tooth effect, generating a vortex at high incidence.

Ahead of the bomb bay on the Victor, an array of vanes were hydraulically extended when the doors opened, shifting the position of the shedding edge and consequently the effective aerodynamic length to depth ratio of the bomb bay. This had a strong effect on the pitching moment induced on the bomb. For obvious reasons, it was necessary to ensure clean separation from the aircraft.

Blue Steel was essentially a small aircraft in its own right and required specialised loading and transport equipment to move around. DS COLLECTION

The Blue Steel assembly line. DS COLLECTION

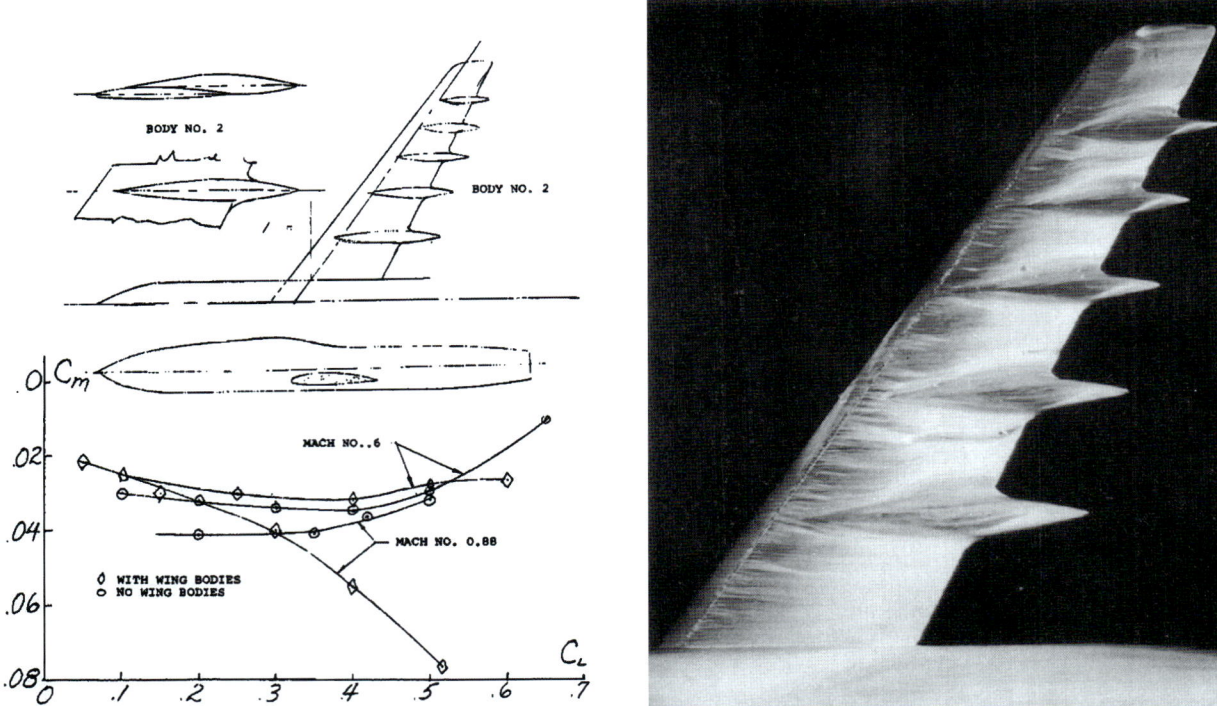

Richard Whitcomb's work on the application of special bodies to reduce the effect of shock waves. The layout of the wing-fuselage model is illustrated, together with a flow visualisation of the wing at M = 0.88, in which the attenuation of the shock was identified, compared to the baseline case.

The Küchemann Carrot fairing on Victor K.2 XH672, mounted on the fixed portion of the trailing-edge outboard of the flap, and helping to reduce the local shock strength at cruising speed (and weight). Its primary purpose however was as a chaff dispensor; it is not clear that they would have been adopted for purely aerodynamic reasons.

SECOND GENERATION: THE VULCAN AND VICTOR MARK 2

Because of the cancellation of the B.Mk.2 series, XM718 would become the final Victor to be built. Operating as an SR.2 reconnaissance aircraft with 543 Squadron when photographed in November 1968, it would be part of a very small group of these jets that flew on after the disbandment of the bomber squadrons, and prior to the K.2 conversion programme. GODFREY MANGION/AVIATIONMT

Turning onto the runway at RAF Luqa in October 1976, XL512 was a relatively newly converted Victor K.Mk.2 at the time. GODFREY MANGION/AVIATIONMT

An early Victor B.Mk.2, XH674 remained operational as an SR.2 with 543 Squadron in the early 1970s, at a time when the majority were stored for tanker conversion. Visible here are the Küchemann pods and underwing fuel tanks, which were not always carried if extreme altitude was required by the mission. GODFREY MANGION/AVIATIONMT

The V-bomber engine with the greatest static thrust was the Rolls-Royce Conway 201 (R.Co.17), providing 20,600lb in the retrofitted Victor B.2. A version with increased bypass ratio was concurrently available for the VC10, but would not have fitted and would likely have traded range for height over target.

11

BOLT FROM THE BLUE

IN JANUARY 1947, the manned bomber aeroplane was the only conceivable method by which a UK nuclear weapon might be delivered to the most likely target: one within the Soviet Union. From the perspective of the Air Ministry and the RAF, it was clear from the content of B35/46 that they perceived a much wider remit than the one-shot, one-way mission that is generally associated with the V-Force. Varied non-nuclear loads were specified—pointing straight back to Trenchard and the RAF's primary strategic role—the projection of air power in a limited war, where conventional weapons and gradual escalation were still highly relevant. For the doctrine that emanated from Cranwell and permeated through High Wycombe and Whitehall, the atomic mission was both opportunity and challenge.

The UK had received a clear warning of what the future held in late 1944, when Wernher von Braun's V-2 rockets began to fall. A delivery system that could barely be tracked, let alone intercepted, offered an assuredness of deterrence that could not be ignored. Britain had not prioritised such weapons during the war, but afterwards had ample access to the German research on rocket propulsion. The frightening nature of the V-2 in context should not subtract from the fact that it was a pioneer, its actual payload-range being very limited. It would be years into the future before the technology would mature.

As with the comparison between the B-52's mission and that of the V-bombers, attacking a target within the Soviet Union would require a much larger and longer-ranged missile if launched from the US than from the UK. Indeed, an intermediate range missile deployed from the UK, thousands of miles closer to Moscow, was compelling for US interests. In April 1954, with prototype Vulcans and Victors flying, but entry into service even of the interim Valiant a year away, an amendment of the Atomic Energy Act allowed both sides to benefit.

Duncan Sandys, as Minister of Supply, entered into agreement with US Secretary of Defense Charles Wilson in the following August, for the joint development of ballistic missile technology. The US would build an intercontinental weapon of 5,000-mile range, while the UK would benefit from US assistance in the design and deployment of the 2,000-mile range intermediate range missile. The latter requirement would ultimately be met in two parts. In the first instance, the RAF would use the Douglas Thor missile, air transportable from the United States but requiring an overseas base due to its limited range. While these were originally intended to be USAF Strategic Air Command weapons, the shock of Sputnik in October 1957 would strengthen Britain's hand: the US needed to be able to get the IRBMs close enough to work, immediately.

However, these missiles were used in partial fulfilment of the RAF's OR.1139, which had called in 1954 for a 1,500 nautical mile range IRBM to be in service within a timeframe that reflected the expected obsolescence of

The deterrent pose of the V-Force. Four of 83 Squadron's Vulcan B.Mk.2 aircraft at RAF Wittering in July 1963, wearing anti-flash white and the antler unit badge on the fin. The first three of the Scampton based jets, in the process of forming the Blue Steel force, are XL427, the future Vulcan Display Flight jet XL426 now preserved at Southend, and XL392. ALISDAIR MACDONALD

the manned bomber. The primary domestic route to fulfilling this objective was the Blue Streak programme, led by de Havilland Propellers and using a liquid-fuelled rocket motor based on the American Rocketdyne S-3D but provided by Rolls-Royce as the RZ.2.

Unfortunately, liquid-fuelled missiles had a major drawback in the deterrent role. The use of cryogenic liquid oxygen meant that the missiles had to be fuelled within a relatively short time prior to launch; the liquid oxygen could not be kept on board indefinitely. For Blue Streak, the LO2 load was 60 tons, clearly not a load that could be provided to the missile within minutes of the attack warning being sounded. The operational scenario consequently drawn up involved two missiles being located together, one being fuelled for ten hours then replaced by the second at the end of this period. The workload and expense associated with this cycling, in order to provide the sub-minute reaction required, was non-trivial to say the least.

Another related problem was the protection of the missiles from a pre-emptive attack. Standing on airfields, the vital and expensive weapons were obviously vulnerable to attack, and it was a British innovation that would provide some mitigation to this: hiding the missiles underground in hardened shelters. Able to withstand the blast of a megaton warhead centred close by, such shelters could maintain the efficacy of the missiles, in the face of a vanishingly small warning period. As life-affirming as this innovation might be, it added another layer of eye-watering expense to the IRBM programme. The freedom-loving, democratic West was also at a disadvantage: just where would it be acceptable to place such structures? In the densely populated UK, lacking both the space for shelters and adequate warning time of impending doom (both somewhat less of an issue for America), these were major problems. The country saw itself as uniquely vulnerable, and the fixed IRBM installations would be a hard sell in terms of making people feel safer, whatever the capability of the missile residing within. Nonetheless, in 1954 and looking to a time a decade ahead, it looked like it was the right thing to do. Blue Streak would replace the V-Bombers when it became available.

Liquid-fuelled rockets were perceived as the more viable option for a long-range delivery system due to the diminishing weight fraction of their propellant tanks and engines as they grew in size. The accurate control of burn time that they conferred was also seen as a non-negotiable feature; even with a thermonuclear warhead it was necessary to be within a sensible distance of the target. Conversely, a shorter-range and smaller rocket would tend to an optimum using solid fuel propellent.

These trends fell down under the tactical and operational

The Bomb. Yellow Sun was the major free-fall weapon for the V-Force in its early to mid-1960s prime. The Green Grass 'physics package' in the Mark 1 version was replaced from 1961 by the American-derived Red Snow in the Mark 2, which was shared with Blue Steel. The stated yield of the latter was 1.1Mt. The flat nose served to rapidly slow the bomb, unlike the preceding Blue Danube casing that featured a streamlined shape.

requirements of the rapid, limited warning launch requirement of the deterrent. It didn't matter how efficient the missile was if it had already been destroyed while being fuelled for launch. Solid fuel propellent could remain onboard throughout, which made up for other disadvantages, particularly when considering the actual reliability problems encountered with the early US weapons. Therefore, a development push saw the start of an American solid fuelled ICBM programme in 1956, with the aim of an operational missile in the early 1960s—likely entering service at a similar time to Blue Streak. Major improvements in the control of burn cutoff and also the size and cost of these missiles were enabling technologies.

One way to address the vulnerability to first strike of the missiles was to mount them on a mobile launch platform of some kind. The US Navy had been sceptical of the wisdom of using large liquid-fuelled rockets aboard ships, being forced to look on as the deterrent function and its associated priority funding was directed towards the USAF. The potential inherent in the smaller, lighter and much safer solid-fuelled rockets led to a vision for an entirely new and, as it would turn out, unassailable deterrent concept. At the end of 1956, the older service was able to drop out of the Army's Jupiter programme, in which it had been directed to build a solid-fuelled version of the baseline liquid-fuelled machine. In its place, a new programme was started for a missile that was compact and safe enough to be launched from a submerged submarine. Coupled with the recently demonstrated nuclear propulsion technology, not only would the launch site be mobile, but it would be almost impossible to find. The new Polaris missile, for which Lockheed was the prime contractor, promised to revolutionise the strategic deterrent and offer a more effective replacement for the manned bomber.

Nonetheless, the bomber aircraft offered a level of flexibility that made it complementary to the new submarine- and ground-launched missiles, not least if it carried stand-off weapons of its own, and even more so if the range of those devices meant it could stay out of the way of interceptor fighters. Even if Polaris was patently more assured of success, a bomber could be launched and recalled could deploy a wider range of weapons (including in conventional roles) and by its nature was a more visible demonstration of the capability of a country's armed forces.

In the same time period therefore, thoughts turned to definite requirements for cruise and ballistic missiles that could be carried by bombers. In the UK, concepts considered by the RAE for powered glide bombs would lead eventually to the placing of a MoS development contract with Avro in March 1956. The Vulcan's manufacturer was well placed to attack the problem, having been working with the RAE since 1954 on a prospective solution to OR.1132, which had been issued in September of that year. This had called for a powered bomb able to be launched from the V-bombers at a distance outside of the defences of the major cities that were the aircraft's targets. Just how far this distance might be was a controversial problem. In 1954, an aeroplane with a maximum speed of Mach 2 (and perhaps a little greater) was well within the bounds of possibility, albeit not operational service. The RAE itself was working on the design of such machines in order to inform decisions on future interceptor fighters. It was clear that the payload of the guided bomb, in the shape of the Green Granite physics package, would be large and heavy. The weapon itself then would need to be of the shape, size and complexity of an aeroplane. Avro's Weapons Research Division recruited R. H. Francis from the RAE to lead the effort, and he later noted that the missile would have to,

> "…accelerate through the transonic speed range and perform various manoeuvres at supersonic speed before reaching its target. It therefore faces most of the problems of the manned high supersonic speed aeroplane; the only ones it avoids are those associated with take-off and landing and with the crew environment, although in place of the latter it has the problem of controlling tightly the environment of its internal equipment."[164]

OR.1132 was not the first time a requirement had been issued for a stand off weapon by the UK. As early as 1950, it had been recognised by the RAE at least that various forms of guided weapon would be a reality in a decade's time. If surface-to-air missiles capable of defeating a high-flying bomber came to fruition, then the efficacy of the V-bombers on which the deterrent was based would lack the necessary assurance of success. To defeat ground-based radar-controlled defensive systems, then the opposite would be required: a bomber that could fly low and fast. Consequently, studies began into a replacement for the still-to-fly B35/46 aircraft, that might become available in time to save this embarrassment.

The answer was OR.314, issued almost exactly four years after the V-Bomber specification, and calling for a dedicated nuclear bomber aircraft that would fly at low level. While it would never come to fruition, at least two specific aspects would kick-start work with future applications. The first was an advanced navigation and attack system, eventually incorporating both sideways-looking radar and visual aspects. This became the starting point of a follow-on requirement, OR.339, which became TSR2. The second was the associated thorny problem of low-level operation: how would the bomber escape its own thermonuclear mushroom cloud? The answer had to be by putting miles between itself and the explosion, which led inevitably to the idea of a powered bomb that would complete the last few miles of the delivery of the warhead.

The low-altitude bomber would deploy a missile named Red Cat, which would have had a range of about 30 miles, using inertial guidance to get to the right place. Assuming that SAM and Mach 2 fighter defences were likely to confront and confound the V-Force from the early 1960s, there remained a credibility gap that would need to bridged somehow, pending replacement. The Red Cat type weapon could offer considerably longer range if launched from the 50,000ft operating altitude of the planned bombers, and that should have meant that the relatively short range of the defensive missiles would mean ingress routes to the expected targets might be safely navigated. OR.1132 assumed that by 1960, the range of such weapons might be 30 miles, advancing to 60 miles by 1964. A stand-off bomb with a range of 100 miles would therefore be a useful countermeasure to this emerging threat, sustaining the capability of the V-bombers until the middle of the 1960s and the introduction of Blue Streak. The operational requirement noted that the bombers,

> "…would not be able to use guided bombs to reduce their vulnerability to fighter attack to any great extent until a guided weapon with a range of 400 to 700 miles became available. This calls up many difficult problems of guidance and target information."[165]

This represented a tacit acceptance that the fighter threat, which was much longer-ranged than the SAM defence, could not be negated by the planned powered bomb. Fundamentally, the OR.1132 weapon would allow some measure of protection from surface-to-air missiles in the period 1960-64, but because deep penetration of Soviet airspace was still needed to reach the targets, there was no substantial improvement in survivability when

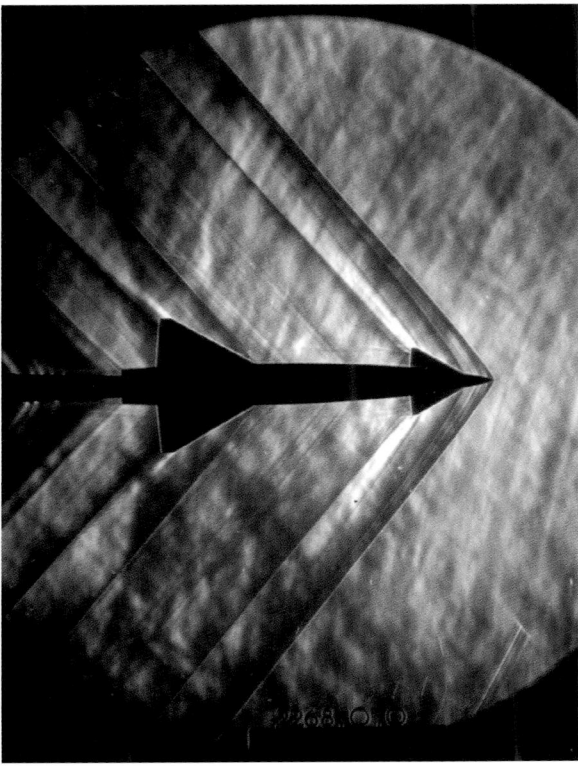

A Schlieren image from wind tunnel testing of the Avro Blue Steel Mk.1 missile. This optical technique allowed the complex shockwave pattern that developed around the missile at Supersonic speed to be visualised. In this case, the missile appears to be flying without yaw or pitch and the pattern around it is symmetrical. AIRFIELD RESEARCH GROUP

challenged by fighter aircraft. Perhaps though, the threat posed by even Mach 2-capable fighters was something the confident V-Force crews felt that they could deal with.

At about the same time that the first Vulcan B.Mk.2 was delivered to RAF Waddington, the first examples of the English Electric Lightning F.Mk.1 to go to an operational squadron were flown to Coltishall in Norfolk. 74 Squadron's previous mount, the Hunter F.Mk.6, could not maintain sonic speed in level flight. The new Lightning was released to service with a limit of M = 1.7. Ed Jarron flew the Vulcan with 35 Squadron, while just to maintain the rivalry, his brother flew the Lancashire-built supersonic fighter. He described an occasion that demonstrated the Vulcan's abundance of power and reserves of lift available at high level, but also clearly indicated the care with which the crew needed to treat the bomber's idiosyncratic transonic characteristics.

"I came across the problem once when carrying out fighter affiliation against a Lightning above 45,000ft. After some energetic manoeuvring on the buffet with full power we managed to get behind him; however, in my enthusiasm I had not noticed that for a couple of seconds he had entered a steep dive.

Keeping him firmly 'in the sights', our speed increased rapidly, and it very shortly became clear that there was no way we could get the nose to come up, so, forget the Lightning, roll wings level, shut the taps and put the brakes out! We rode down in this state for what seemed like a very long time (in reality, only a few seconds) until it finally bit, and we were able to recover. In those few seconds we had reached about Mach 0.98 and lost about 5,000ft. We later heard of an aircraft from one of the Waddington squadrons that lost 20,000ft before the aircraft bit at about 25,000ft with the pilot pulling so hard on the control column that when the Mach number began to decay in at the lower levels, he was unable to prevent an overstress. Happy days!"[166]

That said, Air Chief Marshal Sir John Allison flew the British fighter in the 1960s, prior to a long career associated with the Phantom. In his view,

"The Lightning could not outmanoeuvre the Vulcan at high altitude—nothing beats wing area when the air is thin."[167]

Why could the weapon not offer longer range? A major part of the answer to this lay in the navigation of the missile to the target which, as planned for Red Cat, would use the new technology of inertial navigation. Using accelerometers and a gyroscopically stabilised platform, the system could measure in three dimensions the accelerations experienced by the vehicle it was attached to. Mathematically integrating acceleration with respect to time would give velocity, and in turn integrating velocity with respect to time gave distance. All such measurements are subject to errors, building up with time.

A rule of thumb might be that an early INS system would drift by the order of one nautical mile for every hour of flight; still a relatively accurate system, but one that would require regular correction from other sources. For a crewed aeroplane, that might involve visual sightings of geographical features on the ground, radar mapping or star shots for astro-navigation, all of which provided position fixes. However, a better solution if available was the use of the aircraft's doppler radar, which measured speed-over-ground. This could be used as a regular input to damp the otherwise open-loop (uncorrected) velocity calculations of the INS.

At the missile's launching point, there would be a

Wind tunnel model of the Avro Blue Steel Mk.2 missile, which featured ramjet propulsion and an attached pair of booster rockets. AIRFIELD RESEARCH GROUP

'statement' of the current position from the bomber's navigation system. The missile's INS would compare its calculation with this, aligning its inertial platform and effectively zeroing its calculations, but of course incorporating any uncertainty in the bomber's navigation. To this would be added the increasing divergence from reality of the missile's undamped calculation, as on its own it would have no doppler radar to keep it honest.

In 1961, an American study of such a system estimated a circular error probable (CEP) associated with a 20-minute, Mach 2 flight of a typical missile to be over 4,000ft.[168] In other words, only 50% of those launched could be expected to navigate to within three-quarters of a mile of the target. Even for a thermonuclear weapon, there were limits. This difficulty in accurately navigating to the target and the fact that it became increasing difficult as the time of flight increased, were key drivers towards a shorter-range missile. It would have been plausible to provide longer range than the hundred nautical miles or so that OR.1132 required, but not necessarily to hit the target. For Blue Steel, as the weapon was named, the Elliott guidance system was considered the limiting factor.

What should the priority for development have been? In retrospect, the important aspects were the delivery of a M = 2.5, 100nm range missile to meet the 1960 deadline. Because it was a cruising, winged craft, it was still vulnerable to interception in the same manner as a supersonic bomber might have been. Consequently, the target speed was really the minimum viable, while flight at very high altitudes—perhaps 70,000ft or more—together with low radar cross-section were essential to counter this. It would have been possible, just about, to meet all of these requirements with an airbreathing turbojet powerplant and a largely aluminium alloy airframe. Both technologies were firmly within the state of the art and offered the potential to derisk the weapon, improving the chances of meeting the in-service deadline. Given how short a window that provided prior to expected obsolescence, the ambition and risk that seem obvious in the actual solution pursued deserve comment.

The short range meant that the high specific fuel consumption of a rocket motor could be tolerated. This in turn gave an abundance of thrust, necessary for rapidly climbing to altitude, but also eliminated the need for supersonic intakes and their radar signature. Blue Steel would feature a structure with a very high proportion of stainless steel in order to deal with the uncertainty of aerodynamic heating. These two decisions, taken near

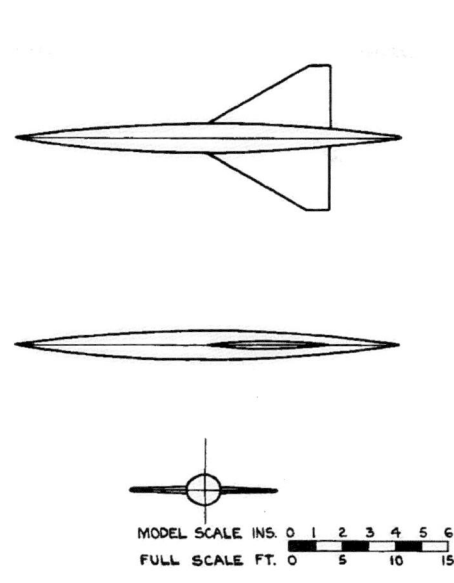
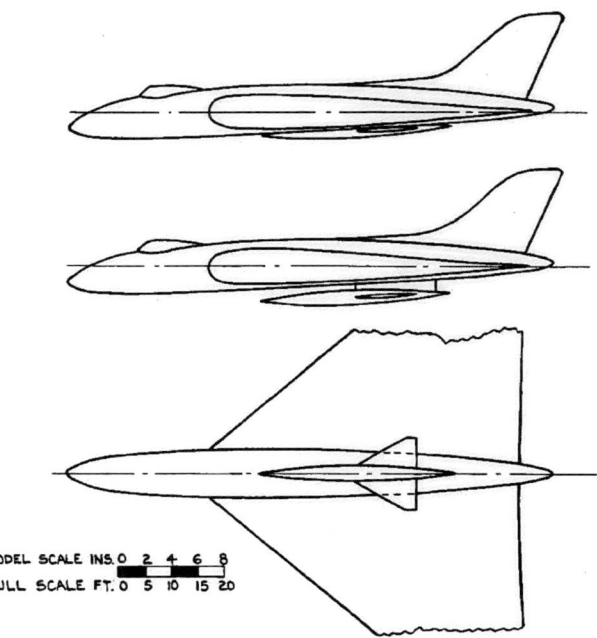

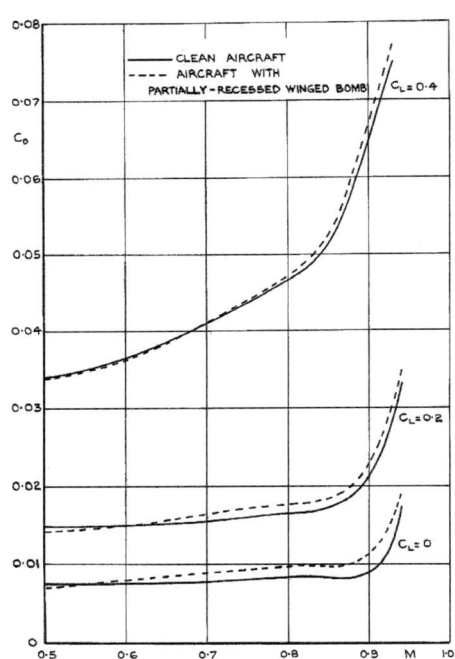

FIG. 8. VARIATION OF C_D WITH MACH NUMBER AIRCRAFT WITH AND WITHOUT PARTIALLY— RECESEED WINGED BOMB.

Results of RAE wind tunnel tests of a Vulcan model equipped with a 'winged bomb', which gives a generic view of the impact on drag of carrying such a weapon on either a pylon or semi-recessed arrangement. In the case of the latter, while not for free, neither was the impact disastrous when considering a typical lift coefficient of about 0.3. The increasing Mach number dependency with lift would be a parameter that could be improved, by detailed design of the missile/lower fuselage geometry.

to the start of the programme, were to prove costly in different ways.

The stainless steel airframe proved extremely difficult to build, requiring research into new methods of manufacturing. Even apparently simple processes like drilling would prove problematic. This undoubtedly caused significant delay, while the initial full-scale test missiles were actually constructed in aluminium alloy to speed up proceedings. While it was barely necessary at the projected speed, an ability to tolerate heating became more important if the missile did need to go faster. It has to be concluded that sticking with the decision to work in steel was a major factor in Avro missing the 1960 deadline, only really justifiable if a follow-on use of the airframe for M = 3+ was being considered. Francis wrote,

> *"We found difficulty, largely due to the time spent in learning the techniques of manufacturing in steel, in getting these trials started and it was therefore decided to insert into the programme a series of full-scale test vehicles made of aluminium alloy instead of steel... The first two of these — inert dummies — were launched in 1958. During 1959 and 1960 a number of powered test vehicles of this series were launched."*[169]

By the required in service date therefore, progress had reached only as far as a test vehicle in a different material, and as it happened a different engine. However, in terms of demonstrating the functionality of the weapon, only the specific internal temperature control and connections

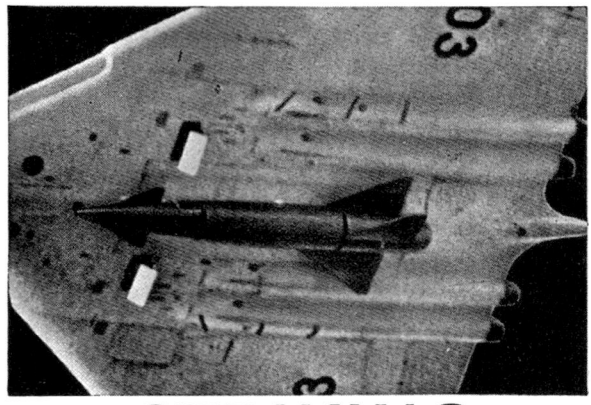

THE VULCAN GETS NEW STRIKING POWER

AVRO STAND-OFF BOMB GIVES V-BOMBERS EXTRA FLEXIBILITY IN ATTACK

Britain's most advanced V-bomber, the new Avro Vulcan B. Mk. 2, due to enter service with RAF Bomber Command next year, will be equipped to carry one of the most powerful weapons ever invented for the preservation of peace. This is the Avro Stand-off bomb, first air-to-ground guided missile to be developed in this country, which is designed to destroy an enemy target without the aircraft having to penetrate the ring of defences. Designed and built by the Weapons Research Division of A. V. Roe, the Stand-off bomb gives new power to the Avro Vulcan which becomes an even more formidable bulwark against aggression.

AVRO VULCAN
POWERED BY BRISTOL OLYMPUS ENGINES

 A. V. ROE & CO. LIMITED, MANCHESTER / *MEMBER OF HAWKER SIDDELEY AVIATION DIVISION*

Avro advertisement showing a test Blue Steel being carried by Vulcan XA903. This would be the only B.Mk.1 to carry the missile, as it was allocated to the test programme. DAVE ROBINSON – AVIATION ANCESTRY

Clearance of the launching characteristics of the Blue Steel/Victor combination, in the ARA wind tunnel. Measurements of the forces and moments on the missile helped to determine its flightpath, which obviously needed to be deconflicted from that of the bomber in this critical phase.

to the bomber associated with the final product had been identified as test points not achieved with the aluminium versions. Almost irrespective of their performance, given the limited SAM-defeating objective, could and should the customer have pushed harder for a missile with a lower thermal tolerance (meaning cruise Mach number), in order to have a useful weapon in service?

The Armstrong Siddeley Stentor rocket motor used kerosene as the fuel and high test peroxide (HTP) as the oxidant. This sadly is where it really went wrong. HTP was a difficult and dangerous chemical to work with and suffered from all of the problems described for the liquid oxygen in Blue Streak: a liquid-fuelled rocket was fundamentally unsuited to rapid generation and the V-Bomber's operational tasking: quick reaction alert in the event of a minimal warning Soviet attack. David Booth, an Elliott Brothers technician involved in the trials of Blue Steel, recalled:

> *"The missile would be gleaming under the glare of the floodlights illuminating the chainmail enclosed loading bay, the ground under the aircraft awash with distilled water. The water was from the protection system that the fuelling team could use if they got HTP on their suits. As you all remember HTP was highly flammable if it touched any clothing or exposed parts of the skin."*[170]

The aerodynamic challenge of the missile should not be overlooked either, as in 1955 when key configuration decisions were made, the world was not awash with test-data on winged craft that could fly at more than twice the speed of sound. The situation was complicated by the need to start the flight in the transonic regime. Although this speed range could rapidly be cleared thanks to the enormous thrust of the rocket, there was a brief period in the seconds after launch where danger might lurk. Flying close to the bomber and on approximately the same heading, there was a logic to ensure that control of the missile was as linear as possible, as it responded to the movement of the aerodynamic centre with Mach number.

What would be beneficial was a configuration that would be stable longitudinally both at launch and supersonically. Avro's studies showed that it would be difficult to combine a just-stable (to minimise trim drag in the cruise) supersonic tail-aft missile with a stable transonic layout, the shift in the aerodynamic centre being too great. A tail first, canard layout could be arranged to have relatively similar positions at launch and cruise, which implied low trim drag and effective launch control authority for the elevators. It was considered that this was only of real importance in the case of autopilot failure, and then only in the case of the first few test launches before the robustness of the system was proven out. Strangely then, at least part of the reason for the complete configuration of the missile was in order to derisk stability and control in the test environment, which was in marked contrast to the approach to materials and manufacturing.

As Handley Page had considered in its extensive tail-first investigations, the canard layout had some advantage of the aft-tail in that both wing and stabiliser surfaces were lifting; together, they could be smaller and lighter than a wing and down-loaded tail when supporting a given weight. For the wing itself, a 60-degree delta was selected, which placed it in the class of geometry where stable vortex lift at moderate incidence could be present. The planform and span were dominated by considerations of the initial transonic launch and low-supersonic pull-up into the climb, as well as, of course, the need to fit under the launching bomber. This latter requirement limited the finesse ratio, or slenderness, that could be achieved; a very important parameter for the wave drag (and hence range) performance of the missile. A minimum body diameter was required in order to package the warhead, while the length was a maximum associated with the space under the carrier aircraft. In the case of the Vulcan, while the weapon would be carried semi-recessed into a special fairing in place of the bomb bay doors, in reality the recess extended forward beyond the original, taking full advantage of the available distance between the nosewheel bay and the aft bulkhead. Blue Steel therefore had a length to maximum width ratio of about eight, while being 35ft long.

Wind tunnel testing of the Vulcan and Victor had been difficult enough, in a Mach number region that was challenging to achieve. Blue Steel obviously needed to be tested at supersonic speeds, and the programme was able to take advantage both of the RAE's new facilities at Bedford, which were coming on line in the mid-1950s, and the brand new ARA transonic/supersonic tunnel, on a different Bedford site. Schlieren imagery, which used the effect of density changes on refraction in order to visualise shock waves, had been utilised while working on the Vulcan's leading edge. Helpful in subsonic flow, this technique that had been so evident at Völkenrode was essential for the analysis of supersonic flows.

One area of concern that Schlieren data would prove useful to analyse was the interaction between the foreplanes and the wing. Specifically, the shocks generated by the foreplane were highly influential to the flow pattern around the fuselage in sideslip, capable of generating asymmetric forces far ahead of the centre of gravity. The tip vortices from the foreplane also passed close to the wing, which of course mounted the ailerons. It was important for autopilot control to simplify any coupled response between pitch and roll control, so these characteristics had to be well understood. The original fin design had similarly sized dorsal and ventral surfaces. It was found that around Mach 1, both fin and aileron effectiveness were severely reduced by a strong interaction between them. While this could be driven through acceptably, and was on the initial test vehicles, for the production missile it was corrected by biasing the area such that the upper fin was considerably smaller, compensated with a longer chord tip and swept trailing edge on the lower. From an installation perspective this was more convenient too, while in the cruise at incidence, the ventral fin was by far the more effective surface for directional stability. Still, it would be nearly at the production stage and following a test programme across nine different wind tunnels before this change was implemented.

The guidance, propulsion and aerodynamics of the missile were all challenging but a further problem was how to minimise the penalty of integration with bombers. Under all circumstances, some penalty in skin friction and profile drag should have been anticipated, given that there was simply more skin overall when the two were combined. For a transonic vehicle subject to compressibility effects, the issues were more complex, leading the RAE to comment that relative to a conventional pylon mounting,

> *"A possible aerodynamic refinement of this mounting at the cost of some structural complexity is to eliminate the strut and recess the lower surface of the aircraft fuselage to take the upper part of the bomb body... In the case of an aircraft having a moderately high aspect ratio wing, mounted in either the mid or shoulder position, it can be assumed that the interference effects between the wing of the aircraft and the wing of the bomb are small. For a delta winged bomber, such as the Avro Vulcan, or an aircraft having a wing set low on the fuselage, the proximity of the bomb to the aircraft wing invalidates such an assumption."*[171]

Victor XL231 landing at RAF Wittering, to join 139 (Jamaica) Squadron in 1962. In its original form prior to the retrofit programme, the aircraft was not yet equipped for Blue Steel or inflight refuelling, and the nose flaps are deployed. DOUG GREAVES

Tests with a representative bomb model, which was of slightly improved slenderness and using solely a (larger) delta wing as a its lifting surface compared to Blue Steel, showed fundamentally that the lifting characteristics of the aircraft were unchanged by the presence of either the semi-recessed or pylon-mounted weapon. At any useful incidence, the flow under the Vulcan was dominated by the shape of the aircraft itself, which supressed the lifting characteristics of the bomb below that found in free air.

The integration difficulties were more apparent when viewed in terms of pitching moment and drag divergence, both with the same root cause. The tests were able to identify an acceleration in the gap between the wings of the bomb and the lower surface of the Vulcan, resulting in a locally forward position of the lower surface shock wave. Under the fuselage therefore, a local supersonic low-pressure region of flow now existed—and because it was behind the centre of gravity it applied a nose up pitching moment. In a related manner, the additional wave drag associated with the new shock characteristics had some effect on drag divergence, which was negligible in the semi recessed configuration below $M = 0.89$ but was both earlier and significantly more detrimental in the case of the pylon-mounting. The pylon itself had the disadvantage of forming its own shock system in conjunction with the fuselage. However, in both cases the RAE was forced to treat the results with caution due to their old problem of Reynolds number similarity.

"At high Reynolds number, the increase of Cm above $M = 0.9$ may not be so great as that shown by the tunnel tests. For the aircraft without bomb, the boundary layer on the lower surface ahead of the shock would be turbulent in flight, but laminar at the Reynolds number of the tunnel tests; the greater thickening of the boundary layer behind the shock at low Reynolds number would be expected to cause a more marked reduction in the nose-up due to the deflected trailing edge, as the lower surface shock moved back, with the bomb in position, however, the boundary layer is probably turbulent ahead of the shock in the tunnel tests (due to the presence of the bomb); hence the conditions are more representative of full scale, and no change would be expected in the effectiveness of the deflected trailing edge due to changes in the lower surface boundary layer conditions caused by difference of scale. Thus, it is seen that the nose-up change of $Cm0$ caused by the bomb above $M = 0.9$ in the tunnel, may be greater than that which would be found in flight."[172]

Once again, the RAE specialists were forced to use their judgement rather than direct reading of the data when making an important assessment. However, it was clear that the data suggested a lower risk to the all-important cruise and height-over-target performance when using the semi-recessed configuration.

The original bombers had been developed over a period of time which largely came to them; there was no choice but to pursue an independent deterrent if there was to be one at all, while the ramping up of East-West tension through the events in Berlin and Korea would concentrate opinion in one direction. The moderate thawing of the atomic freeze between the US and UK of the 1954 Atomic Energy Act was followed by the nadir of relations brought on by the Suez Crisis.

By the time potential cooperation across the Atlantic was again on the cards, the prototypes were already flying and the first production machines taking shape. Programme delays meant the same could not be said of Blue Steel, with the weapon firmly in the development stage at the time of the Sputnik panic of October 1957. A month later, Operation Grapple demonstrated that the UK independent deterrent was built on firm foundations: a true megaton-range hydrogen bomb had been successfully dropped from a V-bomber near to Christmas Island and it had worked as intended. Now, the American public were demanding progress and the UK had proven itself to be much more than a bit player in terms of Cold War capability.

Macmillan was keen to persuade Eisenhower that cooperation on nuclear weapon development would be mutually beneficial, and Sputnik perhaps gave him the opening he needed. The issue would be less one of convincing the president, but instead of him doing the deals with Congress that would enable the required amendments to the McMahon Act. However, the door was gradually opening. The Joint Committee on Atomic Energy was positive about the opportunities that sharing information with a wider range of Allies, including Canada and Australia, might bring. Over the next half a year Congress would gradually move towards deciding in the final amendment that nuclear sharing take place only with allies that had made, 'substantial progress' in the field on their own. Grapple ensured that the evidence was obvious and in the event it was the UK alone that passed the test.

For the first time since the collapse of the Truman, Attlee and King discussions in 1946, the development and sale of fissile material and actual weapons between the two countries was a possibility. A few days after the amendment was passed, the Mutual Defence Agreement (MDA) — a treaty specifically laying out how the technology transfer would work — was signed between Samuel Hood, in his capacity as the de facto deputy ambassador to the United States, and the Secretary of State, John Foster Dulles, on July 3, 1958.

The RAF's greatest difficulty with airborne deterrence — penetration of Soviet airspace protected by effective SAM defences — was of course shared by the USAF. The American strategy was expressed in a subtly different way however: the primary purpose of air-launched cruise missiles was defence suppression, as opposed to being the primary deterrent. In this concept, some of the bombers would carry nuclear missiles and spearhead the attack, firing them at SAM sites and fighter bases, clearing a way for the majority of the force and its freefall bombs. This implied that the missiles needed to be supersonic, not just for survivability but in order to get far enough in front of the bomber force to destroy the defences prior to their arrival. This differentiation made sense for a country that had the resources to tackle the deterrent problem from a number of complementary angles. However, it would have devastating consequences for the UK.

American cruise missile development work had encompassed long-range ground-launched systems, based in part on captured German missile technology. As Avro had described, for stability purposes there were good reasons to move away from a swept wing configuration for these — towards a tail-first layout — and so evolution to the canard-delta had occurred on both sides of the Atlantic. North American Aviation would explore this idea with the X-10, a startlingly advanced unpiloted aeroplane of canard delta layout which also featured F/A-18-like angled twin fins. This 70ft long, turbojet-powered machine was capable of about Mach 2 at altitudes in excess of 40,000ft, which was blistering performance at the time of its first flight in October 1953. It was intended to be a stage in the understanding of the future Navaho ground-launched missile, proving the airframe and guidance system concepts, which over the next few years it successfully did. The Navaho programme itself was cancelled in 1958 as unnecessary, given the success of the Atlas ICBM for which it was effectively the backup.

A viable air-launched cruise missile had been sought by the (then) USAAF even prior to the end of the Second World War. On July 16, 1945, it published the proposed characteristics for a new weapon that might be launched from an unspecified bomber at between 20,000ft and 40,000ft, reach at least 1,200mph and hit a target at a range of at least 100 miles. The acceptable performance penalty to the bomber of carrying the missile was specified

Similar both in size and purpose to Blue Steel, Bell's Rascal weapon was made from aluminium alloy, rather than stainless steel. USAF

as within 2%, implying that it was already understood that the weapon would be too large to be stowed internally.

Three companies were issued initial study contracts but by 1947 Bell Aircraft was the only one still in the running. Subsequent contract changes highlight the confusion of attempting to specify either what was really required (supersonic speed) or what might realistically be achieved, and not for the last time. The B-29 was identified as the candidate carrier, while longer range and a heavier payload were progressively sought. In mid-1947 Air Force Materiel Command (AMC), responsible for the procurement, highlighted that the continuing studies showed,

"A missile carrying a 3,000lb warhead and capable of sustained flight, will not fit in the bomb bay of any operational, experimental, or planned bombardment airplane.

A liquid rocket power plan is the only type suitable for supersonic flight performance and the present time, and probably for five years to come. Fuel requirements… will make it uneconomical and very poor choice for powering any missile designed for ranges in excess of 150 miles".[173]

AMC had therefore already identified flaws in the Blue Steel concept—still a decade in the future. Progress on the Bell missile was slow, as emphasis was placed solely on the complex guidance system for a number of years. However, in February 1950, USAF Headquarters divided the requirement into two parts: an initial Phase I missile would have a range of 1,000nm at up to Mach 2, carrying a 5,000lb atomic warhead and launched from a B-50; a Phase II design would fly at Mach 3 and also be carried by the B-36. These were required in service by January 1954 and July 1955 respectively. If this had come to fruition, then at a similar time to the first Valiants becoming operational with the RAF, Strategic Air Command would have a short-range stand-off capability, albeit provided by their much slower and lower altitude capability piston-engined bombers. In fact, by the end of 1951 SAC had realised that the new swept-wing B-47 would be essential to mitigate the challenges encountered by B-29s over Korea in MiG-15 patrolled skies. They realised that,

"Successful attacks in the future would only be possible if the then existing weapon systems were continuously improved".[174]

General Curtis E. LeMay was by now the commander of SAC and was scathing of the Bell Rascal, as the weapon had now been named. He believed that it would be inferior to other systems in general but in particular, felt that the terminal guidance strategy whereby the missile beamed back a radar picture to the bomber for manual control was untenable. It was likely to be inaccurate, since it required a well-defined radar target, and there were concerns that it would be easily jammed. This caused problems for the relationship between SAC and USAF HQ, who agreed with an alternative assessment that the B-47's performance would not be significantly affected and that Rascal would fill in a gap in the deterrent strategy, for which there was no other option. It was thought that jamming concerns could be addressed in the fullness of time by inertial guidance, while range was still expected to be increased.

Rascal, as eventually service tested, was a substantial 32ft-long craft (Blue Steel was 34ft 11in), carrying a 3,000lb warhead for about 90 nautical miles and with a demonstrated maximum speed of M = 2.95. Despite this, it was built conventionally from aluminium alloy, without the issues of stainless steel that would plague and delayed Blue Steel. Its launch and cruise flight phases were navigated inertially, with the second manually corrected, radar guided system that LeMay had so objected to being used in the terminal dive. It was clear that the pure inertial system was needed quickly, and the missile was actually available for use in its initial form in 1958, at the time of the MDA. By then however, the USAF was absolutely sure it could see better things on the horizon and terminated the programme, with a complete end in December. At a time when Blue Steel was a programme of test shapes and still some years away from the most optimistic estimates for deployment, the Americans had discarded a missile that had a basically similar range and speed performance, packaged in a V-bomber proportioned shape, albeit with a much lower payload.

The USAF's fear that Rascal's range and operating concept left it dead on arrival would lead to a new requirement issued in March 1956, now closely aligned with the start of Blue Steel in the UK. GOR.148 called for a Mach 2 missile with a 350nm range, cruising above 55,000ft and specifically to be launched from a B-52. NAA proposed a turbojet powered, air-launched version of the Navaho, resulting in a contract award in August 1957. Unsurprisingly, it became a very high priority post-Sputnik, with the GAM-77 programme (later redesignated AGM-28) being accelerated in February 1958 and remarkably seeing the first delivery of a production missile in December 1959. In a sign of the times, it was named Hound Dog, and in the primary high level profile had a range of over 650nm. In other words, it would have met the OR.1132 ambition

Dramatic early 1960s take off of a Boeing B-52F Stratofortress, carrying a Hound Dog missile under each wing. These 42ft long Mach 2 cruise missiles were expected to fly ahead of the main bomber force, clearing air defence sites to smooth its passage.

of a missile that could be launched from far enough away to negate the fighter threat, as well as that from the SAMs.

Unlike Rascal, it was fully inertially guided to its target, although it received positional information when mounted on the B-52 from an astro-tracker mounted in the pylon. In fairness, Hound Dog was so long and tall, with the engine podded underneath the fuselage, that compatibility with the Vulcan and Victor would have been difficult to achieve. Nonetheless, the Americans had demonstrated again that a viable missile could be put into service, in a timeframe when it was still a credible deterrent weapon. It was operational in July 1960, when there was no sign of the British missile in the order of battle.

The advent of Polaris (even as an idea) represented a reversal of the situation the USAF and US Navy had found themselves in a decade earlier: now, the more convincing method of providing an assured nuclear attack (and the claiming the funding that went with it) lay at sea. However, both the land-based ballistic missiles and the manned bomber force that it had responsibility for, stood to benefit from the advent of effective solid-fuelled rockets. In the latter case, it became possible to consider an air-launched ballistic missile, achieving hypersonic speeds and consequently as survivable post-launch as any other available system. The manned bomber suffered from the problem that its well-known operating bases might be quickly destroyed by a first strike missile attack, but as a counter could be dispersed to many locations, some of which at least might be beyond missile range.

At the same time, the bomber provided a major flexibility advantage; once in the air, the actual nuclear strike by the missile itself could be retargeted quickly using the onboard systems and real-time understanding of what targets remained. These arguments were undoubtedly motivated by the inter-service rivalry as much as by taking a cold look at the likely success of the resulting deterrent. On the other hand, the capability of the bomber extended well beyond the nuclear mission, and the USAF as an entity greater than SAC would not have wanted to find itself without any platforms suited to conventional war. In this at least, the extraordinary longevity of the B-52 suggests that they were correct.

The response was a study known as WS.199, an umbrella for the ALBM related work undertaken by AMC and a number of contractors. A technology demonstration phase saw Martin design the Bold Orion missile, which in November 1958 was able to show a 250nm range when launched from a B-47. USAF HQ had already seen enough to request a more advanced, two-stage version with a range of 750nm. Progress was again remarkably rapid, with a second launch on December 16, 1958 working well until radar contact was lost soon after second stage ignition. Nonetheless, the impact prediction was calculated as 930nm downrange, and it was clear that

The RAF's new bomber, with new weapons to suit: the Vulcan B.Mk.2, coming soon (in 1960), with Blue Steel and Skybolt.

contemporary technology could deliver the weapon that was required, in terms of a flight vehicle at least.

From a UK perspective, the new possibilities brought about by the NDA in 1958 gave rise to an opportunity. The glacial progress on Blue Steel when compared to Rascal and what would become Hound Dog could be judged against a direct purchase of an American system. A joint USAF and RAF task group for strategic air-to-surface weapons met in Washington in April 1958. Clearly, they would have discussed UK progress and that of WS.199, together with the production systems that must follow. The following month, a new British requirement, OR.1159 was issued, effectively asking for a Blue Steel with a range of at least 600 miles. Avro had looked at modest improvements in the baseline Blue Steel, using combinations of high-energy fuels, booster rockets and external fuel tanks. OR.1159 was to be addressed by a further derivative termed W114 (the original Blue Steel was W100), retaining the hard fought-for steel airframe and finally attempting to justify this through speeds of Mach 3 and beyond. Using a pair of external booster rockets to accelerate to Mach 2, prior to being jettisoned, thrust was then delivered by a battery of four wing-mounted Bristol ramjets. The space vacated by the Stentor rocket engine in the fuselage allowed an on-board doppler radar to be fitted, allowing correction of the INS and hence solving the inherent accuracy problem of the long-range weapon. This solution was different to the astro-navigation system favoured for the American missiles; while it would have been expected to be simpler (especially in daylight) and of lower risk, in operation it was obviously an emitter and hence open to being used by Soviet defences to guide their interception.

The new W114 Blue Steel Phase II was tested, like its precursor, in the supersonic wind tunnels at Bedford and Farnborough. Of particular interest was how the booster rockets would separate safely and it was shown that this could be achieved by careful attention to their pitching characteristics. It is reasonable to assume that the Phase II system could have been made to work, but as it offered little reduction in its own vulnerability compared with the earlier missile. It is therefore hard to make the case that it would have significantly postponed the originally expected obsolescence date of 1965. It survived as a live concept until the end of 1959, by which time the challenge represented by efforts over in the US could no longer be ignored. It fundamentally could not compete with what the RAF now knew the Americans could offer as an air-launched alternative.

The success of the proof-of-concept WS.199 programme was timely; there was post-Sputnik appetite for bolstering the US deterrent and the Air Force latched on to the concept of an ALBM, with enthusiasm at the highest levels. In July 1958, a little later than the commencement of OR.1159 work in the UK, the USAF Chief of Staff General Thomas H. White instructed that the Aircraft and Weapons Board determine the possibilities for an ALBM system which, in no way pre-judging the outcome, he believed should be in service as quickly as possible. The Department of Defense was not automatically sold on the idea, given that funding was in place for a multitude of systems already and that the long-term future of the bomber was in some doubt.

Despite the government reaction remaining lukewarm, by January 1959 the USAF had acted to shift the programme up a gear; firstly, ARDC issued Systems Directive 138A, which initiated the competition for selection of a contractor while defining the priority requirements. The jumping-off point was a missile for the planned B-58B, with subsequent attention being given to the B-52C/D and then the Victor and Vulcan. There should be no doubt then of the intertwined nature of the new missile's development with the strategy of both governments: the weapon was important to both, and the UK was firmly at the table even at the conceptual stage.

A day later, the USAF issued its general operational

Vulcan B.Mk.2 carrying a Blue Steel round.

requirement for an advanced air-to-surface missile, GOR.177. Consistent with the ARDC requests, the stated need was for a rocket-powered air-to-surface missile which would allow its Convair B-58 carrier to stay outside of the range of Soviet defences. And it had to be in service not later than 1963. The B-52 was again stated to be the secondary application, with a helpful imperative coming into play. Because the B-52 squadrons were gradually losing aircraft to depot maintenance for the fitting of the Hound Dog system, it was apparent that time (and cost) would be saved by fitting for the new ASM at the same time. To realise this, progress needed to be made on the new missile immediately.

The Eisenhower administration's policy of massive retaliation, that is, meeting the Soviet nuclear threat with the promise of an overwhelming response, did not mean that the efficacy of potential new weapon systems was not closely scrutinised. One of the many impacts of Sputnik was the strengthening of the defence scientific establishment and its alignment with the nation's priorities. February 1958 saw the creation of the Advanced Research Projects Agency (ARPA), reporting to the Secretary of Defence and tasked with understanding and implementing research and development to address military needs far into the future.

Herbert York was hired as the assistant to the director of ARPA. A renowned nuclear physicist, his tenure in this important role was relatively short, as on December 30, 1958 he was appointed as Director, Defense Research and Engineering, overseeing ARPA and many other organisations, now as the senior engineering advisor to the Secretary of Defense, Neil H. McElroy. It was very much on his desk therefore, in early 1959, that the task of formulating a government response to the Air Force requirement landed. In late March, he advised that he was content to allow a contractor to be selected, but prior to the actual signing of contracts additional understanding of the problems would be required.

Matters were further complicated by the inability of the Joint Chiefs of Staff to come to agreement about whether the weapon was actually required. The Army Chief of Staff was concerned that an effectively artificial dependence on aircraft would result from the allocation of resources to the project, when in reality the day of the manned bomber might already be over. On May 20, York informed the USAF Chief of Staff that five criteria would need to be met by the contractor, prior to a full go-ahead; this project would need to be completed by August 15. These were: operational employment, system optimisation, cost effectiveness, development planning and technical problem areas. He had also established an ad hoc group specifically to look into surface-to-air missiles, recognising just how much of a hot potato the concept would become. Douglas was appointed as the contractor and on July 14, XGAM-87A was allocated as the designation for the system. On January 11, 1960, the name Sky Bolt was announced (later contracted to Skybolt in order to meet a 'one word' popular names directive).

In reality, it would not be until January 1960 that Dr York was fully briefed and subsequently DoD approval given for the project to start. It had been a difficult time for Douglas, forced to stretch out an extremely limited tranche of funding to complete what in retrospect was clearly a weighty block of upfront work. In the second

Vulcan B.Mk.2 XL360 of the Scampton wing taking off from Coltishall in 1969. Towards the end of Blue Steel operations with 617 squadron, the round is visible in its semi-recessed position. XL360 is preserved at the Midland Air Museum, Coventry. FRANK CROOM

half of 1959, York's working group had recommended that the programme be terminated, given the uncertainly over the fifth requirement—the technical problem areas—and presumably its knock on to the third, project cost effectiveness.

The UK involvement moved forward from early March 1960, as the programme headed towards firmer ground, although still without an actual commitment to production and deployment from the DoD. US representatives gave a briefing on the system to officials in London on March 2, followed towards the end of the month by a meeting at Camp David between Macmillan and Eisenhower. The two leaders each agreed to provide the other with something that they wanted, although without quid pro quo. The Americans needed submarine basing facilities in Scotland for their new Polaris fleet, a weapon system that the president would later describe as, "invulnerable". The British received assurances that they would be able to participate in the Skybolt programme, and, if it came to fruition, purchase their own missiles, the use of which would be a UK-only decision. Macmillan would therefore be able to leverage the existing V-bombers, the Mark 2 versions of which were yet to enter service, to provide a credible and crucially independent deterrent beyond 1965.

At a stroke, or so it might have seemed, the difficulties of Blue Streak could be eliminated. The RAF would retain its status as the provider of the UK deterrent, which in contrast to the attitude of the US services largely suited the aspirations of the Royal Navy and British Army. Harold Watkinson, the Minister of Defence put a paper before Cabinet on June 20, giving details of the position. He noted,

"There could as yet be no absolute certainty that SKYBOLT, which was not due to be tested as a complete weapon for about a year, would be successful and it must be recognised that the Americans would not develop it for our use alone. However, the United States authorities were confident that it would be effective and they attached importance to it both for their own Air Force and as a means of prolonging the effectiveness of our V-bombers, which they recognised were an important part of the strategic nuclear deterrent."[175]

The reality was of course quite different to that articulated in the second sentence. If by 'United States authorities' he was referring to the Department of Defense, then it is clear in retrospect that there were grave misgivings about the ability of industry to deliver a weapon which did all that was promised; the challenges of guidance and re-entry inherent in a ballistic missile being compounded by the

Believed to be XL231, this camouflaged Victor B.Mk.2 equipped with Blue Steel awaits at Wittering. DOUG GREAVES

transonic launch from an ill-defined position. On this latter point, the UK government's own scientific advice was hardly supportive. A few years later, the Chief Scientific Advisor to the MoD, Sir Solly Zuckerman, described his then recent recollections.

> *"Almost immediately I started inquiring about Skybolt, I knew that Skybolt was as much nonsense as the manned bomber in a missile era. Furthermore, I didn't believe it could be worked technically, in spite of everything that was being said. I was being provided with information by … the men who had to take the responsibility for providing funds for it in the US defence budget. I wasn't getting information from … Douglas … nor from some Air Force general who wanted [to] … fly Skybolt off a B-52. I was getting the real information from men whose opinion I trusted and who trusted my view."*[176]

Zuckerman was a figure almost without compare; as a world-leading scientist and educator from a completely different field—zoology—he had served his country with distinction throughout the war, generally in operational analysis and assessment and often associated with air power. The future Secretary of Defence Harold Brown, who would follow the same road from the nuclear science of the Livermore Laboratory in California to Washington as had Herbert York, said of him,

> *"During the Eisenhower administration I think Solly felt at home with the PSAC and with Herbert York, the Director of Defence Research and Engineering, but not with the Atomic Energy Commission people or the rest of the Defence Department."*[177]

While Air Marshall Sir Christopher Hartley offered,

> *"By the time I returned to the Air Ministry in June 1961 Solly was Chief Scientist MoD and an object of grave suspicion to the Service Chiefs because of his refusal to accept their statements without query… It was not his fault that he arrived too late to influence some of the major errors of the 1950s, but his scrutiny of the subsequent defence programmes put the country, not to mention the public purse, in his debt."*[178]

If 'Solly' was sceptical of Skybolt, then it must be said that his views on the default post-1965 deterrent, Blue Streak, were no less forthright. Neither to him would represent a viable deterrent. What then might have been the alternative, if there was to be a deterrent at all? At the same cabinet meeting where Watkinson presented the Skybolt plan and walked away with a Cabinet decision to proceed, the only other item on the agenda was the American request for Polaris basing facilities in Scotland. Watkinson's paper on this topic extended beyond the initial discussion between the leaders of the two countries,

Vulcan XL318 was one of 617 Squadron's first B.Mk.2 aircraft and is preserved along with its weapon — Blue Steel.

describing further talks that he had himself held with his own counterpart, Secretary of Defense Thomas Gates. It outlined the significance of the weapon system itself, again using Eisenhower's "invulnerable" description, and that the two to three month long submerged patrols of the submarines would,

"...pose complex communications and other problems, and careful organisation would be required to ensure the efficiency and reliability of their crews. The United States Government would welcome our close association with the United States Navy in this development and they hoped that the United Kingdom Government would feel able to participate, if only on a limited scale, in their POLARIS submarine plans. The use of facilities in Scottish waters would enhance the effectiveness of the POLARIS force by about 30%".[179]

It was not clear where the notion of only 'limited scale' of participation came from, because the feeling of the Cabinet as recorded in the Conclusions was decidedly in favour using the request for basing facilities as a lever for an extraordinary prize — an invulnerable Polaris submarine force of its own. While the UK was indeed in the process of acquiring Skybolt on what would be very favourable terms, given the risk and not least the hundreds of millions of dollars that would be committed on behalf of the US taxpayers for the research and development needed, was not an extension of the capability of Polaris by 30% also of huge significance, that could equally be accounted for in dollars? They certainly thought so, as is recorded.

"As the closest partner of the United States in the Western Alliance, it would be desirable that the United Kingdom should be associated with this form of diversified deterrent, which was well suited to our maritime traditions. It would be desirable that the Americans should accept this as basis of partnership for the facilities they wished to obtain in Scotland and to agree that we had the right to obtain a force of this sort of our own.

...In further discussion it was agreed that an option for the United Kingdom to acquire POLARIS submarines would not involve any commitment for their purchase... It was suggested that our financial resources would not be sufficient to embark on a POLARIS as well as a SKYBOLT programme".[180]

Macmillan summed up the proceedings by declaring that the presence of the US Navy Polaris submarines was in the national interest and could be made justifiable in the opinion of the public, who would be making themselves targets for the Soviets in accepting them, if the effort were a joint venture that could eventually result in a UK (independent) Polaris force. He requested a draft message that might be taken to Eisenhower. The Conclusions do not record which solution, Skybolt or Polaris, was favoured by the government. Just one week after the Cabinet meeting, the first operational example of a Polaris Submarine, the USS *George Washington* (SSBN-598), departed her moorings at Groton and sailed for Cape Canaveral and the loading of a pair of the futuristic weapons. On July 20, her commanding officer signalled directly to Eisenhower: POLARIS—FROM OUT OF THE DEEP TO TARGET. PERFECT. Both missiles had fired from the submerged vessel and impacted 1,100 nautical miles downrange. Skybolt was a complicated theory, Polaris A-1 and its delivery system was a realised, 'invulnerable' deterrent weapon.

12

APOGEE AND RE-ENTRY

The journey from Operational Requirement to Initial Operational Capability for the Vulcan and Victor had taken around a decade. In a world of rapidly improving technology, it would have been surprising indeed if all of the underlying assumptions that had given the make up of B35/46 had still held. The RAE had been sceptical in 1937 about the ability of Volkert's proposed unarmed high-speed bomber to evade defences in the long-term, as the same technology was adopted for fighters. How had the V-bombers fared over this ten-year gestation?

In June 1954, Air Chief Marshal Sir John W. Baker wrote to George Carter, the chief designer (and a board member) of the Gloster Aircraft Company, a member of course of the Hawker Siddeley Group. Carter should perhaps be better known than he is—as the designer of the first British turbojet-powered aircraft, the 'Gloster-Whittle' E28/39, then subsequently the RAF's first operational jet, the Meteor. Indeed, Roy Chadwick had written to him on the occasion,

> "... I am particularly glad to be able to congratulate you on being the first aircraft designer to produce a practical jet-propelled aeroplane."

Baker informed him,

> "... the Air Staff propose to issue a requirement for a new all-weather fighter of advanced design and performance... I thought that you would be particularly interested to have the enclosed RAE/RRE Report No. Aero 2513/G.W. 20 and gain value also from the accompanying ACAS(OR) Paper CMS.1539 ... You will note that the Air Staff and the Establishments' studies point to two different kinds of aircraft ... In view of the important disclosures made in these documents I must ask you to control extremely carefully their circulation."[181]

The Joint RAE/Royal Radar Establishment study referenced was a vast three-part report that covered the radar, aircraft design and armament requirements needed to fulfil the brief espoused by its title: Defence against High Altitude Bombers by Mach 2 fighters. The RAE's proposed design had the appearance of a much larger F-104 Starfighter, perhaps no surprise as it was the layout de jour, the perceived best method of providing Mach 2 speed. The report did describe another solution, with an aggressively swept wing and tail and a single large turbojet. Despite only having one engine, this second aircraft is unmistakably an early form of English Electric's research fighter, which eventually would enter service as the Lightning. The problem to be solved was laid out simply, but is revealing in what it shows of the 'mirror' case: the UK solution to bombing in the face of such a defence.

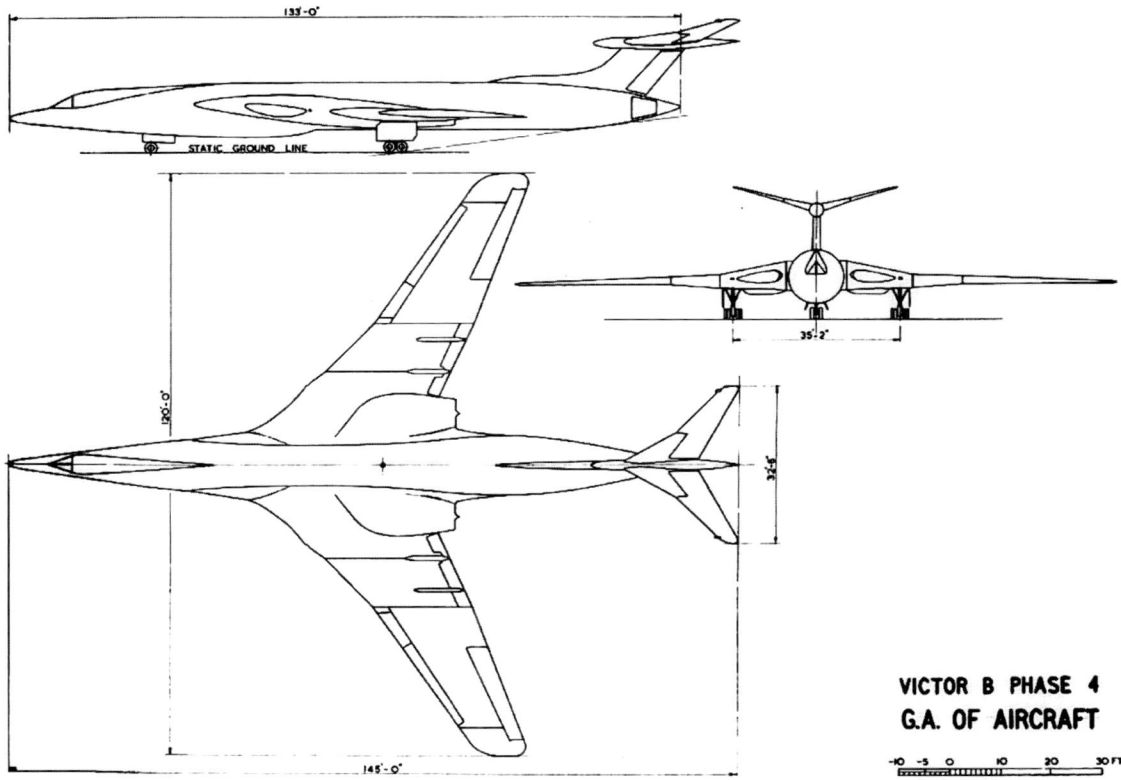

General arrangement drawing of the Victor Phase 4 proposal of 1956, which would have created a bomber with supersonic over-target dash capability, while retaining much of the Phase 2 (B.Mk.2) wing and tail assemblies. VIA BARRY HINCHCLIFFE

"The aircraft is to be capable of intercepting targets flying at a height and speed corresponding roughly to those of the V-class bombers, or plausible developments of these aircraft which might be expected to be in service concurrently with the proposed fighter. The V-class bomber at present cruises at speeds approaching $M = 0.9$ at 50,000ft. Developments might achieve 55,000 or even 60,000ft with more powerful engines. The possibility of a supersonic speed of about 1.3 in the region of the target must not be ignored."[182]

Was a supersonic version of either of the definitive V-bombers a serious proposition? As with the work that led to the Mark 2 versions of both, there was a strong emphasis on continuous development of performance, in the knowledge of just how finely the balance of success or failure against the predicted defences was poised. The rapid drag rise as Mach 1 approached was the major challenge of the original aerodynamic development of all the B35/46 concepts, with its postponement as far as possible being the sought-after prize. Assuming that the inevitable increase in drag caused by the strong normal shocks as sonic velocity was passed through could be coped with by increased thrust, what were the limitations?

On the one hand, a supersonic Vulcan was not on the cards. The problems of longitudinal stability and control had been addressed sufficiently by the autostabilisation systems provided, within the limited working range specified for the aircraft. Somewhere above a true airspeed equivalent to Mach 0.95, the bomber would run out of elevator authority with which to compensate for the rearwards movement of the aerodynamic centre. The obvious solution to this would have been a Javelin-like horizontal tailplane on top of the fin, of likely excessive mass penalty. The development work on the Thin-Wing Javelin had suggested that zero-lift drag coefficient would be of the order of three and a half times that of the subsonic value, at $M = 1.3$. Even if it were only doubled, the increase in thrust that would be needed to balance this was implausible. One would be far better to start again.

In 1956, Handley Page took the modified, extended wing of the planned second-generation Victor and used it for the basis of just such a supersonic machine.[183] Unlike the Vulcan, the Victor did have the ability to provide longitudinal trim independently of the wing geometry, via its imposing tailplane. The new design was described as the Phase 4 Supersonic High-Altitude Bomber and would have been powered by a reheated version of the Conway. While the original Victor had the appearance of a machine designed in conformance with area ruling, this was not

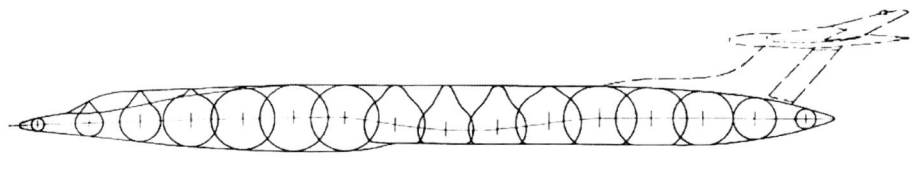
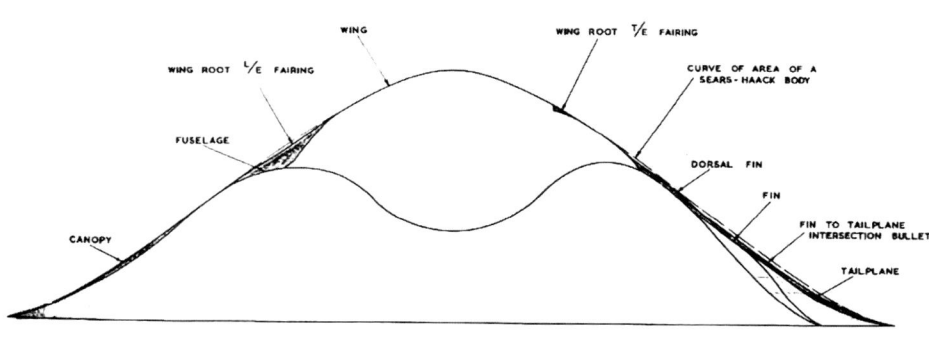

The longitudinal area distribution of the Victor Phase 4 proposal of 1956, with comparison to the 'ideal' Sears-Haack shape. VIA BARRY HINCHCLIFFE

the case and it was to correct this that the major revision in the proposal was made: a completely new fuselage.

The A&AEE trials on the Victor prototypes, which like the Vulcan were subject to engine thrust limitations, had shown that handling at high Mach numbers was good. The aircraft was comfortably flown at M = 0.915 TMN at 46,000ft in level flight, at which condition no marked compressibility effects were found. The specification agreed for the production aircraft dictated a maximum Mach number of 0.95, which had been, "reached with ease" in these early tests, without, "buffeting…wing dropping or longitudinal instability… and all controls were effective." It was noted though that the weight was low and, again like the Vulcan, short period oscillation had been present. Most significantly though, changes in longitudinal trim were small through this transonic phase. The report specifically commented on the opportunity to exploit this very positive outcome.

"It is evident that the maximum Mach number of 0.95 required by Spec. B128P covering production Victor aircraft can be easily exceeded, and it is strongly recommended that the aircraft's capabilities for doing so be investigated as a matter of urgency in order to obtain the obvious advantages of the highest possible Mach number.

"… it should be noted that, as far as could be investigated, there was only a gradual increase in buffet intensity above the threshold… while no buffet occurred… in straight flight at 0.95 M, some buffet is to be expected at 1g at higher weights at the higher altitudes."

Just how slippery the production Victor could be was announced in indisputable terms—with a sonic boom (or 'bang' as would have been known in England at the time) on June 1, 1957. Johnny Allam and his crew exceeded Mach 1 when returning to Radlett from a test flight. In the process, the rear crew members, flight observers Paul Langston and Geoffrey Wass were famously credited with being the first men to exceed the speed of sound backwards. It had been proven therefore that the Victor was statically stable and controllable around sonic speeds, albeit at low lift coefficients that might be unrepresentative of those that would be needed for a high speed, high-altitude dash over the target. Boscombe Down's report implied the importance of this factor,

"[The flight test results] indicate that at the specification cruising speed (500kts TAS, 0.873 M) at 47,500ft and 120,000lb weight (predicted optimum range target conditions with 10,000lb bomb on) the

aircraft should be capable of a turn at 1.3g before buffet commences. It is believed that this... will be adequate for bombing but consideration must be given to... adequacy of the manoeuvrability available for evasive action [and] ability to maintain speed and height when turning at high altitude."

So, the problem was unsurprisingly multidimensional. As Mach number increased, the available lift coefficient prior to buffet would reduce. This was not necessarily disastrous for an attempt to increase maximum speed; the lift force (that had to remain equal to the aircraft's weight for level flight) would remain constant, so long as the product of the square of the velocity and the lift coefficient also remained constant. The former would rise quickly (as a squared term), enabling the latter to reduce, so provided the product could remain under the buffet threshold and the increased (dimensional) drag be balanced by engine thrust, a faster cruise was possible.

The Phase 4 proposal described the new Victor as providing equivalent payload-range performance to the planned B.Mk.2, while cruising at Mach 0.9. Over the target, it was capable of supersonic speed at 65,000ft, subject to the results of,

"An intensive supersonic wind tunnel programme... to check the aerodynamic assumptions underlying this study together with flight tests on the Victor B.Mk.1."

The new fuselage was designed around the concept of the Sears-Haack distribution, a theoretical area distribution that provided the minimum possible wave drag, although notably the underlying assumptions fell down at transonic speeds. In turn, the Sears-Haack body was a developed result of a linearised supersonic flow theory which had been developed in the 1930s by Theodore von Kármán and Norton B. Moore in the United States; the eponymous William Sears had worked with von Kármán at Caltech prior to moving to Northrop and its flying wing programme, while Wolfgang Haack had independently reached similar conclusions in wartime Germany; recurring themes across the V-bomber aerodynamic story.

The von Kármán-Moore thesis postulated that wave drag for an axisymmetric (circular cross-section) body would be proportional to the second derivative of the longitudinal area distribution: the rate of change of area with length had to be smooth and continuous. The Sears-Haack result was generalised, axisymmetric body with an area distribution that mathematically minimised the wave drag contribution for a known volume, which transpired to be a pointed, slender body with maximum thickness at half the length, and symmetrical also about that longitudinal plane. If it were reminiscent of anything to the contemporary British observer, then it was Skylon, the futuristic and apparently unsupported sculpture that was the centrepiece of the 1951 Festival of Britain.

There were really two different 'area rules' that were under consideration in 1956. The first is the best known in the outside world, the transonic area rule of Whitcomb. This was simpler to visualise and with a more direct application than the supersonic area rule, the latter being the general case formulated by another giant of American aerodynamics, R. T. Jones. In describing Whitcomb's work, he commented,

"Recently [he] has shown how the drag at transonic speeds may be reduced to a surprising extent by simply cutting out a portion of the fuselage to compensate for the area blocked by the wing... [This] was based on considerations of stream tube area and the phenomenon of 'choking'—which follow from one-dimensional flow theory... if followed towards the limit as M approaches 1.0 (from above) [linear theory shows that] the wave drag of a system of wings and bodies depends solely on the longitudinal area distribution of the system as a whole".[184]

The largest step in drag coefficient rise occurred around sonic speed, then gradually fell towards a value that, while still appreciably higher than the subsonic value, was also a very worthwhile saving on the peak value. This peak drag that needed to be broken through was a fundamental problem for the low thrust fighter aircraft of the 1950s, and explains why Whitcomb's relatively simple analysis of transonic drag rise and its relationship with area was so important. The best-known application is the F-102 Delta Dagger, an aircraft with roots in Lippisch's DM-1 glider. When modified with a smooth longitudinal area distribution, the reduction in wave drag meant that the J57-powered aircraft could pass through Mach 1 and to the other side. It is perhaps less frequently appreciated that this 'supersonic' aircraft was barely so, with a maximum speed of a little over M = 1.2.

Whitcomb's theory was based in mathematics but used a philosophy extracted from it: the need to minimise the second derivative of area, a smooth longitudinal progression. This could be achieved in many ways, one of which might have been with a high aspect ratio swept wing, as a method of reducing subsonic drag and improving cruise efficiency. For an aircraft with a mission like the

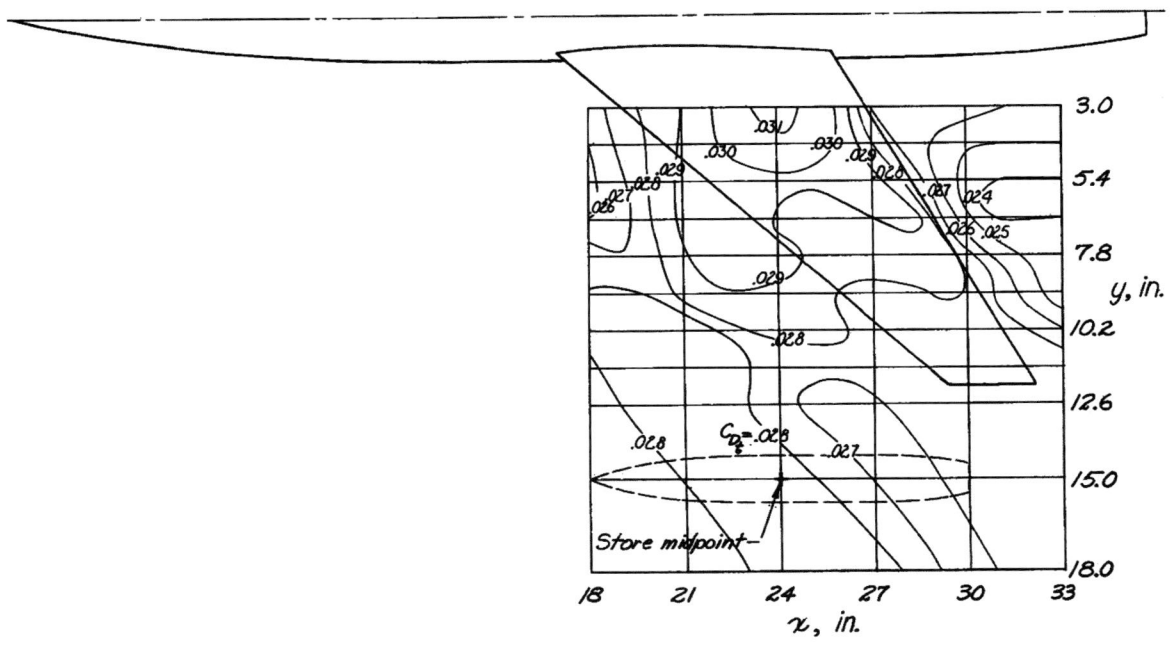

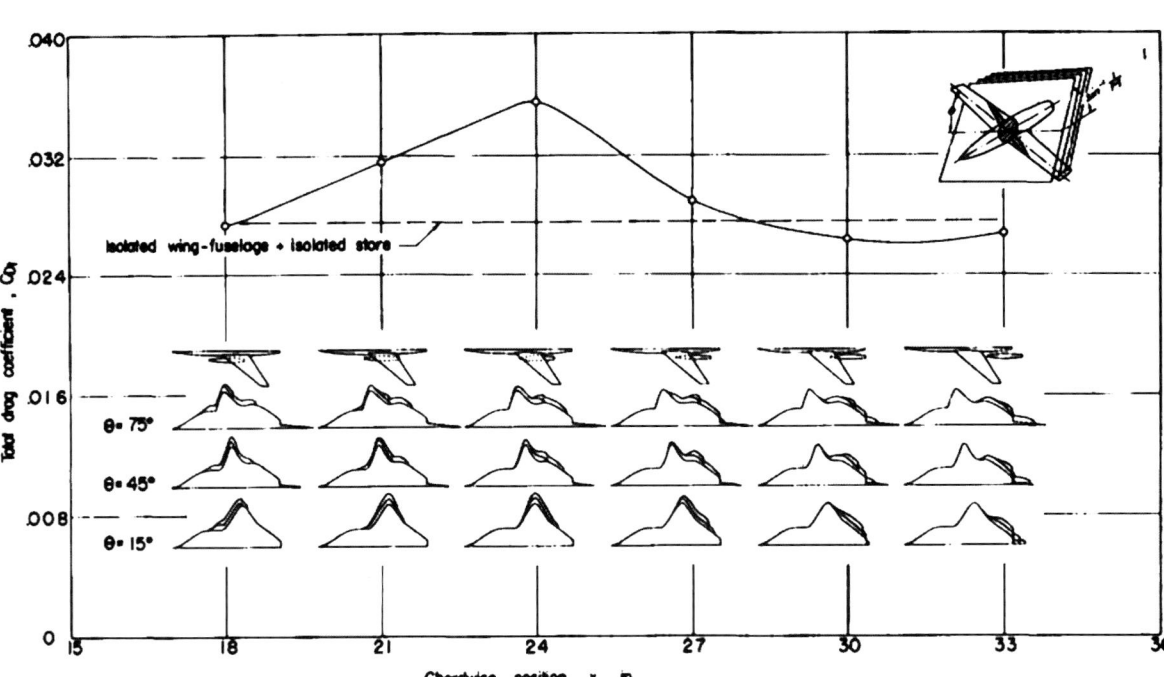

NACA plotted contours of total drag of a wing-body combination at M = 1.6, with an external store at varying position. The sum of the two individual items was Cd = 0.28, so mounting at tip implies no interference drag, while the positions behind the wing showed favourable interaction. The study was unable to correlate its results with those anticipated from the supersonic area rule. It was described as, "simple in concept, but extremely tedious in its application."

Victor Phase 4, in which the supersonic 'dash' was only a small portion, this was surely the type of configuration to pursue. Jones showed however that with a combination of Sears-Haack body (fuselage) and elliptical wings of varying aspect ratio, Whitcomb's modifications reduced the wave drag to that of the body alone at Mach 1, after which the results were very different.

The higher (although still low) aspect ratio geometry studied continued to gain wave drag; at M = 1.2 it surpassed that of the unmodified shape. However, the same body with a wing of about one quarter of the aspect ratio, now spread over a much longer section of the body and—crucially—making the 'waisted' region equally long and progressive, maintained a much lower wave drag until well above M = 1.6. It was concluded,

> "The rapid increase of drag is, of course, the result of the relatively abrupt curvatures introduced into the fuselage lines by the cutout. Such abrupt cutouts are necessarily associated with wings having small fore and aft dimension, that is, unswept wings of high aspect ratio."

Jones's general method recognised that the correct 'area' was not that normal to longitudinal axis for all Mach numbers, but one that was parallel to the specific Mach angle associated with the design condition. For each position, it was necessary to average the complete range of slices that could be defined, if the Mach plane were rotated about the centreline. In so doing, a 'correct' area distribution could be calculated. Bearing in mind that at the time this would be a manual, perhaps graphical process, it is of little wonder that a contemporary NACA report described it as,

> "...simple in concept, but extremely tedious in its application".[185]

The authors had been unimpressed too by the correlation of the theoretical results with those that they found in the wind tunnel. They had been conducting a study of relevance here: the optimal positioning of a large external store on a supersonic heavy bomber. Their final, perhaps exasperated comment after following Whitcomb's lead and studying only three roll angles for area analysis, was,

> "The foregoing analysis thus shows that visual inspection of supersonic area-rule diagrams does not provide correlation with the data obtained on the fuselage-swept wing-store configuration at M = 1.6. It therefore appears that examination of the details of the flow, with particular attention to the effect of the pressure field of one component on another component, may be useful in explaining interference effects. The analysis which follows takes this approach and attempts to provide a more basic understanding of the interference effects encountered."

While the supersonic area rule should have given close guidance for a practical, optimised shape, it was also clear that the analysis capability of the time struggled to implement it. Handley Page's sensible option of approximating as closely as possible the Sears-Haack distribution for the whole aircraft was a pragmatic approach, but it is likely that considerable detailed work would still have been required in the wind tunnel. In one respect, the Victor wing offered a reasonable advantage. While nominally a 'high aspect ratio' design compared to a delta, even in the extended B.Mk.2 form it was less than six, with a very long chord root over which to spread the associated fuselage cut-out, as NACA had found to be important.

Why though, in 1956, with Blue Steel planned as the primary weapon and meaning that the target should not have needed to be overflown, was this hugely challenging supersonic derivative required at all though? The answer, we might postulate, lies in a combination of factors that must surely include,

* The accuracy of a freefall weapon released according to the human-in-the-loop navigation of the bomber, as opposed to the systematic build up errors inherent in a stand-off weapon's automatic systems.
* The general interest in higher-faster-further, and specifically in the case of the Victor, a response to the A&AEE's indication that the basic design had the potential to go far beyond the B35/46 requirements.
* The competitiveness with which HP had to approach the challenge; they alone could provide a (limited) supersonic dash capability, while retaining the expensive and difficult bits of their baseline design.

Assuming that the design had proved plausible, the use of reheated engines for the supersonic dash phase is interesting. While more thrust was essential and the challenge of installing the Conway in the first place indicated that there was no chance of a physically larger engine being accommodated, the cost of the reheat in terms of fuel was equally non-trivial. The brochure offered a hint to the thinking, with the statement,

> "High energy fuels, when available, can be used relatively simply in the afterburners and increase still further the 'supersonic spurt' ranges".

At the time, the USAF was actively studying the problem of the next heavy bomber that could successfully attack the Soviet Union. It had concluded that supersonic flight over the USSR would be a prerequisite, which was extremely problematic given the ranges considered. An active nuclear propulsion programme was underway, but the refinement of chemical fuels represented a technology much nearer to reality. Substituting lithium, beryllium or boron for the carbon in the fuel offered ways of increasing the heat of combustion and providing more energy per unit weight or volume. A 1956 report noted that,

> "The development of boron fuels under the code name of Zip is being actively sponsored by the Department of Defense... If this fuel can be used in place of JP-4, a range increase of 40% will be realised. However, the combustion products, boron oxide, tend to deposit as a solid in the combustor and on the turbine stator blades; intensive research is needed on this problem... current research indicates that combustion of Zip fuel in the afterburner causes less trouble than combustion in the primary combustor ... and will increase range by 25%, for that portion of the flight where afterburner is used."[186]

Perhaps a circular argument developed: new fuels that could be exploited only in afterburning were on their way; the only sensible use of such technology in the V-bombers was to either improve take-off performance, or to provide a short-term improvement in the aircraft's speed, logically over the target. Technology therefore coalesced to suggest a supersonic Victor, whether or not it would be a militarily sensible weapon by the time it might be fielded. The baseline Victor was already proceeding slowly and was still more than a year away from entering service; HP's ability to deliver a supersonic derivative prior to the alternative of the stand-off bomb must have seemed remote.

Blue Steel seems to have been the elephant in the room where the Victor Phase 4 is concerned. If the aircraft had carried it in the same semi-recessed manner as the B.Mk.2 would do, then the Sears-Haack area distribution would be blown. Neither would it need to overfly the target at all, although the missile would have benefitted from a higher altitude launch. Indeed it is not even clear—because it was never on the cards—whether supersonic launch was viable. In the United States, the B-58 Hustler was impressively designed with an external freefall pod, such that either with or without it fitted it conformed to the required area distribution. Although the rigours of the Victor Phase 4 were less, so too were the opportunities to optimise a derivative design. And after all of that, surely its advantages would have disappeared completely with the advent of Skybolt.

In spite of the ongoing work on Rascal and not least Polaris, the same 1956 Air Force Research and Development Command report made no reference to stand-off weapons or missiles of any kind. The aim was still fairly obviously to outrun the defences at high altitude, which was seen to promote improvements in lift-to-drag ratio as the wing became larger to offset the reduction in air density. Notably, the report considered that a supersonic bomber able to complete the intercontinental mission, still to be flown for the most part subsonically, might find an optimum weight up to a staggering 800,000lb. The impact of the supersonic design was severe; in comparing the B-52 which was said to attain the very impressive L/D of 22 (albeit at M = 0.73, an inadequate cruise speed if there were any material threat), the B-58 Mach 2 medium bomber was anticipated to achieve comparative values of 12 subsonically and less than six at supersonic speed. It should be noted that the range efficiency was proportional to the product of Mach number and Lift-to-drag ratio, but the B-52 still comprehensively won as a long-range machine, were speed of no object.

Replacement of the Boeing aircraft by a Zip fuel-burning, Mach 3 dash-capable bomber, specified by the USAF as Weapon System (WS) 110A, was the stated outcome of this and related studies at the time. Over the next two years, technology to an extent came to the aid of the programme. The compression achieved through the slowing of the flow into the engines could be, if well engineered, the provider of most of the aircraft's thrust. For a supersonic cruise aircraft, this would be essential to the machine's fuel economy, but the work involved (as shown by the Lockheed Blackbird family and Concorde in the 1960s) would be formidable. The second major find would be a method known as compression lift—containing the overpressure from the shockwaves under the aircraft as a lift-enhancing strategy. Together, these made the idea of a complete Mach 3 intercontinental mission, as opposed to a short burst on a subsonic journey, a viable prospect. WS.110A was awarded to North American Aviation and would in time result in the XB-70 Valkyrie.

The story of the XB-70 connects closely with Skybolt, and by extension the V-bombers. It was an unavoidably huge radar target, relying on the fact that at a true air speed of more than 1,700 knots and an altitude of 80,000ft, it would still be a staggering challenge for surface-to-air missile engagement, let alone a fighter interception. Although the promise of the ICBM was a powerful driver against a new manned bomber, the immature and unproven nature of those weapons as the 1950s entered their second half was troubling for an absolute guarantee of a nuclear deterrent. This position was held by both

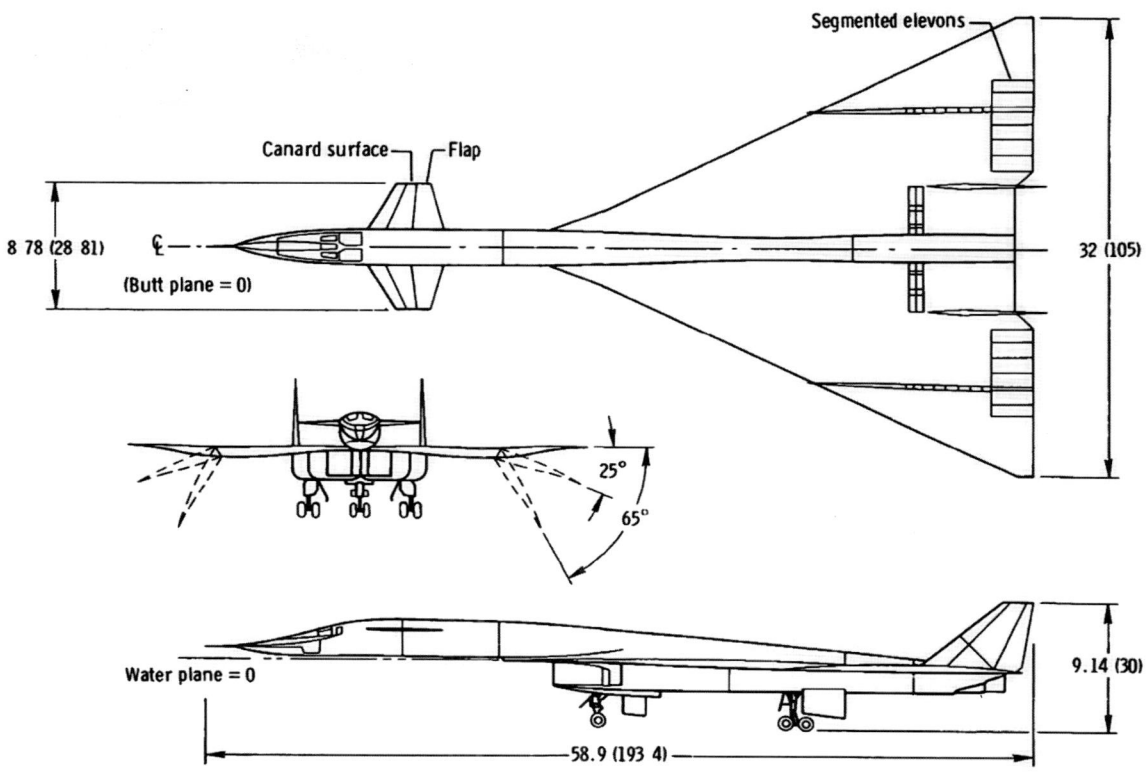

Three-view diagram of the XB-70. The folding wingtips achieved the dual aims of harnessing compression lift and moving the aerodynamic centre forward, to reduce the supersonic trim drag penalty. NASA

the president and Strategic Air Command of the USAF, which at the time was responsible for the procurement of its own systems, rather than the Secretary of Defense. This would change, along with so many other things, when the American public were able to bear witness to the aerospace prowess of Soviet Union in October 1957.

By December, the contract for the new bomber had been placed with NAA, against an assumption of a 250 aircraft programme, to start with a firm order for 12 test jets. Within a year, this was accelerated, with any suggestion of a prototype cancelled and the intention that the test aircraft would eventually find their way to SAC. Each aircraft was expected to cost more than $24m, or in other terms, about three times the price of a B-52. While aircraft cost strongly corelates to weight and the XB-70 was huge in order to carry out its intercontinental mission, it is clear that there was no way the United Kingdom would have been able to justify a recapitalisation of its deterrent force with such an expensive capability. If the combination of deep penetration, target over-flight and freefall weapons were essential than the message that the XB-70 sent was that the UK would have to look for another way, if it wished for a credible deterrent in the late 1960s.

It did not take too long for the cracks to appear. By 1959, it was clear that ICBM progress was exceeding expectations; it was difficult to argue against the practicality of a weapon that could be launched from a submarine. The overall defence budget was becoming unaffordable, which would mean the cancellation that year of the North American F-108 Rapier Mach 3 interceptor. Because this would have shared technology with the B-70, the cost of the bomber escalated accordingly. The Eisenhower administration slowed the programme to a trickle in 1960, down to just two test aircraft and a very limited budget plan for FY1961.

A further consequence of Sputnik was a reorganisation of the Department of Defense, with procurement authority placed into the hands by the Secretary of Defense by law. The XB-70 would become a cause that both the SecDef and Congress would adopt, for opposing reasons. 1960 was an election year, and the Democrat pretender John F. Kennedy was concerned about the missile gap and even more so about the voters of California, where NAA was based. The likely cancellation of their main programme was a threat to livelihoods and predictably, Kennedy therefore spoke up in support of it. Both the Senate and Congress, including the powerful Carl Vinson, were supportive of the XB-70 programme and actually allocated considerably more funding to the bomber than the president had requested.

The spectacular XB-70 first flew in September 1964, although by then its status was purely as an R&D programme. This is the first of the two, known as AV-1.

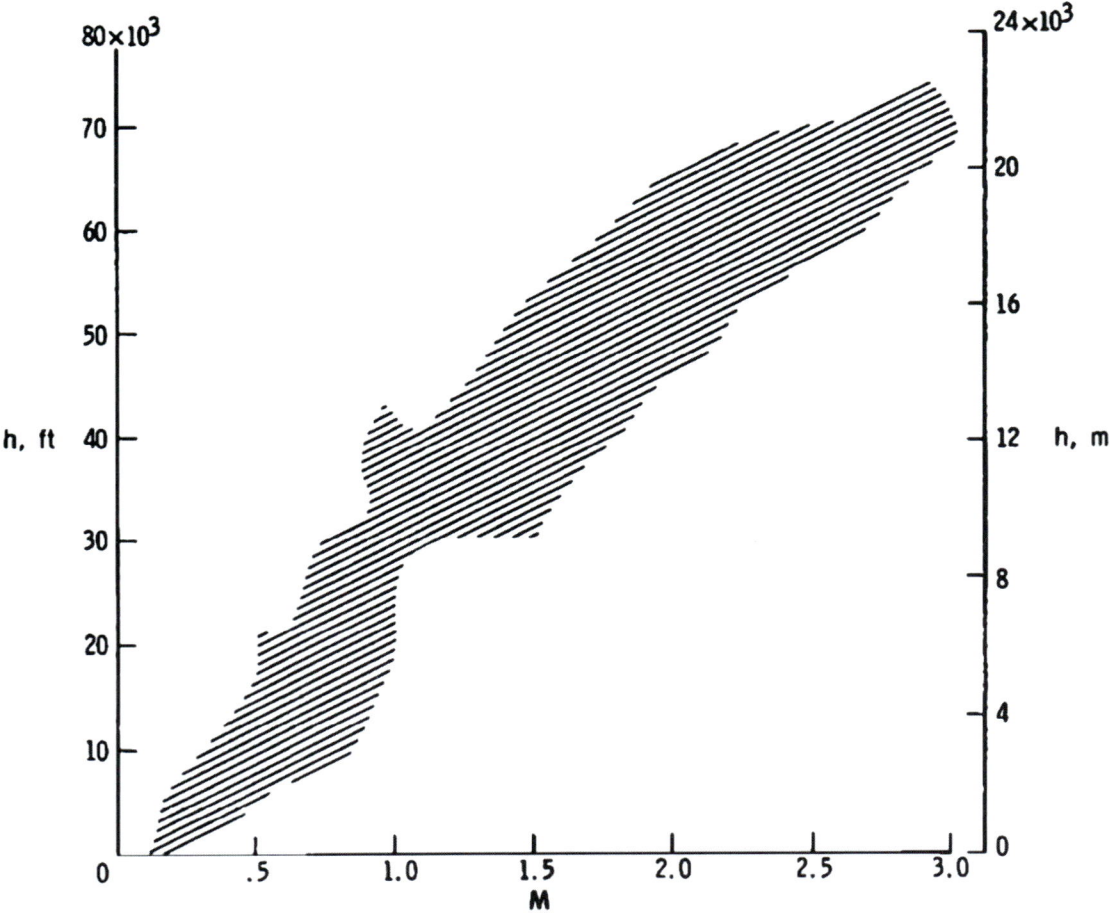

Figure 2.— Altitude–Mach number envelope of XB-70 flights.

Map of the flight envelope of the XB-70. By the end of 1966, sustained Mach 3 cruise for more than half an hour had been demonstrated. NASA

On the day prior to the election, Eisenhower released an additional $155 million to the XB-70 programme, and the Republicans subsequently took California by a wafer-thin margin. The knife-edge result was the story across the country, but the outcome fell of course in the other direction, with eventually far-reaching consequences for the UK deterrent. Nonetheless, as 1961 dawned, the XB-70 remained in the slow lane, the option being kept warm. Kennedy's new administration and particularly the incoming Secretary of Defense Robert McNamara would find a plethora of existing and forthcoming programmes competing for funding and survival. With the XB-70 hovering, the future of the manned bomber in SAC revolved around the B-52, suppressing the defences ranged against it using the forthcoming GAM-87A Skybolt, clearing a path for the accurate dropping of freefall megaton bombs. To McNamara, this was no second-best option: the B-52/Skybolt combination was a more flexible, survivable and crucially cost-effective weapon system.

The problem for the USAF and its supporters in Congress was in formulating a convincing argument that the bombers were required at all. It is often said that it was the U-2 debacle of May 1960, in which Francis Gary Powers was shot down by a Soviet SA-2 missile at 65,000ft, that spelled the end of the Valkyrie. This was clearly not the case, as the bomber was effectively cancelled in the previous year due to the relatively quick progress on ICBM technology undermining the need for the multi-billion dollar aircraft programme. Its election-inspired reinstatement was brief, as Kennedy and McNamara halted the programme again within a month of assuming power. Skybolt was all that remained, and the UK's involvement assumed a new importance for the USAF, who thought it would protect their promising ALBM system from a similar fate. The future of the manned bomber force appeared dependent on GAM-87A, while the assurance of the realisation of the latter programme appeared locked by the Special Relationship.

In February 1960, the plan for GAM-87A was for a first missile launch in August 1961, and a guided launch in January 1962. The first phase of testing by the USAF, in the form of ARDC, would commence in October 1962, while SAC would begin operational testing in August 1963. The test B-52 unit would be the first to become operational, in March 1964. Ten RAF squadrons would be operational, each with eight Skybolt-equipped Vulcans, commencing from the third quarter of 1964. The rationalisation of the types of carrier aircraft allowed the design assumptions to be similarly restricted. Launch would be from the B-52 at Mach 0.8 and 40,000ft, or the Vulcan B.Mk.2 at Mach 0.9 and 50,000ft.

Skybolt had interesting implications for the future of the V-bombers. Watkinson's cabinet paper of June 1960 had been very clear as to how Skybolt would be deployed, and what the future of the RAF's bomber force would look.

"By acquiring 144 missiles, with spares and associated equipment, we should be able in the later 1960s to maintain with the Vulcan Mark 2 bombers a deterrent force equivalent to that previously planned for BLUE STREAK. The Victor Mark 2 bomber would not be adapted for SKYBOLT, but could be a complementary component of the deterrent with a developed version of the British powered-bomb BLUE STEEL, if necessary".[187]

This proposal, agreed in principle by the prime minister and cabinet on June 20, 1960 is extraordinary, as it finally meant that the UK Government made a choice between the Vulcan and Victor. It had little, if anything, to do with the staunch independence of Handley Page in the face of the government's aspirations for industry rationalisation, but everything to do with the independence of the deterrent. The performance of either of the V-Bombers was no longer considered to be in the class that might provide survivability in the second half of the decade; that would need something like a XB-70. Skybolt reduced the need for speed and altitude, and the RAF's first Vulcan B.Mk.2, XH558, had been delivered a few weeks earlier to Waddington. It would make no sense to complicate the procurement process, by making the new missile compatible with two different airframes, and the Victor B.Mk.2 had not yet been delivered in a manner satisfactory for service use. With each bomber carrying a pair of missiles, only half of the previously anticipated force was required. Seventy-two jets made no sense at all, if divided between two types. In a curious and perhaps empty manner, given what was to come, the Vulcan had won.

13

FROM CAMELOT TO NASSAU

By the time of John F. Kennedy's Presidential Inauguration in 1961, Skybolt had supplanted Blue Steel Mark 2 and Blue Streak, together with arguably the XB-70. It nonetheless faced competition from the prospect of a more accurate weapon—Minuteman—which could also more plausibly survive a first strike by the Soviet Union by virtue of its rapid response and nuclear-hardened silo bases. Skybolt was perhaps a more difficult technical challenge, particularly from the perspective of guidance; to an extent, all of Minuteman's problems would be solved on the way to a viable version of Skybolt.

The UK agreement was extremely favourable, in that the enormous R&D cost of the missile itself would be borne completely by the United States. Britain would be responsible for the costs associated with the integration of the weapon on the Vulcan, and of course the missiles themselves, while it would also maintain independence of its deterrent in a fundamental sense by providing its own warheads. Unlike Blue Steel, the ballistic nature of Skybolt meant that the warhead was housed in a re-entry vehicle. The UK had conducted extensive and valuable work of its own in that field, in preparation for the planned introduction of Blue Streak. This developed expertise formed part of the establishment of the UK's status as a country that had made substantial progress in the field of nuclear weapons, and consequently part of the investment that would—unexpectedly—earn it access to the US technology. By any measure, Skybolt was a steal: the UK R&D programme was estimated in November 1960 at an eventual cost of $47 million. This would represent less than 10% of the US investment, all to be spent before any operational missiles would be fielded.

In April 1961, the GAM-87A System Project Office circulated a memo to address a major concern: the deployment of a Soviet Anti-Ballistic Missile (ABM) system. It was thought that a missile system referred to as SA-5 might be operational by late 1964, perhaps with a battery of 12 weapons near a major city such as Moscow.[188] The calculations in the report suggested that an SA-5 would have sufficient time to engage the GAM-87's re-entry vehicle directly above the missile battery, on the basis that the radar cross-section of warhead was so large that it would be detected with sufficient warning. The report proposed that efforts be made to improve the low-observability of the re-entry vehicle, as,

> "Assuming that the Soviets have the SA-5 system in operation, the GAM-87A will be extremely vulnerable to it. It is possible to reduce this vulnerability considerably though not totally. Whether the SA-5 will operate as ATIC estimates is, of course, open to question. However, the design of a missile of the stated capability is not difficult... No real radar sophistication was assumed".

Vulcan XL426 at the 1984 Mildenhall Air Fete.

The third production Vulcan B.2, XH535, unusually photographed flying over the desolate terrain surrounding Edwards AFB, California. The aircraft was detached for trials associated with the Skybolt missile. XH535's career was brief, as it was sadly lost along with four members of the six-man crew, on the May 11, 1964 while conducting low speed tests with the A&AEE.

Decades on, this fear that the Soviet Union might have been able to negate a ballistic missile attack by shooting down space vehicles in 1965 seems pessimistic to say the least. The memo did challenge the assumption that, even if the capability were real, the deployment would be widespread enough to matter.

> *"A full deployment of the SA-5 system for total protection of a target would require nine systems per target… Are the Soviets willing to go to the required expense?"*

An important observation was made in this regard, which would have consequences later.

> *"The GAM-87 is by nature a second wave device. [Is it] probable that Soviet defences, the SA-5 system in particular, will be working at peak effectiveness after the first wave of attack? If they are not, any reduction of GAM-87A radar cross section, though it does not protect it from Soviet peak capabilities, may be sufficient."*

In the same month as the project office memo, an ad hoc task force had been set up at the new president's request to investigate the country's policy towards the NATO alliance, under the former Secretary of State Dean Acheson.[189] The group took input from RAND analysis and proposed the commitment of defence policy to a second-strike counterforce concept, in which the survival of its nuclear arsenal was assured in the face of a first-strike by the Soviet Union. This led on to two other conclusions: the first, that the US could provide a guaranteed, credible umbrella for its allies; secondly, that the nuclear forces should be sized and equipped appropriately for this task.

A third major impact was a realisation that other nuclear arsenals, over which the United States could not exert ultimate control, might be destabilising to the balance struck between the two superpowers. The possession of nuclear weapons by NATO partners was therefore to be discouraged, while a conventional attack in Europe would be subject to a conventional response. Where did Skybolt fit into this doctrine, which was adopted by the Executive and the incoming SecDef, Robert McNamara?

The manned bomber had a fundamental problem if it were to be a second-strike weapon: it was potentially the most difficult option on the menu to protect from destruction in a first-strike. The new Polaris submarines were unproven and their missiles immature, but they were extremely difficult to find and destroy. The liquid-fuelled ICBMs, Titan and Atlas, remained impractical and vulnerable, but their solid-fuelled successor Minuteman I—based in hardened silos—had every chance of surviving to unleash a devastating response on Soviet targets. However, with minimal warning available and operating from well-known and mapped bases, would the B-52 fleet simply be destroyed on the ground? Unfortunately, the implications went further. If the bombers were to give themselves the best chance of survival, were they essentially a first-strike weapon, whose very existence gave the impression to Moscow of an arsenal that had no choice but hit first? McNamara and his team had the responsibility of preparing the plan for the FY1963 budget cycle. He considered the submissions of the USAF in particular, with Skybolt carefully placed as a defence suppression option for the bombers on top of the massive expenditure on Minuteman and the extraordinary XB-70 Valkyrie, excessive on the one hand and ill-suited to the Second-Strike Counterforce doctrine on the other.

On April 4, McNamara appeared for the first time in his new role before the Senate's Armed Services Committee. He was able to set out his stall and to some extent remind them of the more encompassing powers of his office, compared to the pre-Sputnik era. Moreover, he made it clear (in relatively generous terms) that the planned budget generated by the previous administration was indeed a "thorough job" and that it was not plausible to completely change it. Instead, it would be maintained, reviewed and altered only where serious problems or opportunities were obvious.

> *"This is the first time that Department of Defense witnesses are appearing before this committee in support of a request for appropriation authorisation for the procurement of aircraft, missiles and ships as required by Section 412 of the Military Construction Act of 1959. I am sure that we can work out the necessary procedures… No doubt there is room for differences among reasonable men as to what constitutes the optimum combination of programmes for the nation's defence. All I can say is that we have carefully examined all of the principal alternatives and have selected that combination of programmes which we believed will give the nation a fully adequate defence at the least cost, in the light of the threat as we view it today… The people of this country are entitled to a full measure of defence for every dollar spent.*[190]

The world was changing, and with it the technological basis of the weapons that should be procured to meet stated military objectives. This, he said, was happening very quickly.

> "Only a year or so ago the principal general war threat to our security was a surprise attack by large numbers of nuclear-armed manned bombers. A year or two from now our principal concern will be a surprise attack by a large number of nuclear armed ICBMs… Preparations for a surprise manned bomber attack are difficult to conceal: air bases cannot be easily hidden; the staging of large numbers of manned bombers for an attack cannot be readily disguised."[191]

The thrust of McNamara's revised submission was the focus on the right strategic missiles that could be procured for the longer term. The liquid-fuelled Titan and Atlas missiles would need to be replaced, in spite of the huge cost of their acquisition in the first place. He was keen to proceed as quickly as possible to the end-game: the sold-fuelled Minuteman and Polaris missiles. Even in the case of the latter, it was the projected A-3 of more than double the range of the first operation missiles that he wanted. Relative to Eisenhower's plan, the launching of Polaris submarines was to be accelerated, which enabled expensive dead-ends such as surface-ship Polaris launchers to be abandoned. Nevertheless, it was not all about the new systems.

> "We still see the need for a large manned bomber force, at least over the next several years. We now have a force of about 1,500 heavy and medium bombers, together with their accompanying refuelling tankers".[192]

The proportion of bombers on ground and airborne alert would increase, while money would be saved by phasing out the B-47 more quickly than planned. The remaining B-52 force, the largest remaining element, would have help in safely reaching its defended targets.

> "Most of the B-52 heavy bomber squadrons will be equipped to carry the HOUND DOG air-to-ground missile… It will give [them] considerably greater flexibility in attacking targets. Also under development is a more advanced air-to-ground missile, the SKYBOLT. This is a solid-fuel ballistic missile with a design range of about 1,000 miles… Whereas a B-52 can carry only two HOUND DOG missiles, it would be able to carry four SKYBOLT missiles. The successful development of an effective SKYBOLT missile would offset the expected increase in Soviet air defence capabilities by increasing the effective attack radius of the B-52 and the penetration capability of its weapons, as well as increasing its ability to launch 'stand-off' attacks on multiple targets. However, the SKYBOLT is a costly project and presents some unusually complex development problems. Even if the development is successful, it could not be brought to operational status before 1965."[193]

The Secretary of Defense made it plain that now was the time to either do it properly or get out. The plan at that time was to stretch the existing $150m budget to cover work in the following year too. The prime contractor Douglas was adamant that this would not work, and McNamara was inclined to agree. Another $50m was therefore allocated, because in their view,

> "On balance, we fell that the advantages of this weapon system warrant an effective development effort".[194]

Skybolt was now in a curious position, having been subject to persistent attempts of cancellation for more than two years, and generally opposed on both technical and economic grounds by the two administrations under which the programme had been run. Why did McNamara, fundamentally opposed to the manned bomber and looking at what systems would be required towards the end of the decade, at a time when it was supposed that mature solid-fuelled missiles would be common, commit more money to Skybolt? The answer might plausibly have been found in the next part of his discussion, in which he turned his attention to the XB-70. After stating his view that the Mach 3, high-altitude defence was in fact no defence, and that it would instead be forced to proceed subsonically, he offered a further comment that it had not been designed with missiles such as Skybolt and Hound Dog in mind. The conclusion therefore was the full-scale weapon programme should be paused, and the prototypes built to investigate their demanding flight regime instead. This would save well over one billion dollars, albeit there would still be considerable expense in the truncated tests planned.

McNamara almost certainly did not want either of Skybolt or the XB-70 to move forward and consume funding at all. However, he was pragmatic enough to know that he could not win all battles if they were fought simultaneously. The XB-70 was a relatively simple target, and with

Low-speed wind tunnel testing of the Vulcan B.Mk.2 and Skybolt missile combination, in the 9ft x 7ft tunnel at Woodford. These tests probably took place in 1960 and were complemented by high-speed tests in the transonic facilities at RAE Farnborough.

by far the greatest associated short-term expenditure. In his speech, he was able to use the lack of compatibility with Skybolt as evidence of the aircraft's unsuitability for a changing world. The programmes danced around each other, but the game is clear in retrospect: the existence of Skybolt and its proven, in-service B-52 launch platform undermined the case for the XB-70, so to eliminate the latter, the former needed to continue. Eventually, the XB-70 having been dealt with, Polaris, Minuteman and its own complexity could be used to cancel the ALBM programme. In the US, there was no overall gap in capability, but for the UK, cancelling Blue Streak and the final batch of Victor B.Mk.2 in mid-1960, the two possibilities were then either a new home-grown system, or Polaris.

While the political battles around the new missile system were being fought in Washington D.C., the technical problems of integration with the Vulcan were being considered in the UK. From an aerodynamic perspective, the RAE divided the problem into,

"(i) *The loading on the missile during carriage over a range of flight conditions; this was required for assessments of initial motion following release.*

(ii) *The loading on the missile-pylon combination in order to determine the structural loads carried by the aircraft wing and mounting*".[195]

High speed testing, corresponding to the cruise and launch conditions, was once again the domain of Farnborough and the 8ft x 6ft transonic wind tunnel. By this time, the new ARA tunnel would also have been an option. Corresponding low speed testing, which would have been necessary to establish the effect of the missile-pylon-aircraft geometry at high angles of attack and configured for take-off and landing, were within the capabilities of Avro's in-house facilities, in this case the 9ft x 7ft low-speed wind tunnel at Woodford.

At Farnborough, a 1/30 scale Vulcan B.Mk.1 model was used for expediency, fitted with a pylon under one wing only and instrumented with its own five-component strain-gauge balance.[196] The missile-pylon combination was expected to be influenced by the lower surface flow of the wing, with outflow ahead of the leading edge and inflow behind it. As incidence was increased, which at speed would correspond to manoeuvring at altitude perhaps, then the increase in outflow as the undersurface pressurisation became apparent would result in a suppression of the inflow.

When combined with sideslip, the pylon in particular would be subject to a range of angles of onset flow, causing the sideforce it would inevitably generate to vary. For the missile itself, a long cylindrical shape at some onset

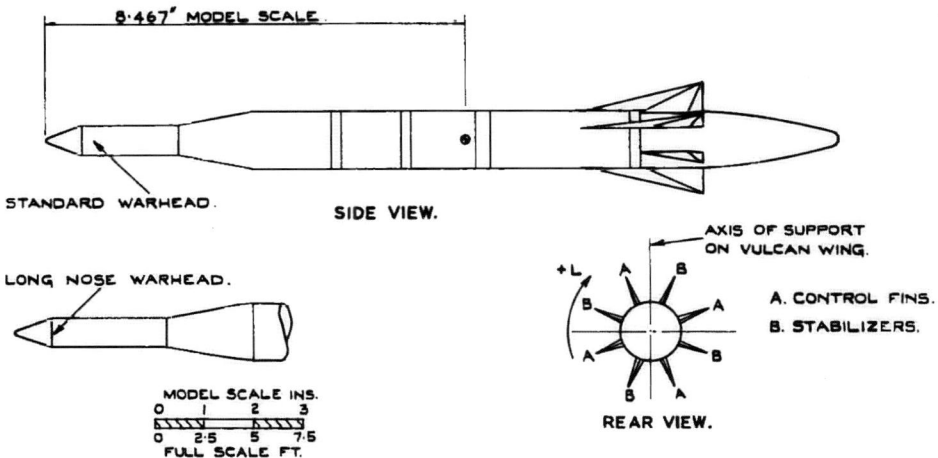

FIG. 3. GENERAL ARRANGEMENT OF SKYBOLT.

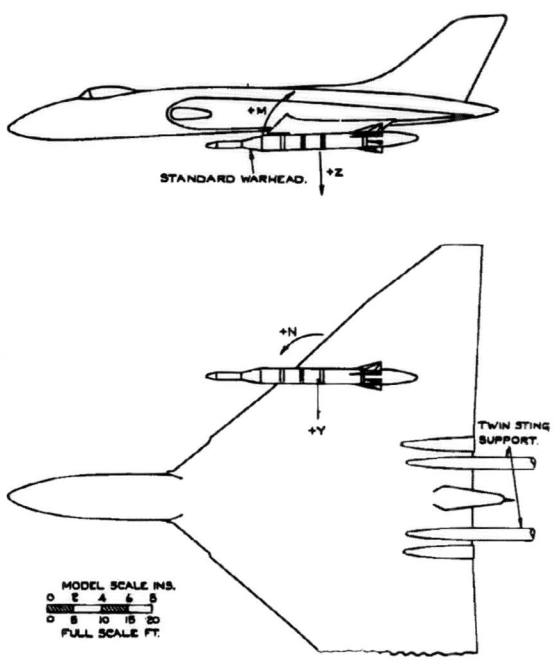

FIG. 2. SKYBOLT INSTALLATION ON VULCAN PORT WING.

The RAE conducted high speed tests on the Vulcan-Skybolt combination at Farnborough. The model used was actually a Vulcan B.Mk.1, for expediency.

angle would create a vortex pair shed from close to the nose, again varying in magnitude and direction. Would these forces be destabilising to the aircraft, or indeed the weapon itself, during the critical launch phase? There is a sense of the tests being perhaps expedited — even rushed — compared to those conducted earlier for Blue Steel. The model represented the launch platform only generically, as there was no prospect of the B.Mk.1 operating with the new missiles and the wing planforms were clearly different.

The post-launch scenario was modelled with a strut connected to the Vulcan model (which itself was not mounted on a balance), as opposed to the twin-sting rig which was used during ARA tests on Blue Steel with the Victor. Nonetheless, the February 1961 transonic programme identified no major concerns, demonstrating instead that the missile naturally developed a nose-down pitching moment and small downwards vertical force when aligned with the local flow. There were subtleties: because of the change in inflow angle with incidence of the aircraft, this alignment varied with load. At the same time, the inflow onto the pylon generated suction above the store, such that its natural download was reduced. In the launch phase, it was obviously critical that the missile rapidly separate from the aircraft and move away from its vicinity. What the wind tunnel tests did not show was the effect of the missile's weight and inertia. Blue Steel for example was found to be slightly unstable in pitch soon after release, but the dominant effect of its 15,500lb mass meant that it would clear this aerodynamic regime long before it threatened the bomber.[197]

Flight test hardware was now becoming a reality on both sides of the Atlantic. From a UK perspective, a short series of six drops by dummy rounds, instrumented to record the appropriate parameters, commenced on December 1, 1961. The Vulcan drops were conducted at the relatively isolated West Freugh test range, near Stranraer in Scotland.

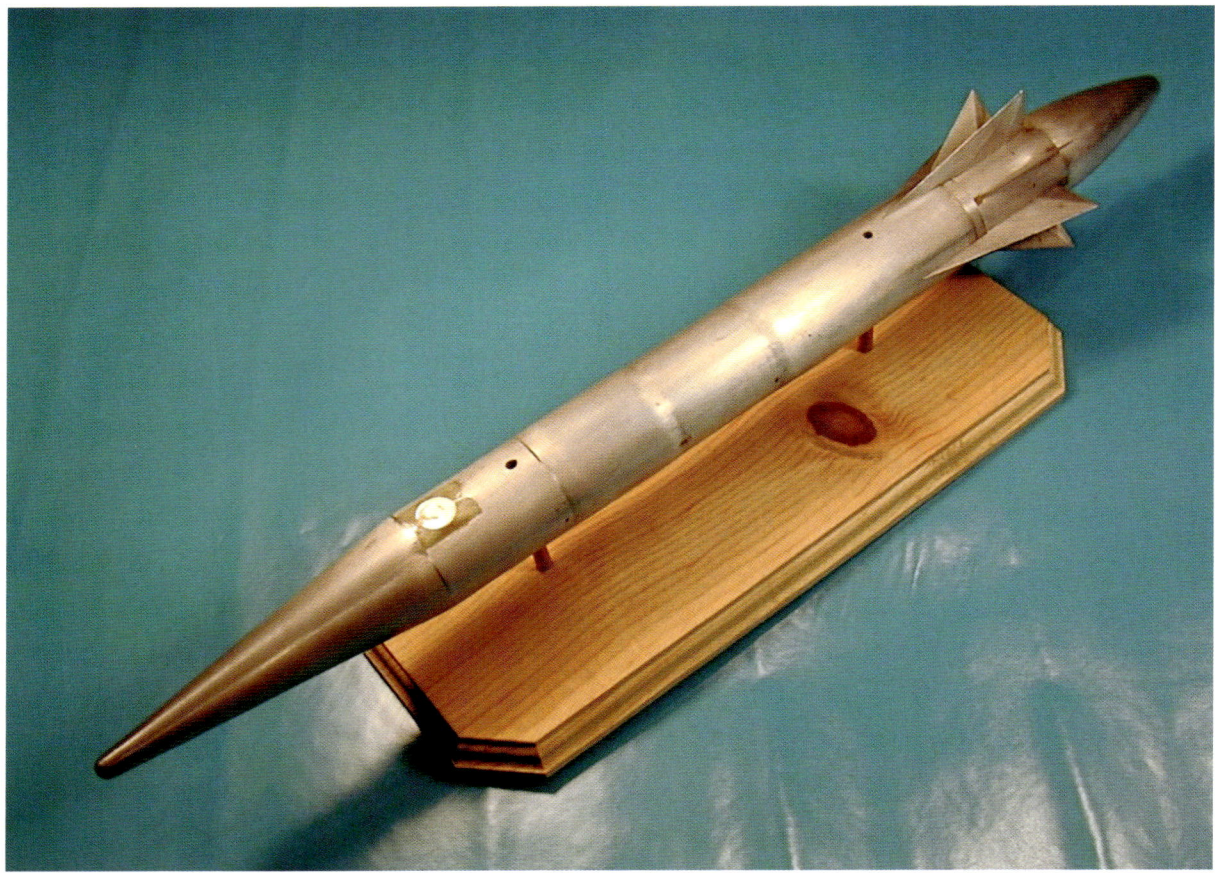

A machined wind tunnel model of the GAM-87A Skybolt missile. RON DOWNEY

The last of these was flown in February 1962. From July 1962, integration of captive mechanical test versions of the GAM-87A on the Vulcan commenced, with the aim of checking out the integration of systems and the missile's operation when connected to the bomber.

Captive flights and dummy drops were possible within the UK, but the launch of a fire-breathing operational Skybolt was not. Neither was it sensible to set up a new test facility half way around the world in Woomera as had been done for Blue Steel, when the Americans were conducting tests of their own at their existing ranges off Florida. A UK test team was therefore established at Eglin AFB, alongside their USAF and contractor colleagues, and given the length of the work planned, their families. A major milestone in the programme was the first powered launch, from a B-52G on April 19. The missile used a test-only guidance system (the operational device was being developed concurrently) and a warhead 'shape' that lacked the ablative coating necessary to survive re-entry. In the event, while the launch from 40,000ft and first stage ignition excitingly proceeded, the second stage ignition failed. Nonetheless, this was an impressive debut for such a complex system.

Four further launches of GAM-87A prototypes from B-52G carrier aircraft would take place before December 1962. All would suffer problems of some kind. The second did not fire at all, while the third and fifth veered out of control soon after launch and had their destruction initiated by the test team. That said, it is not reasonable to describe these tests, as Kennedy and McNamara would publicly do, as failures. The official USAF history of the programme was at pains to point out that,

> *"...the planned test objectives had been completely or partially achieved in most instances... None of the problems encountered during flight testing appeared to be of a fundamental nature but rather in the category of random failures."*[198]

Eighty per cent of the primary objectives had been achieved by the end of flight testing, including all of those associated with the missile's first stage. Flight control instability during the launch phase did require additional wind tunnel testing and restriction of the launch envelope.

The Farnborough tests and the implication of aircraft loading and sideslip were therefore of clear relevance to the successful deployment of the system, but at the same time the identification of the problem meant that fixes

What might have been for SAC: a B-52H carrying a full load of four (mock-up) GAM-87A Skybolt missiles. To reduce the up-front spend, the specific test launches with the turbofan powered B-52H were postponed, with all six using the J57-engined B-52G instead.

could be put in place. Still, the susceptibility of the system to enough random failures would render it impotent; the engineering of sufficient reliability into such a complex system was clearly difficult and expensive, and fed into concerns about the R&D spend that would be necessary. In September 1961, McNamara declared his dissatisfaction at the planned programme, which he did not believe was realistic. The Program Office was instructed to prepare a viable revision, which was submitted to the Office of the Secretary of Defense on October 21. The plan forecast an R&D spend of $492.6m. McNamara then laid down a challenge to the USAF it could have the money, but not a cent more.

> "Over strong opposition of certain of the president's advisors, I have recommended, and the president has approved, the continued development and deployment of Skybolt. Our actions were based on assurances that development would be completed on schedule and at [this] cost... If the Air Force is not prepared to make this commitment today, then I wish to reconsider my recommendation to the president."[199]

The contractors now had sufficient resources to continue on the programme, which would see the weapon ready for deployment in 1964, with production missiles funded from that Fiscal Year onwards. Really for the first time, the programme was able to operate on the basis of assured funding, and a bright future.

Absolutely key to the operation of the missile was the complex guidance system, which like Blue Steel took as an input the accurate position, velocity, altitude and heading of the bomber. The missile carried an INS, aligned prior to launch using its astro-navigation 'star-tracker' system. The interface between the bomber's firmly analogue systems and the digital pre-launch computer required an analogue-to-digital conversion stage as part of what was together termed the AN/ASQ-38(v), provided by IBM.

The four-missile complement of the B-52 used a pair of pre-launch computers, whereas only one was required by the Vulcan. It is interesting that while the Vulcan's higher altitude capability gave better opportunities for star shots, the prospect of a low-level launch had a severe impact on the weapon's accuracy for both carrier aircraft. The actual guidance system was used for the fifth and sixth launches

A contemporary model of the Vulcan B.Mk.2 and its Skybolt installation.

and the pre-launch phase was considered particularly successful; on November 28, it operated for more than four hours, and the two tests together achieved nearly 300 successful star shots, in daylight. This success was achieved after concerns about electro-magnetic interference of the guidance and control systems of the missile from those on the bomber, which were identified by Boeing Wichita in respect of the B-52 installation initially.

The Vulcan's systems were tested at Edwards AFB in June 1961, while a more extensive programme to lay down the fears about the B-52 was conducted at Castle AFB in October and November. It would be June 1962 before a completely satisfactory solution with modified cabling was approved, but there is no reason to believe that the interference problems had not been fully dealt with.

At the end of August, the Controller of the Pentagon, Charles Hitch, together with Harold Brown in his capacity as Director of Defense Research and Engineering, presented a report that McNamara had commissioned from them. Their conclusion boiled down to a comparison of what an admittedly huge tranche of dollars might buy, and they felt that a Skybolt-equipped B-52 force available towards the end of the Sixties (and the twilight of the manned bomber) was a poor investment, when set against the existing, operational Hound Dog and an equivalent buy of Minuteman. The report opined,

> "... the risk that SKYBOLT will fail to work at all is very low; the risk that it will not be a highly reliable... system until the late 1960s is quite large."[200]

Another factor began to play into their hands, as it became apparent that the R&D cost ceiling would be broken. Despite the agreement of the previous October, the USAF was confident that the money would be forthcoming, based on the political landscape it saw. McNamara had expended capital of the cancellation of the RS-70, a frankly ludicrous adaptation of the XB-70 to fit into the second-strike scenario. While that might not have stopped the Secretary of Defense acting, the fact that the UK was embedded in the missile programme, and clearly relying on it, was anticipated to be the clinching factor. What had been agreed with the British? Hitch, Brown and McNamara were under no doubt that the Camp David discussions of

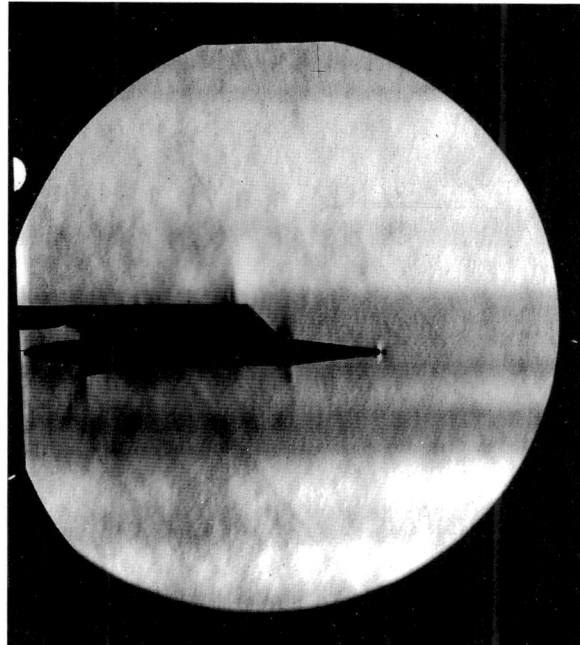

To visualise the formation of shock waves around the missile and pylon combination, a Schlieren system exploits the changes in density and consequently refraction. This shows the missile at the proposed design case for the Vulcan, of $M = 0.90$. Despite the subsonic freestream velocity, shocks are visible around the locations of abrupt changes in shape, at the nose, around the pylon and at the trailing edges of the fins.

March 1960 had committed the United States only to the supply of Skybolt to the UK should it come to fruition. If the programme was cancelled, then — whatever the fallout in foreign relations — the agreement was moot.

Hitch offered a formula: the three men were agreed that Skybolt made no sense and should be cancelled, but to actually carry out the deed it would be prudent to wait until the budget cycle came to its conclusion in January 1963. If the budget were presented to Congress with Skybolt missing and against the predicted scenario of a large deficit, then it would fall to the Air Force to argue for Skybolt to be reinstated, rather the Department of Defense to call for cancellation. The knife would therefore be wielded in December 1962, but secrecy would be maintained as far as possible until that moment. In fact, for the tactic to work, McNamara would need to accede to the Air Force requests for production funding to be released. He did that in September, but on the basis of a month-by-month review.[201] The secrecy extended as far as the foreign partner; a few weeks later, the new Minister of Defence, Peter Thorneycroft, visited Washington for the first time after accepting his portfolio. McNamara told him of his concerns over progress but gave absolutely no explicit indication that the decision had been made to terminate the programme.

Skybolt 20028 was the fourth to be test-fired from Eglin, during the B-52G test programme in 1962. This view shows the twin launcher that would have been used to give the Stratofortress its four-missile load.

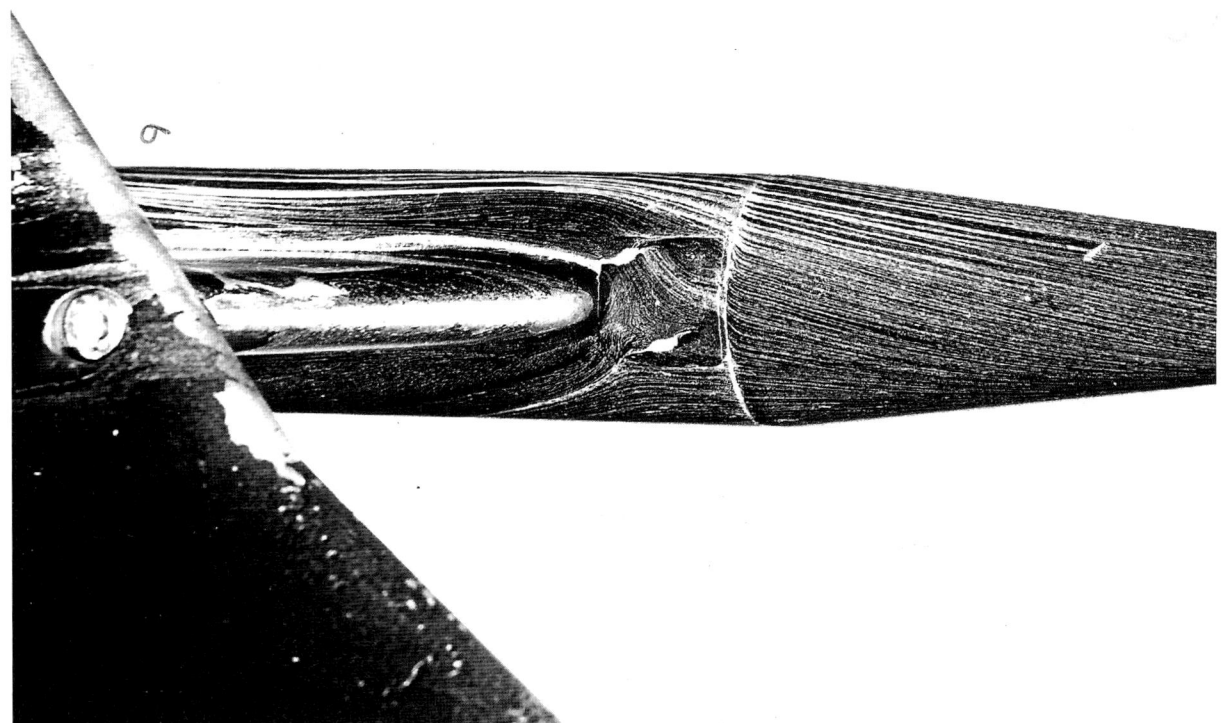

A wind tunnel image of flow visualisation work on the Vulcan B.2 with Skybolt. Luminous particles suspended in kerosene would be painted on the surface, moving with the local flow as and the liquid evaporating in the wind. In this case, the image shows a detail of missile pylon junction, which would be used to identify any local separations that could be corrected and improve stability. RON DOWNEY

Thorneycroft's elevation after Macmillan's Night of the Long Knives restored him to the top tier of British politics. He had met Kennedy many years previously, and the president was clearly comfortable in his company. They journeyed on Air Force One to Cape Canaveral and viewed the progress of American rocketry, during which the subject of Skybolt arose.

> *"...no decision at that stage has been taken to cut it and no indication was really given to me at that stage that it was likely to be cut. I took the opportunity of impressing on [Kennedy] and Bob McNamara that he attached importance to that particular project."*[202]

Outside of defence, the discussions held during the trip were wide ranging, with Kennedy emphasising that European integration was close to his heart and that he believed Britain should be playing an important role. Just a few weeks later and not far at all from the Florida coast where Thorneycroft had enjoyed his time with the charismatic young president, the subject of nuclear missiles was brought into sharp focus by the deployment of Soviet systems to communist Cuba.

Just before lunch on October 21, the British Ambassador Sir David Ormsby-Gore was asked to visit the White House for an audience with Kennedy, for which it had been requested that he was unseen. The president told him that the information he was about to impart was being discussed with nobody else, outside of the US Government. He revealed that U-2 photographic reconnaissance imagery had identified two types of medium range missiles on Cuba, for which it had to be assumed that nuclear warheads were standard. Kennedy had issued a strong statement on September 13 in recognition that this scenario might arise, to the effect that the United States would take whatever action it deemed necessary to ensure the security of its allies and itself.

The president suggested that two courses of action were available. The first was an immediate all-out airstrike on the missile sites. The second was a blockade of the islands, with the aim of stopping any further deployment of Soviet weapons. Ormsby-Gore advised Kennedy, in response to his question, that the first option would be unlikely to gain support outside of the US—irrespective of the risks of escalation—but that the second was one that Britain could support. The president agreed with the assessment and said that his staff had come to the same conclusion. The ambassador immediately reported back to the prime minister and by return, Macmillan offered his support to his 'friend', the president. It was not without concern

The spectacular sight of the Vulcan B.Mk.2 and the twin Skybolt installation. XH537 conducted test drops in the UK, as a precursor to expected live tests from Eglin AFB in Florida.

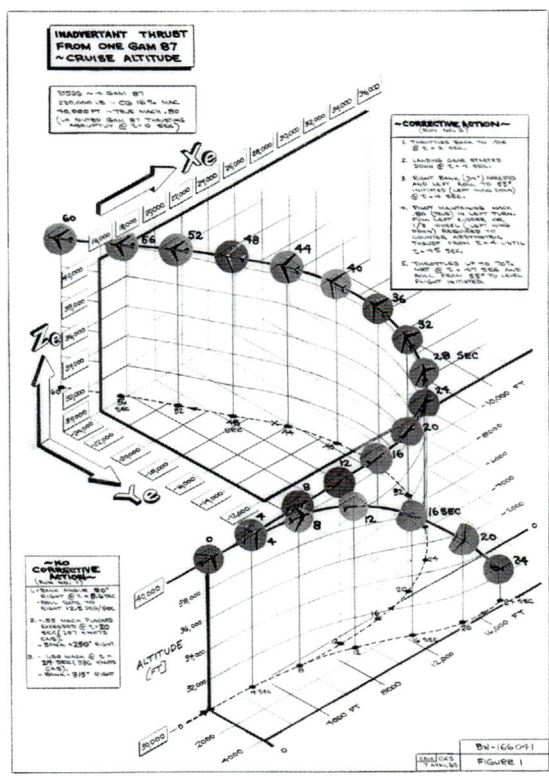

A Boeing study of the effect of inadvertent firing of the motor of one of its Skybolt missiles, on a B-52. The report showed bomber exceeding its maximum Mach number (M = 0.93) after twenty seconds, and four seconds later being supersonic with 315deg bank applied. Corrective action as devised could allow control to be maintained, beyond burn-out of the missile.

though, as the PM wrote to the ambassador,

> *"Since it seemed impossible to stop his action I did not make the effort, although in the course of the day I was in mind to do so. I feel sure that a long period of blockade, and probably Russian reaction in the Caribbean or elsewhere, will lead us nowhere".*[203]

This was a time when the world did indeed hold its breath—October 27, 1962 is often said to be the closest point to all-out thermonuclear exchange of the Cold War. During the process, RAF Bomber Command would also record its highest state of readiness while charged with carrying the United Kingdom's independent deterrent. Michael Robinson (later Air-Vice Marshal) was Officer Commanding 100 Squadron, newly re-equipped with the Victor B.Mk.2. As he described nearly four decades later,

> *"…all available Victor aircraft and crews at RAF Wittering were brought to cockpit Readiness 05. Each aircraft was loaded with one freefall thermonuclear weapon, the crews had their go-bags with all of the necessary route and target information and authorisation codes… Suffice to say that we remained in our cockpits for several hours before being ordered by the bomber controller to revert to Readiness 15. It had been a long afternoon!*
>
> *The fact that we were not ordered into the air says something about the nature of the RAF's deterrence.*

President Kennedy announcing to the US public that Soviet missiles had been seen on Cuba, October 1962.

The more evidently, or visually, efficient and ready we were (as is possible with aircraft if not with submarines) the less likely we were to be committed to war."[204]

But, as we know and not least due to the resolve of the Western forces, an accommodation emerged and, on the evening of the following day Macmillan was able to transmit to Kennedy a message of congratulation.

"It was indeed a trial of wills and yours has prevailed. Whatever dangers and difficulties we may have to face in the future I am proud to feel that I have so resourceful and so firm a comrade".[205]

In return, the president would reply,

"I am grateful for your warm and generous words. Your heartening support publicly expressed and our daily conversations have been of inestimable value in these past days. Many thanks."

On the one hand, these critical events might have represented an opportunity for the GAM-87A, a system of superior potential to those existing, and available to strengthen the US deterrent. Yet this opinion certainly did not prevail at the executive level; the view that Skybolt was not the best way of ensuring safety and spending defence dollars was maintained against the most severe of tests. Perhaps shoring of the tenuous position of the weapon formed part of Macmillan's thinking when just days later, he wrote again to the ambassador, this time with a focus on the future.

"You have several times referred to the president's suggestion that he and I might meet again soon and, in one of the telephone conversations recently, he said that it would be helpful if I could come over when the crisis had ended... If we do have a meeting, I am anxious that it should be on the Bermuda model, a sort of house party, rather than a formal meeting with a great many people present and the constant pressure to make public appearances which always develops in Washington or New York."[206]

The prime minister waited for nearly a week for the response, but it was as positive as he might have hoped.

"As regards a meeting, he says he would be very happy to try and arrange one in December. I emphasised that you would prefer something on the Bermuda model. He said that he quite understood and that would be his own preference. However, he thought the Governor's house had been pretty uncomfortable; apart from anything else, he apparently could never get any hot water. He suggested Jamaica but I pointed out that it no longer belonged to us."[207]

The venue was agreed shortly afterwards as the Bahamian capital, Nassau. Back in Washington, the three defence officials, together with Brown's immediate deputy John H. Rubel were aware of the potential level, if not detail, of the impact of Skybolt's cancellation on the relationship between the USA and the UK. This meant a State Department involvement, and a widening of the circle of knowledge. The 'British Desk' consequently reported back their assessment that,

"Cancellation of SKYBOLT would put in jeopardy not only Bomber Command but a vital element of British defence philosophy... the independent nuclear deterrent.

...we still rely heavily on British real estate all over the world from Christmas Island to Holy Loch. We should carefully consider the consequences of an estrangement.

Kennedy with the British Ambassador to the United States, Sir David Ormsby-Gore, October 1961. The relationship between these two men would be critical to the UK-US deal for Polaris being formulated at Nassau.

In mid-December 1962, Robert McNamara's (left) plan for elimination of Skybolt was coming to fruition, but both he and President Kennedy were becoming attuned to the foreign policy implications, brought starkly in to focus just weeks earlier by the Cuban Missile Crisis.

> *…Assuming that a decision has already been made… [Macmillan] should have as much time as possible to prepare the ground before an announcement".*[208]

Inevitably, there were rumblings about the confidential plan that reached the ears of top UK officials, including Zuckerman. In early November, Rubel flew to the UK for defence discussions and was asked by Thorneycroft to visit him, at home while he recovered from illness. Rubel found himself in a difficult position, ostensibly knowing nothing officially and unable to confirm in either direction, whether the Minister of Defence's concerns about a rumoured cancellation were anchored in fact. Much later, Rubel described that the British man had said,

> *"John, you'll be seeing Bob when you go back, and I would like you to carry a message from me to him. And I want you to tell him that nothing could be worse for British-American relations in the defence field than the cancellation of Skybolt. It would strike at the very heart of our relationship."*[209]

Events however were moving inexorably forward. In early December, McNamara himself, accompanied by Rubel and their staff flew to London for a formal discussion with the Ministry of Defence team. Rubel described how McNamara announced that he would simply read a statement directly from an aide-memoire that had been prepared in advance. Thornycroft, he recalled, looked,

> *"…very cool and self-possessed. He obviously knew what was going to be discussed and had thought about it well in advance. He waited quietly until his turn came. Then he thanked Bob very much for this."*[210]

The Minister was a former Chancellor of the Exchequer and one of the architects of the Conservative party's post-war rebuild. He would have been in no doubt about the seriousness of the situation—the US State Department had identified how closely associated the Tories were with the Independent Deterrent policy—but must also have recognised that he had the opportunity to come out on top. Skybolt had to be painted as essential, it was the first choice of weapon system and the one that had been marketed to the British people as the one that would protect them. Clearly though, as one-third of the US missile triumvirate, it was not the only possibility.

Would McNamara, recognising the difficulty the British found themselves in, hand over Polaris instead? While retaining control of himself and the situation, Thorneycroft suggested that the end of Skybolt was very bad indeed, perhaps serious enough to cause the fall of the government. "We've relied on you Bob, absolutely."

The problem there was that American policy had tied itself in a further knot, while the imprecise and personal nature of the original discussions between Watkinson and

What would have been the two Skybolt carrier aircraft: Vulcan B.Mk.2 XH535 and B-52H 60-0006, over Edwards AFB during compatibility trials.

Gates, then MacMillan and Eisenhower in 1960, left sufficient latitude for misunderstandings to grow. There was confusion on the US side as to whether the Skybolt and Holy Loch agreements were connected; they were not, but it would clearly suit UK interests to suggest that they were. McNamara would make an assumption that allocation of the UK missiles, of any flavour, to NATO was a principle of their supply. It was not, but would cloud the negotiations for both participants. Most crucial however, was the overall US foreign policy towards Europe. Thorneycroft was clear in his own view that Kennedy had an understanding of the continent that was unusual among US politicians and Americans in general, rooted in his time spent overseas with his ambassador father.

The presence of an independent, unilateral nuclear-armed state between his country and the Soviet Union was a destabilising complication in his administration's doctrine; whereas the UK viewed its deterrent as essential to both safety and status. However, Kennedy did want to see a more united Europe as an additional centre of democratic influence. In that scenario, a collective, nuclear-armed force comprised of the combined armed forces of the continent did fit. The US policy that would support this was to promote a combined European Polaris operation, with Britain encouraged to enter the EEC and further cement the relationship. In the way of this stood de Gaulle, with his own concerns about where the UK's priorities would lie, should it come to a choice between the cross-Channel or the transatlantic. If the UK was to have Polaris, then the priority for the US was for this to happen as part of the joint European force. If it were supplied unilaterally, then what signal did that send to de Gaulle, France and the rest of Europe?

McNamara's paper offered the option to the UK of continuing Skybolt alone, or of the supply of the existing Hound Dog. Thorneycroft asked a question that would assume critical importance in the weeks that were to come.

"If you are going to cancel the project, are you going to say it is because it won't work, or because it will cost too much?"

McNamara's response effectively sealed the fate of the weapon completely.

"We won't say that it is impossible, but we will say that technical problems dominate the decision…"[211]

And there lay the crux of the matter. The UK endgame was an intimidating, assured destruction capability with which to face the Soviet Union. Even if the British brought Skybolt to fruition on their own, the Americans would have publicly expressed a view that they, with their vast resources, feared it would not work when required to do so.

President Kennedy observed the firing of a Polaris A-2 missile from the USS Andrew Jackson, *on November 16, 1963.*

What if the Russians took that as their assessment too? The immediate response to the UK stance back in Washington differed between the Pentagon and the State Department. The latter still looked to encourage European integration, British entry into the EEC and by extension the projected Multi-Lateral Force (MLF) project, for a combined deterrent. They could not be seen to be discriminating against France in favour of the UK or any other unilateral player.

At the Department of Defense, a team worked in the period of McNamara's absence under his number two, Roswell Gilpatric, to establish the basis of what they considered a straightforward solution. Although frustrated with what they saw as a lack of British planning for the cancellation of Skybolt, they had no real objection to the supply of Polaris. The London discussions had established that a combined operations plan, as already existed for the two bomber forces, would be agreeable to the UK too in the event that the deterrent was submarine-based. At the DoD, it was clear to the staff that there was no NATO requirement in either case, hence Polaris—provided the UK was willing to pay and could deploy it—was a direct substitute.

After having moved onto further defence discussions in Paris, McNamara's team gradually drifted back over a few days to Washington. Neustadt described how,

> *"Tyler, fresh from Paris, arrived on the fifth day of proceedings. They astonished him. With his ear as sensitive as ever, he had left France feeling that the issue was essentially decided. He found his colleagues talking as though what he just lived through was the future, not the past. Their discussion seemed to him unreal."*[212]

While this was far from a universally prevailing viewpoint, it was almost certainly the return of McNamara and his insight that would drive the process forward. In contrast to the State Department, he had concluded that the MLF concept was dead in the water, because the European nations would not agree to finance both MLF and the large NATO conventional force that was expected. Woolly alternatives that aimed to satisfy the needs of both departments, such as supporting the UK and French bomber fleets in some non-GAM-87 manner, then assisting both with future missiles, were too nebulous to agree quickly, but there was appetite for the idea of a joint US-UK study of future weapons. Discussing all of this with State Department Under Secretary George Ball and McNamara on Sunday, December 16, the president concluded that Polaris should be offered to the UK, bridging any potential gap in understanding of what his country's obligations should be following the end of Blue Streak and Skybolt.[213]

Time was now of the essence, as the planned Nassau talks approached. The two men made appropriate preparations for this. Macmillan had first a planned meeting with de Gaulle, which in contrast to their previous discussion, went badly. It was clear that the French president's position was hardening against the inclusion of the UK, on its terms, into the EEC. His democratic position was stronger than the PM's, who had not seen the expected 'bounce' from his extensive Cabinet changes. Perhaps France simply needed to wait and see what might transpire from an incoming alternative. Nonetheless, Macmillan felt that he should inform de Gaulle of the Skybolt situation, and that he would be discussing Polaris with Kennedy.

For his part, the US president wanted to be personally assured that he was on firm ground. On December 17, the 59th anniversary of the Wright brothers' flight, he took the step of speaking by telephone with his predecessor, Eisenhower. The conversation was cordial and respectful, with Kennedy referring to him as, "General" throughout, and Eisenhower in turn to, "Mr President." The key information that the younger man wanted was whether there had been a true quid pro quo: was the Skybolt agreement part of the price for the Holy Loch Polaris base. Eisenhower assured him that it was not.[214] However, he would depart for Nassau with his team lacking strong documented evidence of this.

The situation then was both unusual and perhaps at odds with the general mythology that has grown since. Macmillan arrived at a low ebb, knowing that his party's reputation for competence in the defence field was being severely challenged, with the way out of the dark woods reliant on his ability to persuade Kennedy that he was left with an obligation to help. On the other side, Kennedy and his team had already decided that Polaris would be made available, should the UK want it. Where then was the complication? To start with, it lay in the fact that British delegation had nailed their colours not just to an independent deterrent, but specifically to GAM-87A as the vehicle that would provide it. This was the PM's status quo scenario—to impress upon the Americans that the missile programme and re-equipment of the V-bombers should continue. On his arrival in the Bahamas though, he learned of comments that Kennedy had made to the press about his belief that Skybolt offered poor value for money, when investments had been made already in Polaris and Minuteman. On top of McNamara's statement to Thorneycroft that it would be the technical basis—its credibility as a deterrent weapon—that would be the reason given for the cancellation, Macmillan now recognised that the missile was tainted. It would have to be Polaris, and he did not know for certain how that would play out.

Meanwhile, on board Air Force One Kennedy engaged with Ormsby-Gore, the British ambassador. The president was finally on the same page—politically—as Macmillan, with an appreciation of the latter's commitments and consequent motivations. In the time of the flight, they concluded a proposal that each felt would satisfy the other. Skybolt would continue, without procurement for the USAF, but with development costs split on a 50:50 basis. Yet again, were this the way the negotiations concluded, the offer was extraordinarily favourable to the UK. There were however two insurmountable problems, one of which was perhaps obvious, the other becoming apparent when Gore was able to speak to Macmillan on arrival. Half of $492m was still an enormous amount of money, well beyond the expected budget for the UK Skybolt programme, and in reality beyond that which the UK could commit to. From an American perspective, an increase in the cost to the UK of its independent posture and subsequent rethink as to its value was absolutely in line with policy. Even were the cost not the problem, Macmillan now believed the missile's reputation was destroyed in Moscow; if it were not bought by the US, especially after the statements in the press, then its credibility was marginal. From the PM's side then, there was a choice: Skybolt was available, but at great expense, or Polaris may be available, if his skill as a negotiator could deliver it.

Thorneycroft was undoubtably impressed once again by Kennedy at Nassau. The discussions, as both the president and prime minister had wanted, were between a small group of top-level participants from both sides.

"...the real thing that struck me was this technique of his, and the relationship between these two men and countries. Whether you could call it a special relationship of not I don't know but what his techniques was—which was unique—was to ask the English some damned awful questions and make them answer it in front of the Americans and turn to the Americans and cross-examine them up hill and down dale and make them answer in front of the English. This was a method which I had never seen adopted between two foreign countries in my life and it was the most refreshing thing I have ever seen.

I remember one question when he knew that a lot of my advisors thought Skybolt a wonderful weapon. He asked—why don't you buy it, why don't you go ahead with it? We will go in it with you. McNamara says if you want to do this thing why don't you go ahead with it? ... If you tell me that this is a good thing we're dropping. Well, I replied that Mr McNamara was one of the best businessmen in the United States and that he just sold these shares. I didn't think it was the moment to buy them."[215]

The president was formally greeted at the airport by Macmillan, prior to the complete party adjourning to the venue for the talks. Walking alone together that evening, the prime minister explained that Skybolt was no longer an option in his view, and that the UK needed the alternative that would still be feared by the Soviets: Polaris. Kennedy had been briefed on the State Department's preferred option, that the talks agree to a joint study of a Skybolt alternative, without commitment to an agreement there and then, but to report back in advance of the end of the Parliamentary recess in mid-January. Meanwhile, Thorneycroft had adopted a militant stance, entirely willing to go home with only a commitment to go it alone on a new system, if Skybolt were not in the offing. He was gradually brought around by the combined efforts of Ormsby-Gore and Duncan Sandys, both of whom recognised that the American 50:50 split was as generous an offer as might reasonably be expected. If it were rejected and toys thrown out of the pram, inevitably the Americans would publish details of their proposed deal in the press, and little sympathy could be expected. Pushing for Polaris, they concluded, was the only way.

The formal talks began the next day, with Macmillan laying out his stall in a coherent fashion, although of course from a particular viewpoint. The UK and the US had together founded the nuclear age and had the shared history of defeating Hitler and establishing the international order. Europe was important to both of them, but in his view, there was no connection between the prospect of EEC membership—which would turn on questions of agriculture—and the substitution of one weapon system that refreshed the status quo, with an alternative. As Neustadt wrote immediately afterwards,

"The president now faced an impassioned older man embodying a valued weaker ally, who invoked in his own person a magnificent war record, an historic friendship, and a claim upon our honour—in Eisenhower's name—to say nothing of one politician's feeling for another. McNamara and Thorneycroft had spoken different languages; these two spoke the same."[216]

Macmillan then played his master stoke: unexpectedly, he handed the US delegation a way to get what they wanted. National Security Advisor McGeorge Bundy made notes of the exchange, recording that the Prime Minister conceded,

"He would be prepared to put in [to NATO] all of his part of a Polaris force provided the Queen had the ultimate power to draw back in case of a dire emergency similar to 1940."[217]

Kennedy's challenge had been to find a solution that precisely mapped to the administration's foreign policy, which an independent Polaris did not. While he may already have decided on a fallback during his December 16 meeting with McNamara, Thorneycroft's view expressed later was that the president came to Nassau with an open mind, ready to listen to the arguments and tailor the generosity accordingly. Macmillan on the other hand was determined that a deal be done there and then; he feared that the conditions would never again be so favourable, with the Americans minded to give the British the extraordinary submarine-launched missile system without further ties that would contradict the 'independent' status. Skybolt was no longer credible, Hound Dog inferior (despite being anticipated to remain in service with SAC) and a joint study a commitment to nothing. At that point, the concept of the deal was approved, and conversation turned to terms.

The Nassau communique issued a day later provided an effective description of the thinking. It went through the Skybolt plan, the decision on its fate and the 50:50 offer, before confirming Polaris would be supplied. The British concessions were framed as the 1960 'spirit of co-operation' idea might have implied; the UK would move towards allocating forces to NATO, which in the short term could include elements of the V-Force. Some clauses, such as vague suggestion of the US and UK Polaris forces acting together within NATO, were sufficiently unclear as to allow appropriate flexibility in the future.

Nassau is significant in the V-bomber story as it represents the final point at which the Vulcan was the future of the deterrent, as well as the present. The challenge of Blue Streak, the long-planned 1965 replacement, had been seen off as its technology became obsolete. In this brief period, starting from the spring of 1960 to the days before Christmas 1962, the Vulcan-Skybolt combination had been painted as the long-term follow-on, a system that would make a slimmed-down V-force sufficiently credible for the remainder of the decade. By the standards of the day, it would have been difficult to contemplate the presence in service of the Vulcan B.Mk.2 much beyond 1970; there was little experience of individual aircraft lasting in the front line for a decade or more. That, of course, would change. As these words were written more than six decades later, two of the actual B-52H Stratofortress bombers that would have carried Skybolt were deployed on NATO training operations to the UK. In the interim, they had carried Hound Dog, gravity bombs of various types, rotary launchers of air-launched cruise missiles and a multitude of other conventional and nuclear stores. Eventually, these jets will be re-engined and will continue to serve into their eighth or ninth decades.

As the American contingent flew on from Nassau, some to Palm Beach, others to Washington, the British hosted a Canadian delegation and all considered that they had achieved positives from the negotiations. Although he would be challenged in the British press and within his party for not returning with Skybolt, the prime minister believed he had secured something much better, while his colleagues that had witnessed the exchange could hardly argue that they might have done better themselves. For McNamara and the Department of Defense, the outcome offered much beyond the direct supply of weapons to an allied nation, an emphasis on NATO that had previously been absent and the security of Holy Loch for the US Navy. The possibility was now open to offer de Gaulle a similar capability, which had never been on the cards with Skybolt. The UK concessions might be the bait that, under those circumstances, could tempt a bigger prize — France as a full NATO member once more.

The participants in the talks were not the only ones to be pleased that day, however. Gilpatric had decided not to instruct the Air Force to cancel the planned sixth and final flight test of GAM-87A, the second with the tactical guidance system. The missile successfully separated from its B-52G platform, fired both stages successively and at commanded cutoff of rocket thrust, the calculated warhead splashdown point would have been 847 nautical miles downrange. At this stage in the tests though, this remained a theoretical, albeit likely to be achievable capability, as there was no re-entry vehicle. The DoD would later argue that this impact point would actually have overshot the target by the order of a hundred miles, while the Air Force press release implied that the range had been physically flown rather than calculated. Still, the objectives of the test had been fulfilled and we are left with the evidence that on at least one occasion, the missile's booster system did what it would have needed to do. Skybolt did not demonstrate its reliability, but did show that it was more than talk.

Macmillan wrote to Kennedy on Christmas Eve, thanking him for a gift initially, but moving on to describe his feelings regarding the events of the past few days.

> *"My dear friend…*
> *…I enjoyed our talks in the Bahamas. Although strenuous, I think they were very rewarding. But it is perplexing that Skybolt should go off so well the next day. It would have been better if it had been a failure. However, those are the chances of life…"*[218]

For the president, the potential impact of the misunderstandings that lay at the heart of the Skybolt affair, including the initial lack of political intelligence as to the effect on the prime minister and his party that the cancellation would cause, left him uneasy. He commissioned his confidant Professor Richard E. Neustadt of Columbia University to review aspects of foreign policy, particularly as they were relevant to the 'Atlantic Alliance'. Neustadt proposed in turn that this focus on the case study of Skybolt, according to the cover note of his eventual report. He delivered it to the White House on November 15, 1963 and it is thought that Kennedy read it over the next days. However, by then he had precious little time before he, his wife and their retinue would emplane on Air Force One, bound for Dallas.

14

BEYOND THE V-FORCE

NEW DEVELOPMENT of the V-bombers in any meaningful sense was stopped by the outcome of the Nassau talks. Blue Steel work remained to be completed, and limited adaption of the bombers and their arsenal to be effective in the face of the Soviet Bloc's countermeasures would of course continue. It would however be brave to suggest that the dark wraparound camouflaged Vulcan B.Mk.2 fleet 'fighting' at Red Flag in 1982, embodied technology very different from that when they had worn anti-flash white 20 years earlier.

Over the Atlantic, the story for the B-52 was very different, because it had fought a hot war and had in some ways been found wanting. The B-52D fleet saw extensive adaptation and upgrading of its electronic warfare systems in particular, in order to survive against a (predominantly) surface-to-air missile threat. By the early 1980s, the prime second generation B-52G and B-52H fleets were cruise missile carriers, ready for future challenges and laughing in the face of attempts to replace them.

Nassau ended the prospect of Skybolt's associated upgrades for the existing Vulcans. Both Avro and Handley Page had been keen to adapt their bomber concepts and make use of their investment, with variants suited to the new role. Of these, the four- and six-Skybolt Phase 6 versions of both jets are familiar. Focusing on the Vulcan, the wing was to have been extended again but retaining a straight leading edge, in planform at least. The drawings we have do not make it clear, but it is likely as we have seen that progressive leading edge droop would have made sense in the finished article. Speed, altitude and even range were now less important than payload and endurance.

The Skybolt carriers might lay on guard on their Operational Readiness Platforms across the country, but they might also, if it could be done efficiently enough, fly long endurance airborne alert missions. In that scenario, being able to carry a very large heavy fuel load and perhaps relief crew members started to look important. Indeed, the cruise Mach number that the wing was designed around was 0.8, with a maximum of 0.87. The manufacturer's literature describes how this gave the freedom to avoid the extent of the B.Mk.2's spanwise tailoring of the lift distribution, which partially explains the simplification of the planform. Of course, as the design Mach number reduced, so did the benefit of these design features, and the cost and engineering complexity they implied.

Avro's Vulcan Phase 6 proposal incorporated a lengthened nose for additional crew, now all given ejector seats, and was stressed for a staggering (considering where development had started) 356,000lb max gross weight. To power the new aircraft (or indeed, as Avro emphasised, converted B.Mk.2 jets), the Olympus 21 (301) would require one of two modifications. Either reheat would be added from nozzles aft of the rear spar, or a shaft-driven aft-fan, which would ingest air from inlets above and below the wing, would be added. In addition, Avro together with Bristol Siddeley proposed the option of using the turbine from the Olympus 22, then under development (as the 22R) for TSR2. The increased temperatures allowable in the new turbine in turn allowed increased thrust, and could be modified as a module swap from the Olympus 301. This engine was eventually termed, by November 1961, the Olympus 23. Avro's numbers suggested that two Skybolts could be carried for about 12 hours, and a maximum battery of six for seven hours. Meanwhile, the strengthening of the B.Mk.2's wing structure, addition of attachment points for the pylons and internal ducting for the missiles' services began to be embodied on those aircraft in production.

We know that Skybolt was cancelled, and with that the requirement for a tailored carrier aircraft. An interesting question though is whether such a machine would have been based on the V-bombers, or if it would have been better to start again? To a great extent, the answer to that question depended on what else was available. If speed and altitude performance were less critical than had been the case previously, was there a better solution than a re-engineered, 350,000lb Vulcan? In fact, the UK government was already invested in an aircraft that was close to this weight and which, in some ways, was more modern and a better choice for the future. However, that aircraft's performance and capability was unlikely to have been available when it was, had it not been for the work that had gone into the Victor and Vulcan. The Vulcan Phase 6 would have been a truly spectacular machine, but a dispassionate examination of the circumstances suggests that it would never have been built, even if Skybolt has come to fruition. The reason for this assertion is the existence of that other aircraft, in 1962 far more advanced down its own path to realisation: the Vickers VC10.

IMMEDIATE MILITARY APPLICATIONS

The Vulcan would likely not have entered service at all in the absence of Herbert Pearcey's insight regarding the transonic leading edge flow over the outboard wing. In his words,

> "Our backroom research received quite a boost [from the Vulcan leading edge work]… It was the understanding … that was to lead us ultimately to the supercritical wing sections that are now commonplace on most civil airliners."[219]

In mid-1955, as results of Newby's work were being disseminated, the UK civil aviation industry had no operational big jets analogous to the V-bombers, but was in the process of developing such types in the expectation of export sales.

The Comet I had never been other than a proof-of-concept that allowed a jet airliner to be demonstrated to the public, validated as viable and—in the limit—exchange its unparalleled speed and comfort for the ticket fares of the few who could consider its cost sensible. To realise its true potential, as had been the case for the Tudor a decade earlier, this remarkable capability needed to be unleashed over the Atlantic. The much enlarged and revised, big-Avon engined Comet 3 would take on this challenge. Examination of its airframe though leads to a difficult conclusion: Britain's fabulous jet transports lagged behind both their military counterparts and, more worryingly, their American competitors, in terms of the high-speed design of the wing. How did the newly proven concept of supersonic expansion and isentropic recompression permeate through industry? Unsurprisingly, it was in a large part used to solve otherwise intractable military problems.

A direct and immediate beneficiary of the Vulcan work was the ambitious Canadian supersonic all-weather fighter, the Avro Canada CF-105 Arrow. This was a large delta design that for part of its gestation was also planned to be powered by the Olympus, albeit the Curtiss-Wright built J67 version. As a part of the same wider industrial group, Avro Canada had access to data on the development of the 707 and 698 (Vulcan) projects. Although the underlying choice of the tailless delta configuration had a top-level similarity; a weight-efficient method of providing low t/c and wing loading, the supersonic requirement drove the actual t/c down to just 3.5% at the root and 3.8% at the tip.[220]

The Arrow demonstrated a reduction in longitudinal stability at moderate angles of attack during early wind tunnel tests at Cornell University. At the same time, there was awareness of Convair's work on conical camber for induced drag reduction and Avro's Vulcan modifications for buffet boundary extension in high subsonic cruise. It was clear that an outboard drooped leading-edge extension might well achieve improvements in all three areas. The eventual 10% chord, drooped extension chosen incorporated an inboard notch for leading edge vortex control (to improve the pitch stability) and at the planned subsonic cruise Mach number of M = 0.925, the droop achieved a significant improvement in the (wind tunnel assessed) buffet CL limit, from CL = 0.26 to CL = 0.41, without a detrimental supersonic drag penalty.

From 1955 onwards, that is concurrently with the Vulcan leading edge implementation, Pearcey's aerofoil work had certainly piqued the interest of another HS group company, the namesake Hawker itself. Concerning himself with the integration of propulsion into supersonic fighter aircraft, work which would eventually result in the P.1154 V/STOL proposal for the RAF and Royal Navy, C. L. Bore wrote many years later that,

> "Pearcey was describing how isentropic supercritical recompressions actually worked, while in other countries people were still arguing whether such behaviour might be possible."[221]

In discussion with Pearcey, Bore suggested that an increase in nose radius from the typical 0.3% chord to 0.8% would increase the extent of the advantageous recompression. This was followed through with geometry developed under the auspices of the Supersonic Air Transport Committee (STAC) and indeed shown to be positive; it might be logical to assume that the CLmax capability was also improved, which was important for the manoeuvrability of the combat aircraft that were Hawker's business. This research provided the basis of the aerofoils for the P.1154 and ultimately the P.1127 (RAF), the first-generation Harrier that entered service in 1969. Hawker would describe the design generically as Blunt Peaky and highlighted the relatively good performance of the Harrier at high lift and high Mach number, compared to the earlier generation Hunter.[222]

Bore's recognition was that the peaky profile was equally applicable to supersonic inlet design, particularly given that some of the NPL shapes were demonstrating shockless behaviour to M = 0.9. Without the ability to sweep a fighter proportioned fuselage inlet in a manner analogous to a wing, a significant pressure recovery and integration efficiency advantage could potentially be realised. Bore applied such shapes to the P.1154 in 1963 and showed remarkable shockless recompression following leading-edge peaks of nearly M = 1.9. The application was cancelled in February 1965 (prior to the perhaps better-known demise of TSR2 in the following April), but the design was transferred to the subsonic strike fighter that rose from its ashes as a development of the P.1127 and Kestrel proof of concept aircraft.

Bore claims the Harrier inlet was the first such design to fly anywhere in the world; he described it as Supercritical rather than peaky, having been using such terminology since 1953. In whatever words might be used, it is clear that the technology had made an appreciable step beyond the Vulcan's leading edge. In designing a conventional high aspect ratio swept wing, which for good reasons was a better overall solution to the problem of a moderately-sized transport aircraft, the need to incorporate a high lift system was paramount. The large nose radius gave sufficient packaging space for a leading edge slat mechanism, as well as making the resulting geometries far more aerodynamically performant than a sharp edged solution. Bore's challenge to Pearcey's concept was clearly a vital building block in the design of future transonic wings.

As an aside, the fixed wing combat aircraft used by the UK in the South Atlantic conflict of 1982 (Harrier, Sea Harrier and Vulcan) can all trace directly substantial parts of their aerodynamic philosophy back to the work of Pearcey in the mid-1950s.

VICKERS V-1000 – BEHIND THE STATE OF THE ART

All three V-bombers had corresponding planned transport versions, as their manufacturers understandably looked to exploit their investment in transonic technology. By definition, these would be large, long-range aircraft; aimed at requirements for BOAC's eventual transatlantic Comet 4 replacement and an RAF intercontinental troop transport. Part of the Vulcan leading-edge study was actually conducted on the wind tunnel model of Avro's contender, the 722 Atlantic.[223] Ultimately, BOAC's enthusiasm was limited and the MoS viewed Vickers as the only contractor with the capacity to take on such a project, which crystallised as the V-1000 long range transport.[224]

The aircraft was certainly reminiscent of the Valiant, but was far from simply a transport version of the bomber. It was much larger, with a wingspan of 140ft as opposed to the 115ft of the earlier machine. The planform strongly indicates the acceptance of the spanwise tailoring concepts espoused by the RAE, with a significant increase in sweep and thickness taper on the inboard region and curved Küchemann tips, all aiming to maximise the extent of the 'sheared wing'-like flow. Excluding the inboard region, the majority of the wing was of 9% thickness Vickers cambered section and 26° sweep. These sections had been finalised by 1954, hence using the state of knowledge of the time.

There is no shortage of opinion on the V-1000 and its fate, including from those involved. Sir George Edwards would state in 1973 that it had been cancelled for, "short term political reasons", and given that he recounted being asked a year later if the programme could be reinstated and subsequently observing BOAC ordering Boeing 707s because "no suitable British aeroplane can be available…", he was doubtless left sore.[225] Dietrich Küchemann would make an almost identical statement, implying a belief among the notables of the sector that a potential lead had been snatched away. However, a rather more prosaic

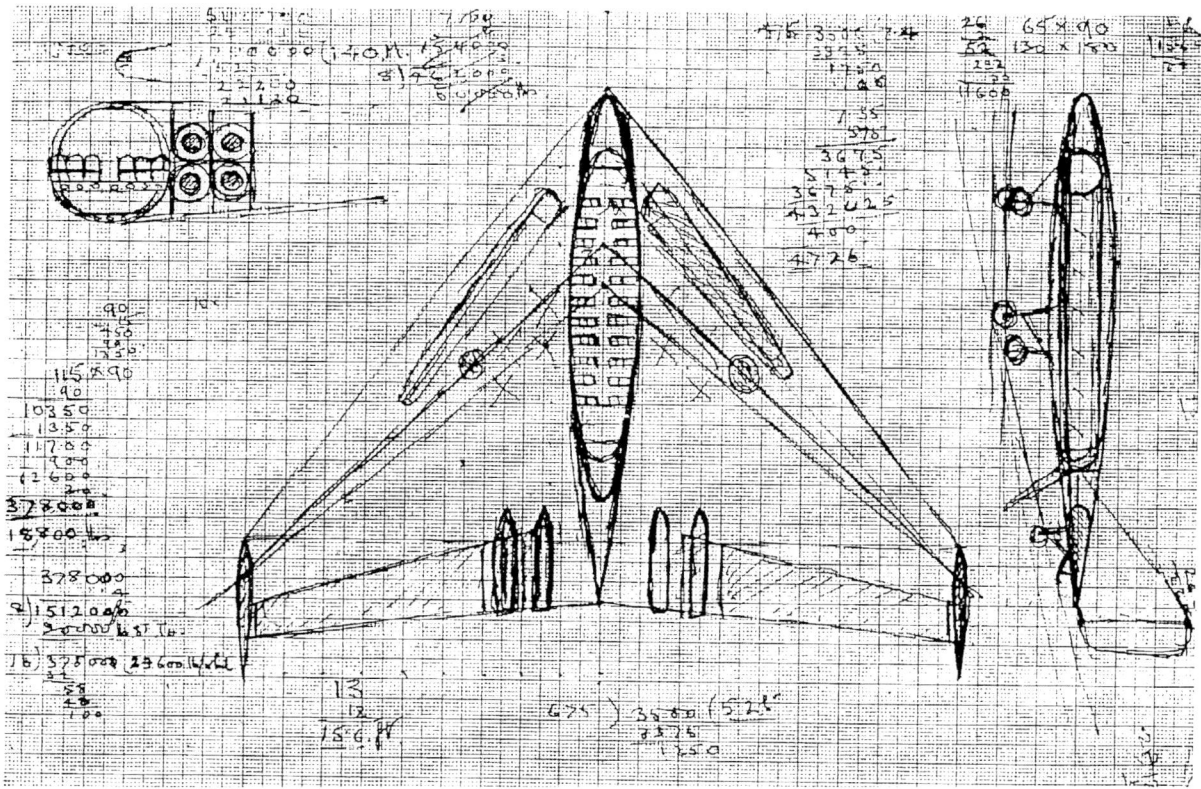

A Roy Chadwick sketch of a potential airliner version of the early 698 concept.

reality has been retrospectively discussed by others taking a less subjective view. BOAC never committed to the aircraft and the requirement for transatlantic range had, by 1954, delayed the entry into service until 1960.[226] As such, the V-1000 would have lost the claimed advantage in lead time over the American jets which were recognised as inevitable (if not precisely defined at the time). Aside from the folklore, it would be surprising if the contemporary development of transonic aerodynamic technology did not have something to offer to the story.

The world that the civil version of the V-1000, the VC7, would have been born into in 1960 had moved on considerably from the assumptions made when assessing the new aircraft's likely competitiveness. In reality, BOAC would that year commence operations with the Boeing 707-436 Intercontinental, powered by a developed version of the same Conway engine. It was however considerably developed beyond the Stratocruiser width, four-abreast cabin version associated with the 1954 launch announcements, or indeed the limited payload-range performance of the initial -120 series. This was a true, fire-breathing non-stop transatlantic machine, cruising at M = 0.83 and with MTOW of the order of 25% greater than planned for the VC7.

Comparing relevant aerodynamic data for the VC7 (planned) and 707-420 (actual) is telling. The physical parameters reveal that the substantially lighter VC7 had a slightly larger wing area and lower aspect ratio, generally indicative of lower overall efficiency. However, the required CL for a typical start of cruise weight was also lower, working in the opposite direction to aspect ratio in terms of reducing induced drag. Significantly, the majority of the span of both wings are of similar t/c ratio, but the sweep of the Vickers design was much less. Without a step in the transonic performance of the aerofoil sections used, the latter would therefore be expected to have a lower drag-rise Mach number and less efficient cruise speed capability.

Reference to available high speed wind tunnel data for the V-1000 was made in section. This was used by Pearcey in a report from 1954 that was classified until the late 1960s due to the sensitivity of the aircraft it described. While pressure plotting data was not available, oil flow and tufts had been used to identify rear shock-induced separation with M, which could then be compared with CL vs. α for a representative M = 0.84 condition. Noting that the calculated value of CL required for the M = 0.78 cruise condition was 0.34, the data showed that this was achieved at an incidence of α ~ 4°.

At M = 0.78, the separation boundary was encountered at α ~ 6°, giving CL = 0.47 and an available margin to buffet of 1.38g. Compared to the JAR requirement of a minimum margin of 1.3g in the cruise, then the

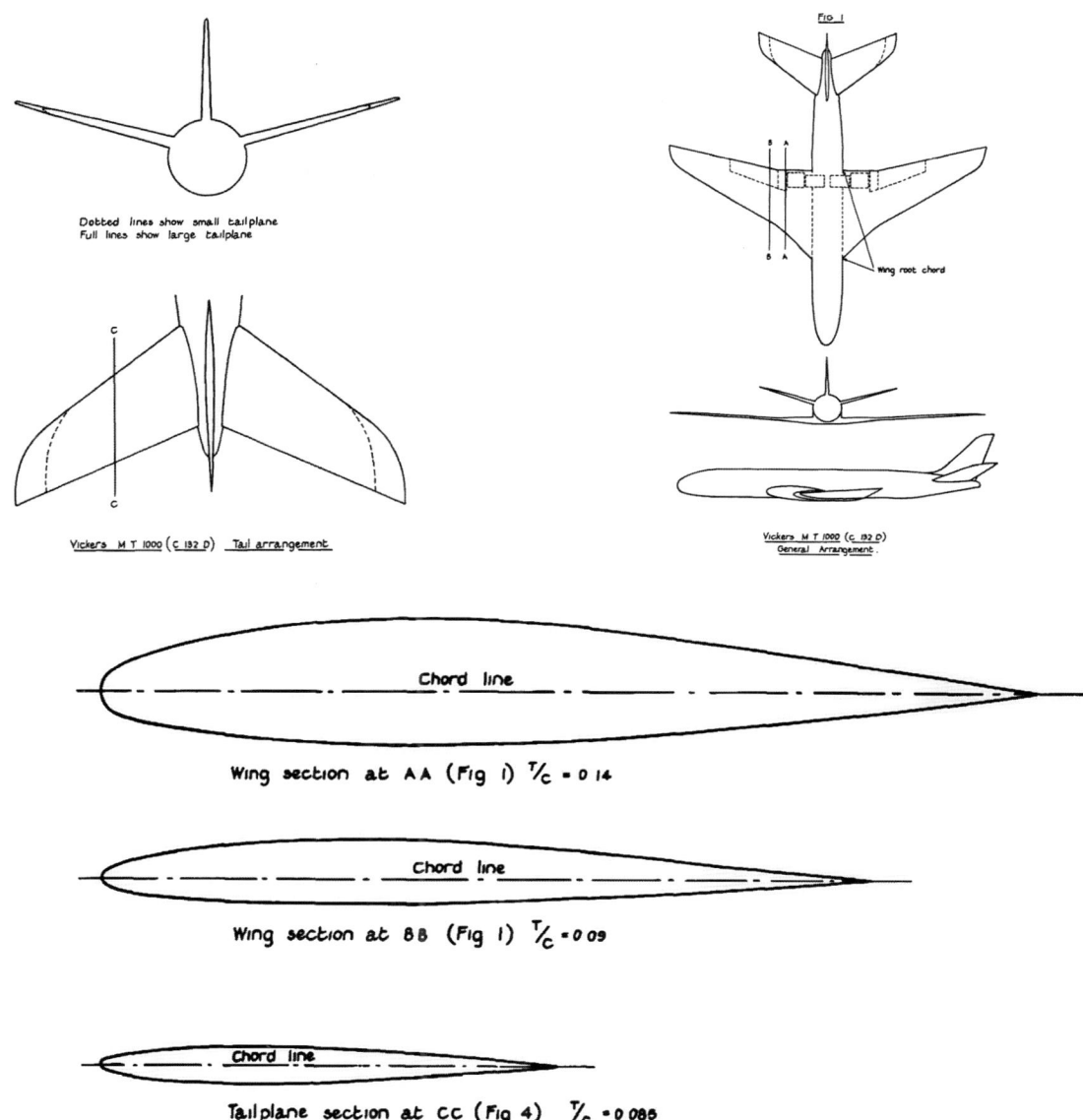

Views of the basic and option configurations of the Vickers M.T.1000 (or V1000) military transport aircraft, as tested in the NPL Compressed Air Tunnel (CAT). Notable features are the relatively large area wing of low sweep, and the correspondingly large tailplane area. The wing sections were entirely conventional. ARC-CP-0485

VC7 in its baseline incarnation, that is without considering weight increase in service to meet future customer requirements, was not generously endowed with growth potential. Had weight grown to any extent, it would not have been certifiable.

The mythology of the V-1000/VC7 rests on the idea that Britain surrendered its stake in the long-range airliner market, for want of completion of this worthwhile competitor. Against this must be set the tenuous operational requirement of the RAF for such an advanced and expensive aircraft, together with the knowledge that its principal domestic customer preferred the 707. The data available would tend to suggest that aircraft as planned would have been limited in at least one aspect of its performance—cruise Mach number—when compared with the Boeing and Douglas designs that would also have been available at the time of its service entry.

Was this important? Boeing's George Schairer documented two relevant experiences from the development of the 707. The first was the decision to offer two very different derivatives of the aircraft (the domestic 707-120 and intercontinental 707-320/-420) to customers, following rejection by Pan American of the original concept that compromised both missions. The second was the desire to operate at the maximum possible cruise Mach number by United Airlines, which resulted in Boeing seeking to offer M = 0.84 rather than the M = 0.80 originally planned, with the aim of beating the M = 0.82 of the DC-8.[227]

Speculating on analogous situations that would likely have arisen had the VC7 come to fruition, then is it plausible that Vickers could have addressed them? The history of the UK manufacturers being able to fund product development in this period suggests that there would have been difficulties. Vickers was developing the Vanguard turboprop as a private venture and would eventually lose £16m on the project; an unhealthy underlying position if it is assumed both aircraft programmes were running concurrently.

Elsewhere, sales of the short- to medium-range Trident would be lost because of the inability of Hawker Siddeley to fund re-engineering of the aircraft beyond its original, tailored specification. Even the mighty Convair in the USA would struggle to crack the market just a few years after the service introduction of the DC-8 and 707, using more modern transonic technology and—on paper at least, offering improved performance. It is hard to conceive of a positive outcome for the VC7 against the already entrenched incumbents; its uncompetitive aerodynamic configuration played its part in this.

MEETING THE TECHNOLOGICAL CHALLENGE – VC10 AND TRIDENT

In the period following the V-1000's cancellation in late 1955, BOAC's initial belief in the combination of the turboprop Britannia and turbojet Comet 3 as competitive equipment for North Atlantic services began to erode.

A year later, it received permission from the Minister of Transport and Civil Aviation, Harold Watkinson, for the purchase of fifteen 707 Intercontinentals with which to plug the gap. This was on the condition that British equipment was bought for the remainder of the network and led to the development of a new Vickers proposal that became the VC10. Watkinson would state in parliament,

> "When the big Britannias start meeting the competition of the Boeing 707s nobody knows which will win … It was right for the corporation to back itself both ways."

Who indeed could say? However, having seemingly not viewed the 12hr Atlantic crossing in the Britannia as a defining factor in the discussion, within minutes he would continue on to say,

> "Beyond that, we must look, for the last big conventional jet aircraft, the VC10 … BOAC regard this aircraft as a world beater. It is faster than the Boeing 707 and has completely different characteristics … It is able to fly the North Atlantic and the Eastern routes, and fly into airports that the Boeing 707 could not use at present."[228]

The Eastern routes, extending through the Middle East, Africa and via the Far-East to Australia, were less developed than the North Atlantic routes in the late 1950s. The shorter runways implied an aircraft with relatively better field performance than the American jets. By the time the aerodynamic concept of the VC10 was formulated, the peaky distribution was available to contribute, with an improved freedom of geometry in terms of leading-edge radius, as Bore's work demonstrated. Describing the design in 1962, Vickers showed data that compared a representative actual VC10 wing section with a rooftop design that had been considered as a competitor.

The pressure distribution of the actual wing at M = 0.74 for example, showed an almost constant supersonic Mach number normal to the local geometric sweep, over the forward part to 25% chord. Because on a tapered planform the local normal tends towards the longitudinal direction with increasing distance along the chord (i.e. the geometric sweep reduces, as seen in extreme form on the Vulcan's delta wing), this represented a gradual reduction in actual streamwise velocity, achieved by isentropic recompression.

The terminating shock resulted in a large step between 30% and 40% chord, but the gradual compression ahead reduced its strength. Compared to the VC7, the achievement of this controlled pressure distribution implied a significantly more advanced understanding of both the recompression effect itself and the three-dimensional tailoring required in practice. Indeed, this certainly was the case and once again would involve the insight of the key RAE personnel associated with the V-bombers.

Johanna Weber is credited with combining the existing linearised theories of wing design to enable thickness, camber, twist and sweep to be analysed together. With her long-term collaborator Küchemann, she had been deeply involved in creating the physical understanding of wing-body and wing tip effects for swept wings. According to Green, Vickers saw Weber's method and Pearcey's peaky pressure distribution and "seized on these two advances," resulting in the VC10's wing being more advanced aerodynamically than any previous civil aircraft.[229]

The aircraft that would result from this process was

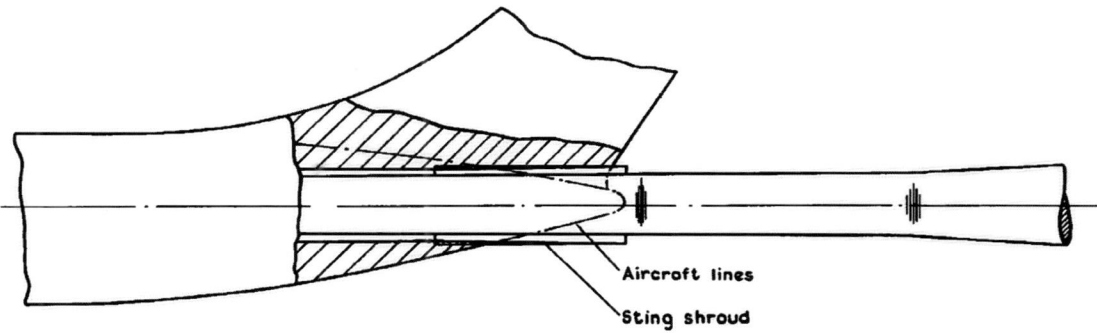

High Speed wind tunnel testing of the Super VC10, with a view of the sting mounting and hence compromised rear fuselage. It was usual to mount the model on a twin sting for further tests, to check the interference cost.

more conceptually like the USA-built competition in terms of cruise design point, typically M = 0.8 and CL = 0.45, but in service often M = 0.866.[230] The wing was swept outboard to 32.5° at the quarter chord (cf. 35° for the 707), of typical outboard thickness of 9.5% and of AR = 7.6, all of which promoted more efficient cruise at a higher Mach number than would have been possible with the V-1000. The contribution to this of the peaky pressure distribution is obvious from examination of the available date; the point of drag rise was delayed by M ~ 0.03.

In 1971, a comprehensive study was undertaken by the Aircraft Research Association to compare transonic wind tunnel data for the VC10 with that from flight test, which had been undertaken by BAC (of which Vickers had become a constituent part) in 1965.[231] The data showed the extent to which the desired two-dimensional 'sheared wing' flow (near constant isobar spacing along the span, when viewed along lines normal to the leading edge) had been achieved, together with the substantial region of supersonic flow, undergoing steady compression ahead of the shock. Flight data representative cruise conditions showed that the isobars in the mid to outboard region were predominantly aligned well with the geometric sweep. The shock was evident as a closely spaced set of isobars, passing through x/c = 0.3 and qualitatively similar to the two-dimensional case. The leading edge suction showed rapid expansion to Cp < -1.0, which had compressed to Cp ~ -0.8 ahead of the shock.

Clearly, the combination of Pearcey's aerofoil function and the Küchemann-Weber planform design insight was effectively implemented on the VC10. Undoubtably, the confidence to develop around the concept and believe that the numbers calculated would be realised, was rubber-stamped by the inadvertent technology demonstrator function of the V-force.

In 1959, de Havilland was developing the Trident airliner for BEA, closely tailored to that airline's requirements of a 1,000nm range and 600mph maximum speed. The resulting 585mph economical cruising speed was equivalent to M = 0.89 in the stratosphere, greater than any other jet at the time.[232] Even with 35° sweep and as seen with the VC10 experience, this specification was a challenge for the contemporary aerofoil and wing design capability. Some years later, a member of the Hatfield team would describe,

"The additional complications associated with flexible swept wings play a large part... since it was due to aeroelastic effects that the derivation of the design loads for an aircraft like the Trident involved so much work.

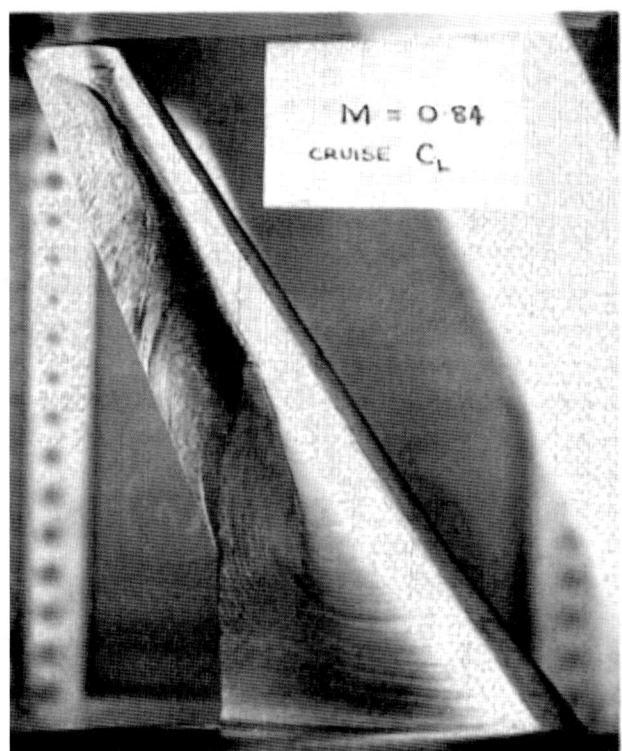 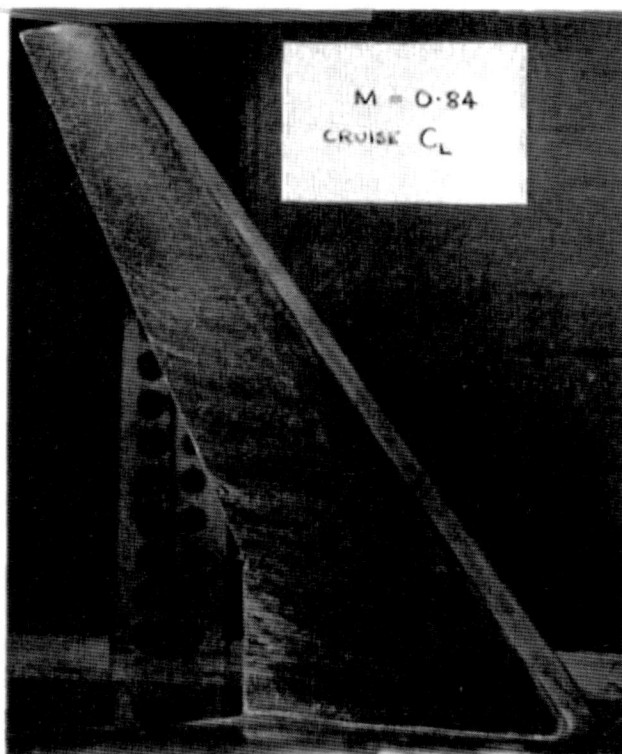

VC10 aerofoil implementation. CD v M with α = 2° for proposed rooftop and actual peaky VC10 aerofoils. The rooftop option is massively separated outboard, while the peaky version remains attached.

... A point which must be emphasised at the outset is the fact that even today, when computers are indispensable to obtaining the design loads... 'engineering judgment' still plays the main part in determining... design loads."[233]

Three concepts that were known then to be critical to a successful design were identified by Dykins et al:
* Using sweep to delay the onset of compressibility effects
* Using sectional design to delay boundary layer separation at high lift
* Using sectional design to carry supersonic flow but avoiding strong shock waves

The latter was of course the rationale behind the 'peaky' aerofoil. De Havilland studied a matrix of three pressure distribution families: triangular (as NACA 0010, the Vulcan baseline); rooftop (as RAE 101, subsequent Vulcan outboard); and rooftop with a supersonic leading edge peak (as 'Newby' modified Vulcan), each with constant spanwise section and an alternative inboard region that promoted isobar sweep. As one might suspect, the combination of 'peaky' section and tailored spanwise shape offered the best performance, showing, "very definite benefits in high Mach number drag characteristics".[234]

As discussed with reference to the development of the Victor, it is the compressed timescales by which new technology was adopted to achieve performance benefit that remains impressive. The Fluid Dynamics Group was inaugurated at de Havilland's Hatfield base specifically for the development of the Trident wing in 1956.[235] The company was confident enough in the basis of its performance estimates (and here it is implicit that the use of advanced aerofoils was inherent), that by September 1957 BEA was requesting permission from HM Treasury to purchase the DH.121 to meet its requirement.[236]

De Havilland had been very recently bailed out by the UK government after the disaster of the early Comet programme, had lost the future long range jet market for BOAC to Vickers and was now essentially beholden to BEA and its changing views on what the new jet should be. Indeed, it may have been DH's willingness to sacrifice the wider commercial prospects of the aircraft and tailor entirely to BEA's very specific (and as it would transpire, misguided) specification that positioned them as the preferred bidder. Under these circumstances, is it plausible that they would have based the fundamental aerodynamic design of the aircraft, around an aerofoil technology unproven by flight test in the relevant Re-M space? Again, the words of Sir George Edwards, bemoaning the lack of model flight tests are appropriate. The

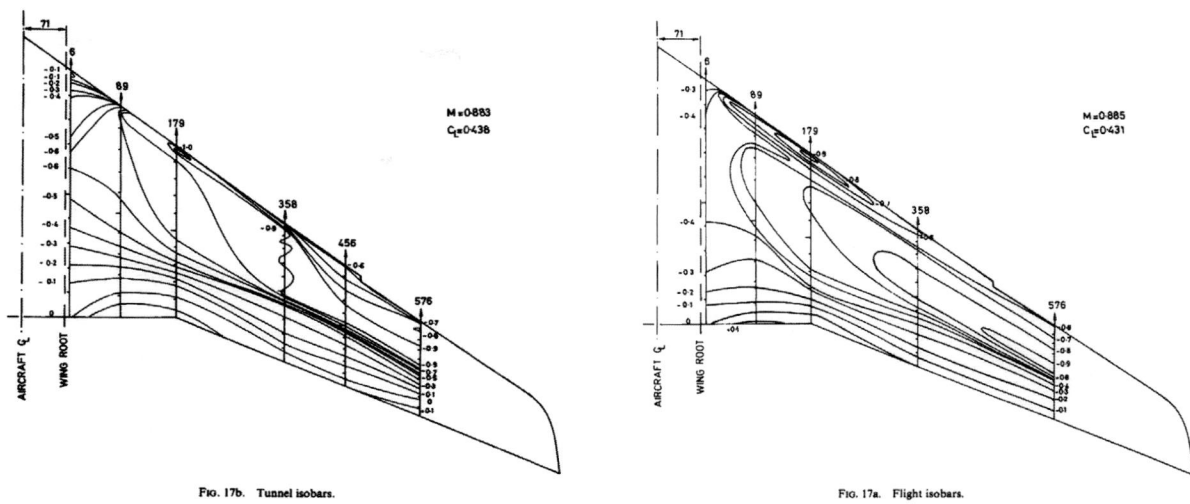

Fig. 17b. Tunnel isobars.   Fig. 17a. Flight isobars.

Cp distribution measured in wind tunnel and flight test on a Super VC10 wing, at M = 0.88.

successful Vulcan implementation must have been, at the very least, of some comfort.

By the end of 1962, two prototype jet airliners that made extensive use of Pearcey's two-dimensional aerofoil design principles to confer improved transonic performance, had made their first flights in the UK. As shown in the contemporary sources, both were biased towards high-speed cruise performance due to the requirements of their respective prime customers, in turn making this technology essential for competitive efficiency. In a 1968 review of the state of the art in transport aircraft wing aerodynamics, Wallace (Boeing Commercial Airplanes) noted the use of peaky aerofoils on the newest US transport aircraft, the Douglas DC-9 and Boeing 747, writing,

"…Pearcey and his co-workers at NPL in England were energetically pursing the optimisation of airfoils with the best possible transonic flow properties that could be developed experimentally. Also, they were working diligently to evolved semi-empirical airfoil design criteria and to determine the basic trades between airfoil nose thickness and camber.

… For the British designs of the BAC 111, the Trident and the VC-10, perhaps much of the consistent improvement must be attributed to Pearcey for his airfoil work and to the Küchemann-Weber development of three-dimensional wing-body theory."[237]

It was therefore clear to (and acknowledged by) the competition, that the UK technology effort with the peaky transonic aerofoil as its two-dimensional basis represented the state of the art for civil transports in this period. But, with the project well under way in 1962, what was the connection between the VC10 and Skybolt?

The UK government had underwritten the development costs of the VC10, on the basis that the technology-driven design would prove attractive on a world market. Indeed, the challenge in terms of a profitable business case did not seem overly ambitious: Vickers anticipated the break-even point being at about 80 aircraft, of which approaching half were accounted for in the requirements of BOAC. The aircraft had a lot of thrust and a wing stressed (and configured) to carry a great deal of weight. The mistake was in using this capability to provide short take-off performance, in order to operate from runways that would be extended anyway to accommodate the ever-increasing 707 and DC-8 fleets that the world's airlines were actually buying. Belatedly, the centre of the programme moved; accepting that Heathrow and the destination airports that would make money actually had long runways and planning to make use of them, the fuselage length and capacity were increased in the Super VC10 version. This variant made full use of more powerful Conway engines, structural reinforcement and aerodynamic advances that a brief increase in development time allowed. As part of the reconfiguring of BOAC's original massive order, the RAF was allocated 14 hybrid versions for transport, essentially standard length Super VC10s.

It did not take a huge leap of imagination to consider further development of the aircraft as a Skybolt carrier, now that the performance requirements of such a military machine had reduced and so crossed over with those of the civil machine. The limited changes required would likely compare favourably with the additional work proposed for a very limited Vulcan Phase 6 production run, while being able to take advantage into the future of a large installed

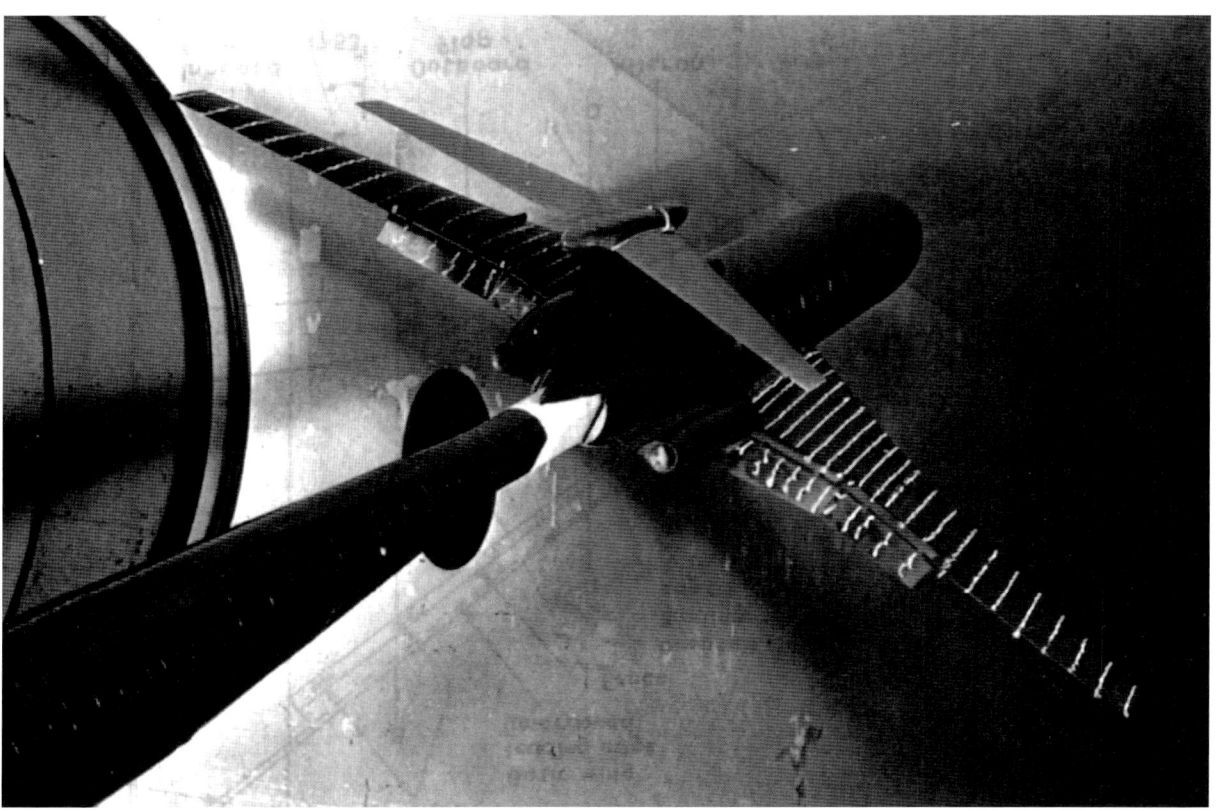

Like the VC10, the near-contemporary Trident made good use of the technology developed for the V-bomber programmes, including three-dimensional aerodynamic design techniques. This image shows the Trident 1C undergoing testing at the RAE's 8ft x 8ft tunnel at Bedford.

base fleet of similar aircraft in the civil world. The Vulcan and Victor had sowed the seeds of their own demise, but of course, that's not how it worked out.

EXPLOITING THE LEAD IN EUROPE

The prospect of European cooperation in the design of civil transport aircraft was perceived as an important strategy in several quarters by the middle of the 1960s. One of the findings of the 1965 Plowden Report on the aerospace industry immediately emphasised by the UK government was the need to promote European collaborative projects; this also sat well with the concurrent policy that pursued membership of the EEC.[238] Both BAC and Hawker Siddeley were keen to develop future subsonic aircraft, initially as outgrowths of the existing VC10 and Trident designs, while in France Sud Aviation (BAC's partner on Concorde) was seeking government launch aid for its proposed Caravelle replacement. A memorandum of understanding (MoU) that would establish a committee to investigate further the market and technical requirements for such a collaborative aircraft was signed by the UK and French governments in May 1965, while Germany would join the effort in February 1966.[239]

Essentially on an independent basis, aircraft companies opened negotiations across borders to form groups that would submit proposals to this committee. In September 1967, the three governments signed a further MoU for a 12-month project definition phase, still without launching the then-proposed 267 passenger A300 but committing funding to development. In contrast to the final situation however, there was a strong emphasis on the UK involvement in the bespoke engine programme, with a 75% workshare being allocated to the UK (with Rolls-Royce as prime contractor) for development of the RB.207, compared to 37.5% for the airframe under sub-contract to Sud Aviation

BAC would eventually withdraw from negotiations with Dassault and Sud Aviation, pursuing an alternative strategy involving UK-only projects. Hawker Siddeley responded to the collaborative challenge by partnering initially with Breguet and Nord, resulting in a proposal aircraft called HBN-100.[240] However, the French government preferred Sud Aviation to act as its project lead, resulting in HS forming an alternative alliance, focused on its acknowledged expertise in wing design. By June 1968, the result of the project definition phase was the baseline A300 design, now with over 300 seats and with thrust requirements beginning to present a challenge for

Another ex-airline military VC10 was XX914, used for research by RAE Bedford after a career with British United Airways. This view shows the tailoring of the inboard wing sections and the large fence and the absence of dihedral on the tailplane. STEVE FITZGERALD

even the 'paper' RB.207 design, by some margin the highest thrust turbofan available in Europe.

The relevance of this early history is in what happened immediately afterwards. Rolls-Royce was already committed to the RB.211 for the Lockheed TriStar and could not allocate the necessary resources to two programmes, leading to abandonment of the RB.207. Without an engine in this thrust class, the A300 as envisaged was technically dead. Even had that not been the case, an MoU requirement for the end of the project definition phase was 75 firm orders from the respective national airlines. With the concept of government-owned flag carriers still highly relevant to the future sales prospects of a civil aircraft at the time, the ambivalence of BEA and Lufthansa regarding the proposal added further difficulty to the situation. Fundamentally, the programme was offering the airlines an aircraft that was too big for their requirements and which was not technically feasible in the absence of the RB.207 engine.

Post-Trident 1, the Hawker Siddeley team at Hatfield (formerly de Havilland) had contributed to improved developments of that design (long range Trident 2 and high capacity Trident 3), the HS.125 executive jet and the HS.681 military transport. Progressively, the aerofoils used on these aircraft incorporated nose droop and rear loading, innovations that improved the low- and high-speed performance respectively. These concepts were also being pursued at NPL and the RAE, in support of the various military and civil programmes for which transonic aerofoil design remained a critical technology.[241]

NPL assisted HS in the wind tunnel-led development of controlled increases in rear loading by extending the region of 'sonic rooftop' on the upper surface, thereby increasing the pressure delta between this and the positively loaded lower surface. By 1965, NPL had tested an aerofoil that incorporated the combined aerodynamic concepts of peaky leading edge, sonic rooftop and rear loading, while a year later HS would commence the design along these lines for the HBN-100; ordinates being passed on to the RAE in July 1966. By the time of the September 1967 MoU therefore, Hawker Siddeley could demonstrate an appreciation of the state of the art in transonic transport wing design, coupled with access to the complementary UK government research agency knowledge and (also via ARA) transonic wind tunnel capacity, that was likely challenged in Europe only by BAC.

The aerodynamic development of what would become the wing for the A300B was discussed in a contemporary account by D. M. McRae, which showed the inputs to the aerodynamic philosophy. We shall consider only the

A unique VC10: ex-airliner ZA141 was modified to K.2 tanker standard for the Royal Air Force in the early 1980s, and wore the camouflage scheme then used by the Victor K.2 force. It is seen here at Farnborough in 1982, perhaps a hint of what a Skybolt-carrying VC10 might have looked like. DICK GILBERT

third column, which traces a direct line from Pearcey's NPL work on the peaky aerofoil (of which he has stated the Vulcan application was the validation), through the Trident to the A300B concept. As with the VC10 and Trident applications, the supersonic expansion-isentropic compression phenomenon was the vital physical mechanism that reduced the strength of the inevitable upper surface shock. The adaption of additional mechanisms (most importantly the rear loading) improved cruise efficiency considerably compared to the earlier aircraft, but also opened up the design space at aircraft level.

The A300 and A300B aircraft were not designed to operate at such high cruise Mach numbers as the VC10 and Trident, so that the improved transonic aerodynamics could be 'cashed in' as reduced wing sweep and increased thickness, both of which reduced wing structural weight. This was illustrated by Dykins et al, showing that the step from Comet to the initial peaky section and three-dimensional design geometries of the VC10 and Trident, then the subsequent step to the A300B, were each worth of the order of 2% in equivalent thickness to chord ratio.

The decision by Rolls-Royce to abandon the RB.207 created an engineering necessity to reassess the size of the new aircraft, in the direction that the airlines wanted. Now a 250-seat aircraft designated A300B, it could use the same powerplants as already planned for the 747 and DC-10. To the UK government and BEA, there were complicating factors. The 75% engine workshare had evaporated, with the return on investment now hinging on involvement in the airframe. The nationalised airline clearly preferred the UK-only BAC 3-11, powered by the RB.211, the latter already subject to launch aid but with the risk spread by its exclusivity on the TriStar.

Consequently, it is easy to understand the scepticism with which the UK government viewed the European Airbus programme as 1968 drew to a close. It had already heavily committed to one airliner programme (Concorde) that was struggling for firm orders and was now in danger of adding a second to its portfolio. The choice was between supporting a pan-European design with significant UK workshare (albeit without engine exclusivity), a completely UK design that was approximately one year behind and would need to compete with the Airbus, or none of the above and require industry to support itself commercially. As is now well known, the latter was the chosen course of action, although it was well into 1970 before the final decision on launch aid for the BAC 3-11 was announced.

The A300B programme still needed expertise in advanced transonic wing design and Hawker Siddeley remained keen to provide it. In a remarkable turn of events, HS invested £35m of its own money and obtained launch aid not from its own government, but from that of the Federal Republic of Germany, taking a strategic long-view on how to rebuild its aerospace industry. While clearly not the only factor, the stature of HS in this field was such that, on behalf of its taxpayers, a foreign government underwrote the original UK involvement what would shortly become Airbus Industrie.

As it came to pass and with the disappearance of the combined airframe-engine programme, the position in which the UK finds itself today, as a major constituent and beneficiary of involvement in Airbus, grew from the proven transonic wing design expertise of Hawker Siddeley. While the aerodynamic prowess so deeply intertwined with the UK Independent Deterrent policy played its part in de Gaulle's post-Polaris veto of the UK's EEC membership, that same technology delivered the country's participation in a vital, pan-European project.

THE CONNECTION TO SUPERCRITICAL AEROFOILS

It could perhaps be considered a missed opportunity that the 'Peaky' tag was used by Pearcey and the wider UK aerodynamic community, rather than following Bore's advice and hence 'Supercritical'. The latter has come to be associated with the parallel effort in the USA by Richard T. Whitcomb at NASA, to exploit the wave drag benefit of near-shockless aerofoil concepts, for which the review by Harris[242] provides a comprehensive summary. Harris noted the large leading-edge radius and flat suction surface of the Whitcomb aerofoil, designed to create supersonic expansion and limited recompression, commenting further:

> "... these two concepts are consistent with the work done by Pearcey, when he demonstrated that the essential geometric feature of sections designed to exploit the isentropic compression due to waves reflected from the sonic line is an abrupt change on the upper surface from the relatively high curvature of the leading edge to a relatively low curvature downstream and that this can be provided with a large leading-edge radius."

Whitcomb himself discussed the early NASA work on supercritical aerofoils in a 1974 ICAS paper, again linking the pioneering work of Pearcey and colleagues to the new concepts:

> "The first airfoils designed specifically to delay drag rise by improving the supercritical flow above the upper surfaces were the 'peaky' type airfoils of Pearcey. They provide an isentropic recompression of the supersonic flow ahead of the shock wave located on the forward region of the airfoil... These airfoils or their derivatives were used on the second generation of the subsonic jet aircraft designed in the USA."[243]

While Harris mentioned Whitcomb's, "intuitive reasoning and substantiating experimentation", the published material from the NPL team on the subject meant that the US effort did not commence from a cold start. The first Supercritical aerofoil incorporated a slot at around 75% chord, with the specific aim of suppressing boundary layer separation following the rear shock. After this investigation in 1964, a non-slotted version was developed incorporating the features now recognised; data on this geometry was circulated internally at NASA Langley in 1967, with the general concept classified. However, as alluded to earlier, this effort was paralleled independently by the work at NPL and subsequently RAE, with a recognisable UK Supercritical aerofoil being wind tunnel tested in 1965.

Over the next decade, both organisations would continue to refine and selectively disseminate information on their work. A major topic that can be extracted is the need to operate effectively off-design, with the recognition that near shockless flow might be achieved within a very narrow operating window, but that this was not practical for a real aircraft and that the aim had to be effectiveness and efficiency in the presence of weak rear shock.

The ultimate outcome was an understanding of the flow physics of the Supercritical aerofoil, that meant it was available to be incorporated in the design (using also the huge advances in computing in the intervening period to create the specific three-dimensional geometry) of third-generation subsonic transports such as the Airbus A310 and Boeing 757. The major review works referenced above notably all discuss the variation in leading edge radii to ensure that the supersonic expansion-isentropic compression phenomena occurred, as described by Pearcey widely at the start of the 1960s and in his communication with Robert Pleming.

15

UNDER THE RADAR

THE SECOND World War had brought into focus the importance of effective reconnaissance aircraft, which by the nature of their role had no choice but to overfly dangerous places. Taking this seriously meant altitude performance and work done by early Spitfire conversions had convinced the Royal Air Force to invest. In some ways the development work that had culminated in 1945 with the Spitfire PR.XIX and Mosquito PR.XVI foreshadowed the advances of the V-Force.

Able to cruise above 40,000ft, with crews accommodated in the relative comfort of pressure cabins, they were certainly not invincible but they had every chance. This philosophy continued beyond the war; what was true for the bomber in turn mapped to a surveillance machine. The Mosquito's de facto replacement, the English Electric Canberra, was similarly procured with reconnaissance equivalents: the PR.3 and PR.7 corresponding to the major bomber types, the B.2 and B.6.

Since the Canberra had been licenced to Martin in the United States for the B-57 contract, the sequence was followed there too and a profusion of RB-57 aircraft were procured or created through modification. The RAF's definitive type took the vastly improved Avon 200 series engines, effectively non-reheated versions of those used in the Lightning, and combined them with a longer chord inboard wing and extended tips. The resulting PR.9 also incorporated a new nose with an offset fighter-type canopy based on the B(I).8 low-level bomber version, and proved such a successful asset that a dwindling number would survive to support the UK armed forces in Afghanistan as late as 2006.

The RB-57 was also extensively modified from its baseline state, with the RB-57D featuring greatly extended wings of exceptionally light construction and incorporating integral fuel tanks. The engines were changed to the ubiquitous P&W J57, with the package aimed at flight up to 65,000ft. Entering service at the end of 1956, the new machines could (and did) overfly the Soviet Union from forward bases, confident in their immunity from interception, although of course the overflights were not authorised. The RB-57D predated and anticipated the career of the Lockheed U-2, as it turned out, in more ways than one.

Beginning in 1958, the capabilities of the RB-57D were utilised by the Central Intelligence Agency in support of another thorn in the side of global communism: Taiwan. A number of Taiwanese pilots were taught to fly the B-57, prior to a small number of the early, single-seat photo-reconnaissance RB-57Ds being transferred. Wearing the markings of the Republic of China, they began to overfly their vast communist neighbour in 1959, identifying military installations and capabilities. On October 7, Captain Ying-Chin Wang was piloting just such a mission, which was identified and engaged by the PLA's 2nd Independent AAA Battalion, based at Tongzhou, in the south-east of Beijing.

The RAF's high altitude reconnaissance version of the Canberra, the PR.9.

The unit was, unfortunately for the RoCAF pilot, one of a small number equipped with a relatively new weapon—the Soviet supplied, solid-rocket booster-propelled SA-2 missile system. It would be a disservice to refer only to the rocket itself however, as the complete battery included the advanced radar systems necessary for guidance. It was an integrated air defence system that had the potential to be effective at even the extreme altitudes where the RB-57D would fly, although that was all but on the limit. Its presence was also unknown to the Taiwanese.

At 12:04 hours local time, the first of a salvo of three missiles was fired. The burn of the first stage was short, under ten seconds, before the liquid-propellant second stage fired and accelerated the weapon to about Mach 3. It was about 40 seconds post-launch when the missile hit the RB-57D; its fragmentation warhead peppering the aircraft with shrapnel. At that altitude, it was absolutely essential that the pilot be in a pressurised atmosphere of some kind, whether that be a pressure cabin or pressure suit. Ying-Chin Wang did not survive what was, historically, the very first successful downing of an aircraft by a surface-to-air missile. It had happened at a higher altitude than any of the PLA's fighter aircraft could reach, and in fact they had been directed to clear the area. A new era had dawned, perhaps a little more rapidly than had been anticipated by advocates of the high-altitude bomber.

The RB-57D flew faster than—and nearly as high as—the aircraft that had been directly procured by the CIA for a similar purpose, in this pre-satellite era. The Lockheed U-2 is much better known today, for reasons that are simple to understand. It forms a point on an arc of aircraft development that is undeniably romantic; the elite team of the Skunk Works under the talismanic Kelly Johnson, putting their radical ideas into sculpted titanium reality. It featured prominently on the day that the Doomsday Clock stuttered towards midnight, during the Cuban Missile Crisis of October 1962. But, more than any other event, it is associated with the mishandled embarrassment of May 1960, when the CIA specified one in a series of clandestine overflights of the Soviet Union, originating in Pakistan.

Francis Gary Powers' U-2 mission is often cited as the end of the high-altitude idea, with profound consequences stemming from the shooting down of his jet by an SA-2 battery, at an altitude that was in fact only a few thousand feet higher than the earlier RB-57D incident. Doubtless, the fact that Powers survived and found himself at the centre of a newsworthy and politically damaging espionage scandal is part of the key to the event's legend. What is absolutely clear though, is that it did not cause an immediate change in the Western bomber strategy on the following day. The task of SAC and Bomber Command was more difficult but not impossible, while the procurement of still faster and higher-flying jets was pursued.

The RAF's first Mark 2 V-bomber of any kind to enter service, Vulcan XH558, had not even flown when Powers escaped the U-2's shattered fuselage. The USSR was vast, and the cost and time required to establish a defensive ring of SAM systems around a definable target like a major city equally so. It had taken double-digit launches of missiles to successfully intercept the U-2 and even much later, in Vietnam during the 1970s, only 10% of SA-2 missiles fired at B-52 bombers actually resulted in a shoot-down.

It was reasonable to suppose that a well-prepared, well-trained V-bomber fleet, specifically warned to the SA-2 threat and using tuned electronic countermeasures, might continue to successfully penetrate Soviet airspace if the ultimate drastic step was needed. As has been described, in May 1960 the future of the airborne nuclear deterrent was expected to sidestep this problem by the use of 1,000-mile range stand-off missiles, themselves difficult or perhaps impossible to intercept. In the event that that target did need to be overflown—a design case for a Skybolt that failed to fire—then the surest way to avoid detection from ground-based radars was well known; hide from them by using the unevenness of the Earth's surface or more readily the very curvature itself.

The move to low-level operations for the V-Force was another implied consequence of Nassau. There was now a clear plan for the future—Polaris would enter service with the Royal Navy when it became available as a British weapon system towards the end of the decade. Until such time as it could assume the deterrent responsibility, then the V-Force would continue. Any credibility gap would need to be bridged, and that would in turn need to be done using the nuclear weapons that were already available, with one notable exception. The free-fall bombs and Blue Steel, which was on the cusp of service entry, would now need to be launched in the face of the Warsaw Pact's defences in the period beyond 1965, a time when it had been assumed for many years that the V-Force would be obsolete, replaced by Blue Streak in the first instance.

The first step after the end of Skybolt, in January 1963, was an assessment of just what could be done to maintain credibility and effectiveness. It was concluded that low-level penetration of Soviet airspace would be the answer, with a series of consequences that flowed from this inevitable decision. Money would have to be spent on adapting the bombers, including strengthening of the Victor B.Mk.2, adapting Blue Steel for low level launch, and a variety of systems upgrades including new navigation systems and terrain-following radar (TFR).

The V-bombers had been designed to achieve a maximum realistic speed at high altitude, a condition where density and the speed of sound were low. It followed that the dynamic pressure seen by the aircraft, and consequently the Indicated airspeed shown to the pilot, were also low compared to sea level values. A Vulcan or Victor was not therefore likely to be troubled by a limiting dynamic pressure in the high-altitude cruise; instead, the limits were to do with the changing topology and surface static pressure distribution caused by the development of shock waves. The limit was one of Mach number, depending on the particular aircraft, of a value greater than $M = 0.9$.

At low level, the opposite would be true. The bringing to rest of air that would be more than six times as dense as that at 50,000ft above the Earth would be a challenge for the static loading of the aircraft's structure. The bombers had been specified for a maximum dive speed of 415kts, which at sea level equated to a Mach number of 0.63. Ignoring any considerations of thrust and drag, the maximum allowable speed at low level would be in a range below that where compressibility would be a problem. It would be all about structure, but not in a simple manner.

Below an altitude of about 1,000ft, flight takes place in what is effectively the Earth's boundary layer. The actual thickness is affected by the terrain, in the same manner that a wing is affected by surface roughness. In a situation where a bomber was attacking an overland target at perhaps 500ft or lower, then the assumption had to be that the 'boundary layer' would be turbulent, with random variations in velocity due to gusts. These of course could occur anywhere in the atmosphere, but it was the relatively short length scales and frequency of occurrence, associated with shearing viscous flow and turbulent mixing, that would make the ride rough. At high altitude, there might be steady winds and rare fluctuations, but an aircraft sufficiently close to the Earth's surface would feel an atmosphere as the people walking on it did, subject to unsteadiness and the effects of the varying landscape.

In 1961, the RAE published design data including a chart that it suggested be considered for correcting the standard spectrum for gusts to different altitudes. The data included showed that above 40,000ft, the frequency of encountering a gust of a particular velocity was of the order of 1,000 times less than at sea level. This might cause a ride that was less than ideal for the crew, but would have far-reaching consequences when the aerodynamic responses of the aircraft were considered.

Understanding the effect of turbulence on the fatigue life of the bomber fleet was an essential concern, especially in the shadow of the Comet disasters. The comparison of conditions at low and high altitudes was highlighted by a description from a 1958 test report on the Valiant.

"Records of flight in turbulence were obtained when flying straight and level at airspeeds of 260 knots and 320 knots EAS. Measurements were taken at the two airspeeds in three flight bands: 50-100ft above ground, 2,000-7,000ft and 14,000-20,000ft above mean sea level. The third and highest height band was intended to cover turbulence above 25,000ft; but unfortunately, during the period of the flight tests, no turbulence was found above 20,000ft. Most of the turbulence in the two upper height bands was found by flying in or just below small to medium cumulus clouds. Turbulence encountered in the lowest height band was the results of several effects which included convective air currents and light to medium winds."[244]

Viewed another way, the contrast of turbulence occurrence between the high- and low-level cruise was stark. In the stratosphere, it could not be found; just above sea level it could not be avoided. The study had specifically instrumented the aircraft's nose undercarriage and mainplane rear spar. The results showed that—unsurprisingly, the main source of cyclic loading for the former was during taxying, but for the rear spar it was surely concerning that the effect of turbulence was greater than any other flight regime.

The findings of the study should not have been, and perhaps were not, a great surprise. In April 1957, the second prototype Valiant, WB215, was being used for trials with rocket-assisted take-off gear (RATOG), a system which mounted a de Havilland rocket pack under each of the pairs of engines, subsequently jettisoning them when their work was done. The rear spar structure of this machine had been upgraded to represent that of the full production standard Valiant, although it was not the same. As the test pilot brought the Valiant in a gentle bank to align with the drop area for the rocket packs, a bang strong enough to convince the crew initially that they had collided with their chase aircraft, a Meteor, was felt and heard.

Thankfully, they were able to recover back to base, where it was discovered that the rear spar had indeed failed, initiated by a fatigue crack.[245] Of course, it is simple to conflate this event with the problems that the Valiant would encounter later. At the time, it might well have seemed that a nonstandard aircraft, in a test regime at maximum all-up weight, had suffered a failure that might never be seen again in the production standard fleet. Perhaps it was related to the cyclic loading, or even acoustic signature, caused by the rocket packs? In any event, as a service aircraft the Valiant would continue. Just over two weeks later, on May 15, 49 Squadron's XD818 would make history by dropping the first UK thermonuclear weapon, as part of the Operation Grapple tests in the South Pacific.

In the early 1950s, the NACA created a new standardised description for the gust response of an aircraft. This took into account the size, wing loading and lift curve slope, giving a comparative, normalised number that aimed to characterise the response to a sinusoidal vertical gust. Remarkably, the Victor B.Mk.2 and Vulcan B.Mk.2 at the same weight had almost the same value when assessed by these criteria, which implied that their rigid body movement due to a change in vertical wind would be similar.

From the received wisdom, this was at odds with how the two types performed at low altitude. The Victor's greater lift curve slope meant a change in the angle of onset flow would cause more of a change in lift, but the Vulcan's smaller wing loading had a compensating effect. While this simple assessment might provide a first pass for design load criteria, the dynamic response of the structure was more complicated and the resulting strains very important. The advent of flexible, swept-wing aircraft would require an equivalent step change in the assessment, if they were to be assured as safe in gusty conditions. NACA explained the problem as,

"The response of swept-wing airplanes in rough air involves a number of complications not present in the case of unswept-wing airplanes. These complications are due principally to the increased importance of torsion for swept-wing airplanes. This torsion in turn results in significant effects on both the airplane aerodynamics and stability. In addition, the airplane vibratory modes may no longer be approximated by simple beam-bending theory but may require consideration of coupled bending-torsion modes. Few experimental data exist on the character and magnitude of these problems."

As speed (and so dynamic pressure) increased, a realistically constructed thin, high-aspect ratio swept wing would inevitably twist along its length. As the aerodynamic moment would be nose down, then the outboard part of the wing would reduce incidence and so unload. This static aeroelastic deformation had the positive effect of moving the aerodynamic centre inboard and so reducing the root bending moment. This could be calculated and understood relatively easily, but more concerning was how the rapid changes in twist, and hence strains causing stresses, might affect the structure. Fatigue being the response to cyclic loading, the big question was how the dynamic modes of the aircraft might be excited by the frequency content of gusty air. Although the peak loading might have been well below the maximum the structure was designed for, if it were made to oscillate in bending for example, applying loading cycles to the root structure, failure could occur at much lower loads due to the accumulated damage.

The NACA's experiments involved flying their B-47 through clear air turbulence at about 5,000ft. The response of the structure was measured using strain gauges at many positions all around the jet; the deflections measured then allowing calculation of local stresses. In order to remove the effect of static aeroelastic deflections, the wing twist due to dynamic pressure in particular, another set of manoeuvres were flown, this time steady pull ups under similar conditions. Varying dynamic pressure allowed these results to be extrapolated back to a theoretical zero-airspeed case, and so an equally theoretical rigid aircraft of the same shape. The results were certainly interesting, but complicated. Fundamentally, the static deflection alleviated some of the effect of the dynamic strains, such that it could be judged that the overall twisting of the outboard wing was a beneficial effect. The results suggested that the bending of the root was less of an issue than excitation of the dynamic modes outboard.

"...a detailed analysis of the aeroelastic effects are required for the successful prediction of gust strains. With regard to the prediction of bending strains, it should be noted that at the root stations the strain records are of an essentially low-frequency nature and thereby reflect largely rigid airplane motions and bending in the first mode, which is at approximately 1.5 cycles per second. At the outboard stations, the strain records indicate considerably more evidence of the higher frequencies, suggesting that the higher vibrational modes become more important in regard to strains at these locations in the present case."[246]

What this meant was that the flexible wing of the B-47 would oscillate at a very low frequency in bending from the root, but responded to the higher frequency excitations outboard, that a stiffer wing might have been expected to damp out. To properly sign off the structure, complicated calculations followed by expensive testing would need to take place. As the 1960s dawned, improvements in computing technology would mean that the flexible wing and engine pylon configuration would become a more plausible design concept, but in the 1950s it might need to be designed in a conservative manner in order to achieve sufficient life.

In 1980, the USAF looked back on the structural aspects of the B-47 programme, to understand what lessons could be learned. Describing the process in the early 1950s as it had entered service,

"The structural analysis of the B-47, including the standard static test and the abbreviated flight load survey, proved that the item under test would support at least 150% of its design limit load. However, it provided no assurance that the test item would survive smaller cyclic loads in the order in which actual flight imposed them. This, repeated cycles of less than maximum loads, including warping, twisting and bending motions, might do more damage than the direct application of much larger loads. Absence of precise information concerning these inflight loads, theoretical or actual, explained how an unexpected fatigue problem could suddenly threaten the life of the B-47 aircraft".[247]

The final line was a reference to what had happened to a number of B-47s, numerically by far the most important nuclear bomber in SAC, and scheduled to remain in service until 1965. From 1957, low level operations became an aspect of training, but just a few months later in March 1958, the jets began to suffer catastrophic in-flight break ups, traced to fatigue of the inboard wing and the fuselage junction. The fix was the reaming out of attachments and various stations, for the insertion of oversized bolts that resembled milk bottles, which would give the programme its name.

"The ultimate 'fix' for the B-47 wing was incorporated in Technical Order 1B-47-1019, which appeared on May 29, 1958, along with the kits required to reinforce the wing root… The work called for in these three technical orders (1019, 1020, and 1022) comprised the phase of the B-47 rescue work identified as Project Milk Bottle. This endeavour

eventually encompassed structural modification of 1,622 B-47 aircraft. The first half of May was a build-up period, the project crested in August and by January 1, 1959, all B-47s had been inspected and reworked at least once".[248]

Low level operations triggered a comprehensive test programme for the Vulcan in particular. A representative aircraft, XM596, was taken from the production line and installed as a new fatigue test specimen. Operating under a constant regime of applied loading to mimic those that would be seen in the new flight regime, it would highlight where fatigue damage would cause structural failure and allow the subsequent preventative measures and repairs to be defined. An investigation was flown using an A&AEE test Vulcan, XH539, in which it was instrumented and tested along a number of low-level routes over Northern England and the North Sea. These flights were conducted with an RAE Canberra B.6, that itself was generally involved in work measuring and establishing gust response.

"On the curve for the Vulcan the peaks occur at frequencies which are in close agreement with those of the first four symmetric natural modes established in a ground resonance test; these were 3.42 c/s (fundamental bending), 5.79 c/s (fuselage torsion-wing bending), 7.59 c/s (second wing bending), and 9.58 c/s wing torsion...

The percentage contribution of the flexibility of the structure to the mean square aircraft normal acceleration has been calculated for the Canberra up to 6 c/s and for the Vulcan up to 4 c/s... For the Canberra the modes at 2.8 c/s and 4.5 c/s make a contribution of 6.4% and for the Vulcan the mode at 2.8 c/s makes a contribution of only 0.89%. Modes at higher frequencies have not been considered because their overall contribution is small and difficult to determine... The calculations give a percentage increase in No [zero crossings per unit distance] due to these modes of 29.5% in the case of the Canberra and 8.0% in the case of the Vulcan. The higher values of No for the Canberra compared with those for the Vulcan... are consistent with the rigid body responses being such that the ratio of high frequency energy to low frequency energy is greater for the Canberra than the Vulcan."[249]

The Vulcan's fundamental first bending mode occurred at more than double the frequency of that for the B-47, while its response to higher frequency excitation was significantly damped. Notably, the Canberra had a lift curve slope greater than that for the Vulcan, although less than that for the Victor. What was important was that the Vulcan's dynamic response had a relatively tiny impact on wing strain (and hence stress), with the report noting that,

"...the calculated contributions of the flexibility of the structure to the wing strains is 20% for the Canberra and 4.5% for the Vulcan..."

As an early machine, it is unlikely that XH539 had received the wing strengthening modifications for Skybolt at the time of these tests. The effect was to further stiffen and strengthen the wing, both properties that would globally help with the problems of structural life, although they may also have introduced local problems that would need to be dealt with in time. Once again, the vagaries of chance had served the Vulcan's cause, but nonetheless it should be noted that the aircraft's structural dynamics were a feature of its inherently stiff configuration, while the delta provided the aerodynamic combination of low wing loading and lift curve slope together. The former conferred the requisite high-altitude performance, while the latter suppressed gust response at low-level, ensuring that the huge area was adaptable.

The most obvious change to the V-bombers as they moved to low-level life was their plumage: the upper surfaces were rapidly painted in the RAF's Medium Sea Grey and Dark Green disruptive camouflage scheme. Anthony Wright was a Navigator Radar on Valiants with 148 Squadron, and participated in what *Flight* magazine identified as the first public appearance of a V-bomber in such attire, when his aircraft was displayed at North Weald, Hucknall and Church Fenton in May 1964.

A month earlier, he had deployed to Canada along with a second Valiant and crew from the squadron, in order to pioneer the new low-level routes that would simulate the Russian Steppe, being flown from Goose Bay. This formed a key part of V-Force crew training for the rest of the aircraft's career. With the benefit of a long-range, high-level transit from Lincolnshire to add to the value. In some ways, the transatlantic journey of over 2,000 nautical miles was an extremely relevant preparation for the Vulcan's new Hi-Lo-Hi mission profile, as Martin Withers who flew the Vulcan as a pilot from the 1970s illustrated.

"With the 300 series engines, we would typically transit across 'The Pond' at FL410/430 above all civil traffic straight from take-off and cruise at .85 IMN at less than 80% rpm... we would always

Part of the role of 27 Squadron's Vulcan B.Mk.2(MRR) force was the monitoring of nuclear testing around the world. XH558 is seen landing at Midway Island in the mid-1970s, resplendent in glossy camouflage.

take as much fuel as possible on operational flights to enable us to reach a (very theoretical) diversion airfield post-strike. Even the Blue Steel aircraft had specially designed saddle tanks in the bomb bay around the missile".[250]

Ed Jarron concurred that the fuel load operationally would have been,

"…just as much as possible. Once we had the much smaller WE177 weapon we were also able to put 16,000lb in drum tanks in the bomb bay. [The initial cruise altitude was] usually high 40ks. We had the ability to get all the way up to 60k."[251]

Norman Bonnor was a Navigator Radar on 100 Squadron. He described having been selected in 1965 to take the formidable Victor B.Mk.2 to the SAC bombing competition in the United States, only for the participation to be cancelled suddenly by Bomber Command.

"The Victor fatigue specimen at HAS Woodford had shown some cracks that could mean structural weakness… Although the specimen was many hours and cycles ahead of even the highest hours flown by a Victor 2, caution was undoubtably the sensible option. Our training targets were dramatically changed from the original ones based primarily on hours and sorties to new definitions based strictly on training value."[252]

Concern about the condition of the Victors had the effect of reducing flying hours that were allowed, but Bonnor related that it did serve to switch the emphasis towards gaining as much value as possible from every mission, making each one stimulating. This might involve fighter affiliation and both high- and low-level routes, with the aircraft now always equipped with a Blue Steel round, allowing navigation in the same manner as if going to war. In August 1964, Valiant WP217 suffered a failure of its starboard rear spar, which precipitated a temporary grounding of the fleet after inspections showed the problem to be far from limited to one airframe.

Two things were already known to the manufacturer and HQ Bomber Command, although not necessarily to the operational squadrons. The first was that the extensive use of DTD683 alloy was always likely to have been life limiting; it had done a great job as a material with desirable properties in the short-lived Lancasters and Halifaxes of

20 years earlier, less so in the case of the V-bombers. The second was that fatigue at low level would hasten the first, rapidly using up the aircraft's life.

DTD683 could cause failure by the mechanism of stress corrosion cracking, and given that the material was found deep in the structure of the bombers, inspection was difficult. The solution to this problem was one of the responsibilities of the RAF's Central Servicing Development Establishment (CSDE), and from the 1950s onwards it had progressively introduced non-destructive testing radiography methodologies based on radio-isotopes and X-rays. These were problematic in the sense that access to both sides of the part to be inspected was required, and so the introduction of ultrasonic testing in the early 1960s was a welcome (and necessary) innovation.

In common with standard practice at the time, the V-bombers were designed on safe-life principles, meaning that their structures were expected to fail at some point in the future, but that a finite life in terms of flight cycles and hours would be guaranteed by calculation, testing and the inspection regime. The CSDE began working on structural problems with the Vulcan in the early 1960s, generally using X-rays to examine the intakes, fin and rudder. In their view,

"The area of greatest concern was the risk of a failure of one or more of the 52 'finger brackets' attaching the rear pressure bulkhead to the surrounding structure".[253]

The CSDE ran courses attended by those in industry, and in 1964 a Vickers employee realised that they may be able to help analyse a failure on a Viscount spar, where a crack had been discovered around a bolt hole. Michael Murden, then OC Development Flight CSDE, recalled that,

"Both Viscount and Valiant used DTD683 and the design of both spars was similar so it was hardly surprising that there was much anxiety at the manufacturers. We had already begun our investigation when the first Valiant spar failure occurred at Gaydon on August 6, 1964".

Murden travelled to Marham to look at the Valiants there and try to understand how the very difficult inspection of the in-situ spars might take place. Ultrasonics were the only viable answer, but even then a custom holder for the transducer would need to be manufactured and positioned.

"At that time all RAF probes were being manufactured by a very small company in Brixton so, accompanied by the chief technician in charge of the project, I went to see them, taking the section of Viscount spar. I asked the designers whether they could make a suitable probe for our use explaining the need for a smaller diameter, ideally about half the one-inch size then currently in RAF use. I could not mention our worries about the Valiant for security reasons, so I told them that the RAF was considering buying a Viscount for Boscombe Down. A few days later, the chief technician came to see me and explained that he had trapped his middle right finger in a door and would be unable to hold a probe until it had healed. The MO had given him a finger stall to protect the injury. It consisted of a leather middle finger taken from a glove and held in place by elastic around the wrist. As we joked about his injury we suddenly realised that if the ultrasonic probe could be attached to the finger stall it might make the difficult inspection possible".

The special probe did indeed enable an inspection of the fleet, rapidly identifying that the problem was widespread. The 61 aircraft then in service were categorised such that some could continue to fly, while at the opposite end of the scale, others were perilously close to failure and were grounded until such time as a repair scheme could be developed.

One of the first to be signed off at Marham in later September, was the notable Grapple machine, XD818. Others were dismantled to allow deeper metallurgical assessment of the rear spars. To assist with the repair plan, XD816 was flown back to Vickers and over the coming months would be re-sparred as the prototype for the required modifications. It was not to be however, as it was decided that the cost of refitting the ageing aircraft was excessive, given the few years that they might continue afterwards, and outside of the core deterrent requirement.

The whole Valiant fleet was therefore withdrawn from service in January 1965. It had, of course, done everything that had been asked of it: entered service on time (and using an amount of American funding), demonstrated a limited capability in the Suez crisis, but more convincingly undertaken the absolutely vital Grapple trials a year later, and finally lasted as a nuclear bomber until the projected Blue Streak in-service date. The argument can be made in many directions, but consider this: would the RAF have thought the Vulcan mature enough in 1957 to have carried out the vital nuclear tests, a race against the incoming

A Victor K.1 of 214 Squadron, near Gan on July 1, 1970. MIKE GRIERSON

Three Victor K.1 tankers of 214 Squadron await a night tasking at RAF Masirah, Oman in May 1970. MIKE GRIERSON

The Vulcans of 617 Squadron on the Operational Readiness Platform (ORP) at RAF Cottesmore, in the mid-1970s. ANDY LEITCH

Victor K.2 XL512 looks as menacing as an aircraft might do, as it turns onto the Luqa runway. GODFREY MANGION/AVIATIONMT

An unidentified 55 Squadron Victor K.2 in 1979, receiving fuel in the United States.

A 57 Squadron Victor K.2 in the mid-1980s, showing its spectacular lines. DAVID HENDERSON

test-ban treaty and essential in convincing both the Soviets and Americans that the UK could do it alone?

The movement of the strategic elements of the V-force to low-level operations, concentrated on the Vulcan and Victor, is sometimes painted as an overnight event. In fact, it was more gradual than is perhaps appreciated. The Waddington wing, equipped with the B.Mk.1A, started to train in March 1963. Their aircraft were not refitted for the new role in the way that the B.Mk.2 would be as time went on. Indeed, even the provision of the newer variant's windscreen heating system, that allowed the material properties of the transparencies to be retained in lower ambient temperatures and so confer bird-strike resistance to higher speed, did not transfer over to the earlier aircraft.[254]

In the following month, the Victor B.Mk.1As of 55 and 57 Squadrons at Honington, together with 10 and XV at Cottesmore, began their gradual progress. However, the

Vulcan XL386 of 44 Squadron at RAF Goose Bay, on a Western Ranger flight. IAN PERRY

A Victor K.2 on the frozen ramp at RAF Goose Bay, Canada. DAVID FRANCIS CLARKE

two Cottesmore units would disband in 1964, partially as the new B.Mk.2 entered service, but more importantly as the Victor 1 would become a tanker. Early in 1964, the free-fall Vulcan B.Mk.2 wing at Coningsby began its own low-level training, while in the spring the Blue Steel Vulcan wing at Scampton also joined in. As with their counterpart Victors at Wittering, the work to introduce the new missile was probably a key player in the late move to low altitude, but even as they did, there was still no procedure for low level launch. That was to some extent academic, as the weapon's initial employment was in an unpowered condition, dropped in a pop-up attack like the Waddington wing's Yellow Sun 2 free-fall bombs.

Viewing the potential of the V-Force aircraft to operate as low-level bombers, perceived as essential to their credibility in the second half of the 1960s, it is clear that the coupling of their aerodynamic characteristics with those of their structures was very important. Subject to the same gusty environment, this aeroelastic response produced a quasi-static distortion of the structure, according to its stiffness. The bending and torsion of the wings in particular caused local strains (changes in dimensions) that set up stresses in physical parts.

The Vulcan's low aspect ratio and dimensionally deep inboard geometry helped to reduce the level of both bending and torsion seen globally under these conditions, compared to the Victor. On top of this structural deformation, the effect of the gusts was to apply changes in the aerodynamic loading with time; the stresses were cycled. The effect at low level (where gusts occurred much more often per unit distance, and because of this had a frequency associated with them that could lock into the aircraft's natural modes) had the potential to be severe.

Many more cycles were applied every mile compared to what happened at high level, and as was seen in the B-47 case, flexibility had the effect of amplifying the cyclic loading at high frequencies, which would have been rapidly damped out on a stiffer aircraft, or one with a lower response to changes in angle of attack, like the Vulcan. This was tempered to some extent by the positive effect of the static distortion, but overall, a flexible structure was going to vibrate a lot more. This made fatigue, the wearing out of metal parts with use at stress levels below their maximum, a real problem. It was particularly acute in the 'safe-life' V-bombers, for which physical modelling using a static test specimen was essential. It would allow the life calculations to be validated and the safety margins approved.

For the Valiant it hastened an inevitable demise, long after the jet had completed its essential purpose. For the Victor, the picture is more complicated. Long before low-level missions were considered, the B.Mk.1 fleet was already scheduled for transfer to the tanker role, which was a high-altitude mission. The B.Mk.2 force would have seen its life limited by fatigue, but might well have been easily repaired or adequately strengthened in the first place to avoid this.

For the Vulcan, the picture was not completely rosy either. The B.Mk.1 fleet was also on the verge of becoming superfluous, and less capable as a tanker compared to the Victor. It is unsurprising that it was not adapted to any great extent for low-level, but the still-in-production, Skybolt-strengthened B.Mk.2 was a different story. Its aerodynamic and aeroelastic characteristics made it the most suitable for low-level operations; if there were to be a V-Force at all post-1965, it would have to rely on the highly developed Avro delta. There were simply not enough Victor B.Mk.2s to make any other choice viable. The Vulcan had won again, but partially because a commitment had already been made: it had to be made to work, and so it was.

Vulcans of 44(R) Squadron at their base of RAF Waddington, in the early 1980s. XM603 and XM655 wear the medium sea grey and dark green upper surface camouflage, over light aircraft grey undersurfaces and with toned down insignia, that was typical from the mid 1970s onwards. XM652 however has the final wraparound scheme with dark sea grey, which was adopted based on experience at Red Flag and worn by most RAF combat jets throughout the 1980s. IAN PERRY

16

IN THE RECKONING

Procuring two expensive bombers to meet the B35/46 requirement can be viewed in many ways. The mission had to be fulfilled, and the British record of getting the answer right at the first time of asking was not entirely promising. The new layer of complexity that transonic flight added muddied the situation further—the consequences being Hunter and Swift, Vixen and Javelin, Vulcan and Victor. In 1948, the Royal Aircraft Establishment said that it could not advise as which one would work best nor whether any of them would work at all. Under those circumstances, the decision to build types based on two fundamentally different conceptual designs had a rational basis. What then can we say about the types that emerged, in comparison with each other?

Firstly, there can be no direct apples-vs-apples comparison of the actual aircraft as built, because the goalposts were ultimately moved to suit the programme and—it is alleged—the aircraft were intentionally different. The spectre of the RAE's suggestion of an 'all-out' high-risk aircraft can be applied not to the tailless aircraft that was anticipated, but to the Victor as it was implemented. Compared to the Vulcan, it consistently offered greater maximum take-off weight, altitude at constant thrust, maximum usable Mach number and longer range. These characteristics were demonstrated by the two prototypes, which validated the concepts and flew within a matter of months of each other.

In 1953 therefore, it was known to the Ministry of Supply and the Royal Air Force that both were sensible flying machines, and assuming that no dragons lurked in unidentified places, both could form the basis of a B35/46 service aircraft. What was also clear from wind tunnel tests, even if the performance envelopes of the prototypes themselves were very limited at the time, was that the longitudinal trim available on the tailless delta design would limit the maximum Mach number.

The Victor would inherently be able to achieve a greater speed without putting itself in an unrecoverable dive, due to the trim that its huge tailplane could provide. It could and did safely fly supersonically. There can be no suggestion, however, that this was done in any way efficiently. At any sensible cruise weight, at altitude and with lift equalling weight, the onset of wave drag effects at speeds much above $M = 0.9$ were the true limiting factor. In practical terms, both aircraft were well suited to the 50,000ft altitude, $M = 0.875$ design case; both had margin, considerably greater in the case of the Victor, but both suffered similarly from drag rise and its effect on range.

The difference in engine specification between the two types is also worthy of note. The Olympus used in the Vulcan heralded a new technology—the twin-shaft engine—and allowed rapid thrust growth in the initial stages of the B.Mk.1 development. The Vulcan was therefore able to take advantage of additional thrust to compensate for any performance deficit, and also to counter the

The appearance of the Victor remains otherworldly from most angles. The earliest B.Mk.2 to reach the RAF was XH669, seen here as a K.2 tanker with 57 Squadron, in 1986. Four years earlier, it had been captained by Steve Biglands as the long slot tanker on Black Buck 1.

The final flight of Vulcan XM605, landing after being gifted to the Castle AFB museum, in 1981.

Turning onto the runway at RAF Luqa in October 1976, XL512 was a relatively newly converted Victor K.Mk.2 at the time. GODFREY MANGION/AVIATIONMT

effects of increased weight as items including the ECM tailcone were added. A more rapid climb to a higher altitude had a positive effect on range performance, assuming that the cruise specific fuel consumption was similar.

The use of the Sapphire in the Victor, an engine that was adequate initially but at the wrong end of its development curve, was more problematic. The Victor B.Mk.1 always struggled for thrust on take-off, and it is even more of a tribute to the outstanding aerodynamic design that its cruise performance was so good. The inboard wing region had been expensively and time-consumingly redesigned to just provide sufficient thickness to accommodate the engines. To use new engines that were physically larger and needed a greater mass flow from the intakes would need even more. A clear advantage of the Vulcan design therefore was that, of the two, it could more simply and quickly take advantage of more powerful engines as they became available. The Conway as used on the Victor 2 had a greater roll-off of thrust with altitude than the Olympus, but significantly better cruise fuel consumption. Avro were able to solve engine integration problems with the Olympus 301 and Vulcan much more rapidly than Handley Page did on even the lower thrust Conways used initially.

Perhaps this is reminiscent of the Spitfire to some extent; on the one hand, it 'cheated' compared with the Mustang in that the later versions achieved their greater top speeds through the high power of the Griffon. On the other hand, this was a pragmatic solution to getting things done quickly. We do not know what would have happened if the high-thrust Conway 201 had been installed in the Vulcan, in terms of the work required to make it safe, nor whether the Victor and Olympus 21 would have worked well together, although on paper the altitude capability would have made this a formidable, if shorter ranged asset.

For decades, the comparison between the Vulcan and B-47 has been a classic of aircraft design textbooks. These two completely different designs, more specifically with very different aspect ratios, both appeared to be competitive in their superficially similar roles. The Victor can of course be thrown into the mix, forming an intermediate step in broad terms. The first point of confusion stems from the assumption that aspect ratio is a surrogate for efficiency; that for two well-designed aircraft of similar technology level, it would provide a direct indication of range performance. In turn, this is probably because aircraft design is often taught by using variations on the established, tube and wing, civil transport configuration. If one knows roughly what a competitive aircraft looks like and has a bank of regression data to hand from existing designs, then wing area becomes essentially a fixed parameter around which the other design variables move. The key to understanding the three bomber configurations is that wing area was not fixed, hence aspect ratio did not sufficiently characterise the performance.

Doing what Vulcans did: XH558 executing a steep climb out, during its days with the RAF's Vulcan Display Flight. DICK GILBERT

As has been described, a key technical difference between the Vulcan and Victor was in the choice of cruise lift coefficient, or to put it another way, how hard to work the wing. There was no such thing then as a transonic aerofoil, designed to exploit the new flow regime by working with the supersonic expansion. Instead, Avro concluded that a lift coefficient of about 0.2 was a sensible limit, beyond which the effects of wave drag and unpredictable behaviour would be excessive. Handley Page worked with a more demanding value of 0.3, which if achievable implied that the wing area would reduce in proportion, to just two thirds of that required by the Vulcan, for the same lift force at the same speed. Both were subject to the 100,000lb all-up weight target right at the start of development, then Avro seized upon the structural and volumetric possibilities of the delta. The approximate size of the bomb being known, along with obviously that of the crew and to an extent the engines and systems, it was immediately apparent that a wing of smaller area could not possibly accommodate them internally. There would have to be distinct fuselage and wing 'organs'.

Handley Page was robust in the technical marketing of its bomber concept, with Godfrey Lee himself presenting to the Royal Aeronautical Society, with coverage in Flight magazine, on the likely performance of the three configurations.[255] His opening assertion was that it be assumed that all three aircraft had similar spans, which would mean that their induced drags would be similar. Operating, he assumed, at a condition close to the maximum lift-to-drag ratio (which was on the one hand valid as it would give the greatest altitude, but on the other hand not the greatest range), then a basic aerodynamic model of any aircraft implies that the induced and profile drags would be of similar magnitude to each too.

The analysis assumed a wing area for the crescent layout of one and a half times that of the swept, while the delta would be two and a half times. These numbers do not appear to have been plucked out of the air; the proportion of area for the Vulcan B.1 compared to the B-47 was indeed 2.5, while that for the Victor B.1 as built would be 1.68. Lee's analysis summed the individual drags of the constituent parts of the aircraft in dimensional form, which if taken at face value showed that the profile drag, i.e. that attributable to skin friction and pressure losses, but not compressibility, was 20% higher for the delta than the other two, which returned similar values. On the assumption that the induced drags were the same, then the delta could be expected to have 10% more drag in the cruise, and hence a range and altitude penalty.

Lee pointed out that delta might be oversized, so reducing the span to 90% of that of the crescent (again, the actual proportion of Vulcan to Victor span), then the

Victor K.2 XH673 on display at Mildenhall, wearing the hemp colour scheme and the markings of 57 Squadron.

The RAF's retained display Vulcan, XH558, flying at Mildenhall Air Fete in 1988.

An unidentified Victor K.2 being serviced in a Bahrain. The hemp colour scheme was adopted by the RAF for the camouflage of large aircraft on airfields in the 1980s, but came in useful for the Gulf conflict in 1991 and subsequent no-fly-zone duties.

Victor K.2 XM717, for many years the youngest surviving Victor, immediately after the Gulf War. The 'Lucky Lou' nose art is just visible.

Vulcan XH558 of the RAF's Vulcan Display Flight, 1990. DICK GILBERT

profile drag would be approximately the same, but because of the reduction in span, induced drag would rise by 20% and the situation would be back where it had started. This fundamental analysis appeared compelling indeed, especially when it suggested that the rib weight of the deep delta wing would contribute to making it the heaviest of the three options too, although it was conceded that overall aircraft weight was unlikely to be very different, whichever layout was chosen.

Lee's analysis is open to challenge on several counts, but fundamentally Farnborough's wind tunnel tests on the developed Victor and Vulcan configurations showed that the induced drag factor for the latter was superior to the former. Firstly, the spanwise lift distribution of the Vulcan was closer to the elliptical optimum (based purely on aerodynamic considerations), which provided some compensation for the smaller span. Secondly, induced drag could indeed be expressed as being inversely proportional to the aspect ratio, but it was also directly proportional to the square of the lift coefficient. Here, the Vulcan's design criteria of a lift coefficient just two thirds of that of the Victor came into its own.

The third challenge to Lee's analysis is that the maximum range for a turbojet powered aircraft occurs not at (L/D)max, but at (M.L/D)max, according to the usual first order assumptions applied to the Breguet range equation for an aircraft with constant Thrust Specific Fuel Consumption (TSFC). In any case, V_{md} (at which (L/D)max occurs) is a condition of neutral speed stability, and so it would be expected that typical cruise velocity would be faster to ensure positive stability and reasonable pilot workload. The velocity to achieve (M.L/D)max is greater ($1.316 V_{md}$, for the parabolic model), and consequently the zero-lift drag is proportionately greater than the induced drag.

From the classical, academically taught model therefore, it would be anticipated that cruise velocity lay somewhere between V_{md}, at which $C_{d0} = C_{di}$, and $1.316 V_{md}$, at which $C_{d0} = 3 C_{di}$, with zero lift drag becoming rapidly dominant. If the 90% span version of the delta had indeed brought the profile drag in line with that of the crescent, then this would become more important with speed, bringing them closer together. Lee does not explicitly state what effect on area the span change caused, but if we assume that the area too was reduced to 90% of the previous, then that would map very closely to the actual Victor vs Vulcan proportions (anchored to those of the B-47), as built rather than from Lee's assumptions.

It is striking and perhaps humbling to realise that the design of the V-bombers took place at a time before these theoretical values had been validated in depth. Long-range, high-speed and high-altitude cruising turbojet aircraft barely existed. In mid-1952, a series of trials for just this purpose, using the first Avro Ashton (WB790) were undertaken.[256] A straight-winged research aircraft derived from the Avro Tudor airliner but powered by four Rolls-Royce

XL426 was one of the final Vulcan B.Mk.2s in service with the RAF, supplementing the six tanker aircraft with 50 Squadron, until the latter were withdrawn in March 1984. For the next two years, it served as a display aircraft and is seen here at RAF Mildenhall.

Nene centrifugal flow turbojets, it was recognised to be limited in its representation of the range characteristics of the advanced jet bombers that were planned. Nonetheless, aside from the Comet 1 that was just entering service, it was clearly one of the only platforms available to fly stabilised high-altitude sectors of 1,000 miles or more, even if its firmly subsonic aerodynamics philosophy meant that maximum specific air range was achieved well below drag divergence.

The latter point is also important when assessing Lee's thoughts. He considered only profile and induced drag, with the implicit assumption that compressibility effects were not important at the cruise (best range) Mach number. This was of the order of M = 0.84, which available analysis of Vulcan B.Mk.2 cruise performance, and the data from Victor B.Mk.2 XH669's flight tests, suggests was just into the drag rise for both types. This is unsurprising in itself, as the most efficient cruise is generally achieved by flying a little faster than drag divergence, as 'M' continues to increase even if 'L/D' is reduced by the compressibility effects, meaning there is a flat region of the curve on which higher speed has a low penalty. Wind tunnel tests for both B.Mk.1 versions of the respective types also implied that the wave drag penalty of the specified M = 0.875 'over the target' speed was reasonable.

The actual developments of both types favoured the Vulcan, as the average thickness of the wing dropped dramatically, which would have been expected to confer a very favourable efficiency step on the B.Mk.2 of itself, while the inboard region of maximum thickness changed little. This could be cashed in different ways; the sensible cruise Mach number could increase, but so could the lift coefficient before shock induced separation, implying a gain in altitude would be possible. Matters were less rosy for the Victor, which had a thicker inboard wing and relatively small extension of its already thin outboard region. This also fits with anecdotal evidence that the B.Mk.2 was less slippery than the earlier versions at altitude, and the worsened drag rise character from the wind tunnel.

Is this implication that the Vulcan's development potential was greater than that of the Victor fair? In a word, no. The story of the definitive V-bombers is one of choices being made, the making of which colours our view of them. The Victor's innovative construction was a necessary evil, given its inherent flexibility. One choice that could have been made would have been to apply this to the Vulcan, as Avro's Phase 3 development proposed. Equally, a drastic change to the Victor at approximately the same time would have seen a much greater wingspan increase than that actually adopted, comparing the proposed Phase 3 and actual Phase 2A designs. Harry Fraser-Mitchell analysed the actual payload range performance of the Vulcan and Victor, concluding that only the Victor B.Mk.2 could achieve the 3,000 nautical mile range requirement of B35/46 (which corresponded roughly to Lincolnshire-Moscow-Lincolnshire), and then only with underwing fuel tanks. This leads to another choice; what if the latter were equally possible to apply to the Vulcan, and a higher weight cleared for service? But such an arrangement was not used. We find ourselves once again in an apples vs oranges situation, with both designs having evolved from their starting points and with their differing strengths exploited. It also tells us something chilling about the post-bombing phase of the V-bomber crews' mission: they were not coming back.

One interesting counter factual would have been a situation whereby the Victor entered service when the Vulcan did in reality, and the Vulcan entered service when the Victor did. If the Victor had been established in service, would the Vulcan have been persevered with at all—what did it offer that the Victor did not, and would that have allowed a definitive choice to be made prior to committing to the B.Mk.2 versions of both? The evident superiority of the Victor prototype when compared to that of the Vulcan, judged purely against the cruise and altitude requirements, surely would have counted heavily in its favour if it had been leading the way. However, there were other criteria including manoeuvrability that were already seen as positive attributes of the Avro bomber, out of the box.

The better question really is to ask why the Victor programme proceeded at a relatively slow pace. Its design was inherently more complicated. The radical spot-welded construction has been touched upon, but the difficulties associated with flutter testing in general and the tail in particular exacerbated this. The loss of the first prototype was a direct result of the complex nature of the beast; for a critical few months there were no flying Victors in the flight test programme at all. Conversely, even at the late stage of Vulcan's high Mach number buffet problems being identified, it was still possible to simply incorporate a physically substantial fix, embodied quickly on production machines that were by then becoming available.

There may have been differences between HP and Avro's respective design, production and project management processes which contributed to the delay, but it must be the case that the Victor would have taken longer whichever firm had built it. If Avro had known they could deliver the first production Vulcan a year or so later than they did, how much more refined might it have been? Still flying in its Mach number, longitudinal stability-limited box no doubt, but perhaps lighter, stronger and longer-ranged.

The advent of Blue Steel and then Skybolt changed the rules. The former benefited from being launched as high and as fast as possible, while giving another example of a choice that was made. Both bombers could use the spare volume in their bomb bays for auxiliary fuel tanks. The Victor however was cleared to a greater maximum take-off weight, so was better able to use this capability. In a war-emergency situation, the Vulcan B.Mk.2 would have taken off at 210,000lb, some 6,000lb in excess of its basic limit. This was still insufficient however to fill all of the tanks and carry the missile. The Victor B.Mk.2 was cleared to a higher weight, 224,000lb, hence again could fly further by virtue of its additional fuel mass.

Many reasons have been postulated for the decision to not allocate Skybolt to the Victor, typically referring either to the desire to see HP as part of a larger group, or to the technical challenge of the aircraft's low ground clearance. We should be in no doubt that, had the Victor been the only bomber available, it would have been carrying Skybolt. In 1960, the B.Mk.2 was only just entering the flight test phase and the first machine had been lost in unexplained circumstances. Major problems with the aircraft's engine integration were apparent, and were not being solved quickly. Again, the Avro bomber's Mark 2 version was well ahead, even if its capability in range terms was less. It would not need to fly so far or so high, while only half the number of airframes were needed. The Vulcan did everything required, had proven itself sound and was rolling down the production line without apparent problems. The Victor's complexity had finally caught up with it, and its advantages had become less relevant.

The move to low level operations suited the Vulcan much more than the Victor, although in a reversal of the situation to that point, it was the Avro bomber that was invested in, while the Victor might have had its problems solved if the will (and money) had been there. The shadow of Skybolt and the diminished number of the second-generation Handley Page jets that had been the consequence, was the real factor behind this. The Victor was a superb technical achievement, but given the additional time and effort spent on it, so it should have been. Both of these aircraft, as hopefully is by now obvious, were attempts to engineer beyond the established state of the art. The comparison to the B-47 is of interest but also lacks an element: that aircraft's limiting Mach number was similar to the best cruise Mach of the Vulcan and Victor, and its ceiling was not far beyond 40,000ft. It carried a larger payload further, but the estimate of the Royal Aircraft Establishment was that a machine in its performance class would be overmatched by the likely missile defences it would have to face.

This book has been the story of the engineering of the Vulcan in a competitive environment, for which it was necessary to understand the drive to build the Victor also. Engineering is the application of science to solve real-world problems; in the ashes of Europe at the end of the Second World War, that problem was how to make sure nothing similar could happen again. The great war leader Churchill had spoken of how the Soviet Union respected only strength, and so it was ostentatious might that was presented to them. The commitment of the people and the organisations involved in this: the government, the national scientific establishments, the Royal Air Force and the aerospace firms, was total. They delivered, almost, what had been asked for, even though it looked to the experts that it might be impossible.

In 1942, Avro had started to deliver Lancasters to RAF Bomber Command, a military unit that was gathering strength and capability, but which still struggled to place high explosive bombs within the bounds of a city, using the unpressurised, ill-defended machines at its disposal. Twenty years later, in 1962, Bomber Command was in the process of equipping a shining white fleet of transonic, stratospheric jets with their new weapon for the next ten years and beyond: a sub-orbital spacecraft that would place a megaton-range, thermonuclear warhead within the walls of a small town if required to do so, from 1,000 miles away. Here was a demonstration of strength and resolve indeed, and when tested in the heat of Nassau, the versatile aircraft were able to adapt to another situation, another show of a commitment to stand up for peace.

Each Vulcan and Victor had a crew of five men; in the period from Valiant's introduction in 1955, to the end of the strategic bomber squadrons in December 1982, this encompassed hundreds of people. They trained knowing that no good would come of them, their aircraft or their mission, if they were called upon to go. What would be left afterwards? Would they have gone if it had been for real? Clearly, for the deterrent to work, the enemy had to be convinced that enough of them would make the distance.

BLACK BUCK

It is April 30, 1982. Beginning at 22:50 hours, the Victors begin to roll on the long runway of Wideawake Air Force Base, Ascension Island. They head south, loosely associated and with anti-collision lights on—ten of just 37 Victor Mark 2 jets that had ever been built. Now, their formidable load-carrying capability has been harnessed to tank the fuel needed to project air power forward. Rebuilt over

A Victor K.Mk.2 landing at Wideawake AFB, Ascension Island, in 1984. PAUL VINCENT

the previous decade at the home of their rival, Hawker Siddeley at Woodford, they have been transformed into Victor K.Mk.2 tankers.

Eleventh in the stream is a bomber, the instantly recognisable shape of the Vulcan B.Mk.2. Captained by Squadron Leader John Reeve, XM598 had been laid down for Skybolt the best part of two decades earlier. Tonight, she carries a full conventional load of 21 1,000lb bombs, together with all of the fuel that could be physically accommodated in the volume available. More than 4,000lb over maximum take-off weight, 598 is going to war. After her, two reserve aircraft, a final Victor and then Vulcan XM607 in hands of Flight Lieutenant Martin Withers, take their turns to blast down the runway and into the night.

Approaching 40 degrees south, XM607 slots in behind Victor XL189 and her vastly experienced tanker crew led by Bob Tuxford. Eighty feet of hose trail from the hose drum unit aft of the former bomb bay, the vast space centred on the long span of Johanna Weber's mathematical realisation of Lee and Lachmann's vision, the jet's constant critical Mach number crescent wing. The huge underwing fuel tanks mounted, now permanently, are just inboard of the homage to Küchemann, the 'window boxes' adding to the almost organic complexity of the purposeful machine.

Tuxford's men know their work is nearly done, just the refuelling of the third jet on the formation, XH669 captained by Steve Biglands to complete and then set course north for the bar. XH669 is no stranger to difficult situations, as just the second of the capable Mark 2 Victors built, in 1959 she assumed full responsibility for the flight test programme, after the tragic disappearance of XH668. She has been a test machine and a full-fat nuclear bomber, stowing Francis's mighty Blue Steel below her bomb bay. But tonight, the growing turbulence and dancing St Elmo's fire are hinting at danger to come. Fighting to make contact and then keep it as the electrical storm begins to bite, XH669's probe tip severs and not just that contact, but any possibility of receiving more fuel is gone. A wise man prioritising his own safety would surely turn for home now, but Biglands, Tuxford and their crews are made of something else. Without enough fuel in the formation to fulfil the plan, knowing that running out is certain death in the freezing waters of the Atlantic, they change places. Biglands will give the fuel back before turning for Ascension, an action itself costly in their limited energy, but too much and they will perish. Any thoughts of home must be postponed for XL189's crew; they are the long slot tanker, going all the way.

Just over an hour later, the two lonely aircraft, hours from their temporary base and half a world from home, move into position to conduct their final refuelling. Because they have to maintain radio silence, Tuxford has to rely on the insight of Dick Russell, an experienced Victor pilot attached as a sixth member of XM607's crew, to understand and articulate that they will not be able to pass enough fuel. The Vulcan will go into the attack somewhere between seven- and 8,000lb down on the plan, in terms of fuel on board. They will need to assume that the single Victor meeting them on their return, off the coast of Rio, will be able to transfer enough to get them back on track.

On board XL189, the situation is even more bleak. To make the mission work, they will be leaving the refuelling bracket more than nine tons short of the plan, and their normal reserves of safety. If nothing changes, they will run out more than an hour before reaching Wideawake. XM607's probe makes contact with XL189's basket. Kerosene flows into the Vulcan's tanks, seven per side in Lindley and Chadwick's delta, realised under the leadership of Stuart Davies. When the lights go on to indicate the end of refuelling, Russell and Withers are confused. The numbers do not add up. All these years later, it is a simple thing to say that they must have had faith that it would be alright. They might just as easily have considered the return flight an idea of purely academic interest, in the unlikely event that they survived the Mirages and radar-guided flak. They go on.

With a few hundred miles left to run, the navigation challenge over the featureless ocean is vivid; the pair of Carousel INS systems, borrowed from British Airways and its modern aircraft and strapped into the Vulcan, agree to nothing better than being within 30 miles of each other, and somewhere entirely different to where the bomber's own system and Nav Plotter Gordon Graham have calculated through their own dead reckoning and star shots. Pragmatically, the Vulcan way, they split the difference and enter the averaged position into the NBS. The jet descends to 300ft, drawing on the two decades of low-level experience that the V-force has amassed.

In the end, just as it would have been with Yellow Sun, Blue Steel or WE.177 on board, the work of the engineers and their slide rules and wind tunnels can do no more. It is down to the tip of their finely crafted spear and entirely different set of skills: lone V-bomber with the tenacity of Withers and Peter Taylor in their Martin-Baker seats, the AEO Hugh Prior, Nav Plotter Gordon Graham and Nav Radar Bob Wright on lower deck. For the last-mentioned pair, there is an obvious question. Have they managed to guide the crew to the island?

Wright fires up the NBS system's H2S Mk.9A radar, the product of Sir Bernard Lovell's team and built on the foundations of the first airborne navigation radar ever fielded, used by the Lancasters flown over Berlin nearly 40 years earlier by their boss, the CAS Air Chief Marshal Sir Michael Beetham.

At first Wright sees nothing on the analogue oscilloscope-like display. XM607 gains a little height and then echoes of Mount Usborne, a peak of more than 2,000ft on East Falkland, named for a crewman on HMS *Beagle*, appear strongly to the relieved navigator. In the dark, without sight of land and after eight hours over the ocean, they have navigated to within a mile or so of their planned position. The data can now be fed into the NBS for a theoretically automatic navigation to the target, much as might have been done for a Blue Steel launch from low level a decade and a half earlier. They are instead taking a route reminiscent of the Waddington wing's contemporaneous weapon, the free-fall Yellow Sun Mk.2.

Systems set, XM607 accelerates to 350kts and on Graham's signal, the four Olympus engines, perhaps handled with more care than Stanley Hooker had displayed on the test stand, go to maximum thrust. The jet climbs to its attack altitude of 10,000ft, selected at the last minute to give a little more clearance to the predicted engagement envelope of the anti-aircraft defences. Wright carefully positions his radar markers onto the green, ghostly impressions of three identifiable headlands close to Stanley and—with remarkable functionality for a system of such age—the NBS indicates to the pilots their track to the target.

Brief flashes of the lights of Stanley are all that the crew will see of their target. Prior activates the newly attached Dash Ten ECM pod upon recognising the signature of a gunnery control radar; the aircraft has to be straight and level to make an accurate attack. The NBS would have opened the bomb doors automatically, but the crew do this themselves to make sure. Graham, focused on the output of the NBS, counts down to bomb release and—finally—the first of the 21 drop away. All those years earlier, the engineers in Woodford and Farnborough had shown the change in longitudinal trim as the centre of gravity of the Vulcan moved through its range. Now, XM607 unloads nearly ten tons over a few seconds, and Withers has to compensate for the jet's desire to both heave and pitch, instead holding her steady. With the last one gone, XM607 does what Vulcans do best, full power once more and a 2g climbing turn, putting as much distance as possible between the crew and the defences. She is gone into the night from which she came.

If it happens at all, it will still be some hours before XM607 sees a tanker. On board Tuxford's Victor, steadily heading north, the five men go through the abandonment procedure and consider how best to stay together if forced to bale out of an empty jet. The system works though, with both of the jets meeting their respective tankers, hurrying south. Guided with the help of a Nimrod, XM607's attached refuelling expert Dick Russell is suitably delighted by the sight of the K.2 rolling out in front of them, in a textbook rendezvous. Plugged in but with fuel leaking, it could still go wrong. XM607 takes on 36,000lb, at a rate of more than two tons a minute. For the first time since they have set off down south for real, Martin Withers can relax and know that all he has to do is fly the familiar jet home.

Black Buck, flown by bombers about to be scrapped or that had been turned into tankers because they were otherwise obsolete, was at the time the longest ranged bombing raid in history. It is often misunderstood, but the nominally tactical attack on a runway was of course a strategic invitation to the enemy to sit up and take notice. A circle drawn on a map, with Ascension Island at the centre and Stanley on the edge, would contain many other potential targets. The Argentinian leadership would need to be very sure that it could accept the consequence of another raid by a Vulcan, an aircraft with the day job of strategic nuclear bomber, on one of these.

The following day, May 2, the nuclear attack submarine HMS *Conqueror* positioned itself such that a three-torpedo salvo firing solution was created on the pride of the Argentinian Navy's fleet, the ARA *Belgrano*. Within these two days, the pattern of deployment of the occupier's military assets would substantially change. The potent Mirage III force was retained to defend the mainland, while the only other large naval unit, an aircraft carrier, proceeded back to the safety of the littoral waters. Neither of these actions won a war; both of them were substantial contributions to containing the threat, and with it the potential casualties. The leadership demonstrated by Martin Withers and Bob Tuxford in the command of their aircraft and crews, saw them awarded the DFC and AFC respectively.

Be in no doubt, they would have gone.

BIBLIOGRAPHY

Armstrong, F.W. 1976. The Aero Engine and its Progress—Fifty Years After Griffith. The Aeronautical Journal 80 (792): 499-520.

Baldock, J.C.A. 1958. The Determination of the Flutter Speed of a T-Tail Unit by Calculations, Model Tests and Flight Flutter Tests (Report 221). AGARD.

Barnett, U. R., and R. H. Lange. 1950. Low-Speed Pressure-Distribution Measurements at Reynolds Number of 3.5 x 10(exp 6) on a Wing with Leading-Edge Sweepback Decreasing from 45 Degrees at the Root to 20 Degrees at the Tip (NACA-RM-L23A50a). NACA.

Becker, J. V. 1980. The High-Speed Frontier: Case Histories of Four NACA Programs, 1920-1950. NASA.

Blackman, Tony. 2012. The Avro Vulcan—Making it Work. Journal of the Royal Air Force Historical Society 53: 66-78.

Blackman, Tony, and Garry O'Keefe. 2012. Victor Boys. Grub Street.

Bore, C L. 1993. Propulsion Streamtubes in Supersonic Flow and Supercritical Intake Cowl. The Aeronautical Journal 97 (967): 257-259.

Bowes, G. M. 1974. Aircraft Lift and Drag Prediction and Measurement. In Prediction Methods for Aircraft Aerodynamic Characteristics (AGARD-LS-67). AGARD.

Brebner, G. G. 1953. The Calculation of the Loading and Pressure Distribution on Cranked Wings (ARC-R&M-2947). Aeronautical Research Council.

Bridle, E.A. 1948. Assessment of the Relative Performance of the Bypass Engine and the Orthodox Double Compound Jet Engine (ARC-R&M-2862). Aeronautical Research Council.

Brookes, Roger. 2006. The Handley Page Victor: The History & Development of a Classic Jet.

Brown, D G. 1980. Airbus Industrie—Past, Present and Future. The Aeronautical Journal 84 (839): 395-407.

Browne, G. C., T. E. B. Bateman, M. Pavitt, and A. B. Haines. 1972. A Comparison of Wing Pressure Distributions Measured in Flight and on a Windtunnel Model of the Super VC.10. Aeronautical Research Council.

Collingbourne, J. R. 1951. High Speed Wind Tunnel Tests on Half Models of Preliminary Handley Page B.35/46 (H.P80) Wing Designs; RAE TN Aero 2103. Royal Aircraft Establishment.

Collingbourne, J. R., and P. H. Cook. 1954. Appendix IV: Layout and Performance of an Intercepter Fighter Aircraft to Operate at a Mach Number of 2; RAE Aero Report 2513B.

Collingbourne, J. R., and Prior B. J. 1951. Wind Tunnel Tests at High Subsonic Speed on a Half-Model of the Prototype Handley-Page (Victor) B.35/46 Wing and Fuselage; RAE TN Aero 2103. Royal Aircraft Establishment.

Constant, H. 1958. Pyestock's Contribution to Propulsion. The Journal of the Royal Aeronautical Society 62 (568): 257-267.

Courtney, A. L., and T. V. Somerville. 1950. The Use of

High Wingloading in a Medium/Low Altitude Night and Bad-Visibility Day Fighter; RAE R. Aero 2373. Royal Aircraft Establishment.

Curran, J. K. 1973. Comparative Turbulence for a Canberra and a Vulcan Flying Together at Low Altitude; ARC-CP-1244.

Davies, S. D. 1969. The History of the Avro Vulcan. The Aeronautical Journal 74: 350-364.

Driggs, I. 1950. Aircraft Design Analysis. The Aeronautical Journal 54 (470): 65-116.

Dykins, D. H., J. A. Jupp, and D. M. McRae. 1988. Esso Energy Award Lecture, 1987. Application of Aerodynamic Research and Development to Civil Aircraft Wing Design. Proceedings of the Royal Society of London. Series A, Mathematical and Physical Sciences 416 (1850): 43-62.

Edwards, G. 1974. Looking ahead with hindsight. The Aeronautical Journal 78 (760): 134-146.

Ewans, J R. 1951. The Aerodynamics of the Delta. *Flight*, 10 August.

Eyre, K. 1952. Flight Measurements of the Pressure Distribution on a Tempest Wing up to a Mach Number of 0.8, ARC R&M 2489. Aeronautical Research Council.

Farren, W. S. 1956. The Aerodynamic Art. The 44th Wilbur Wright Memorial lecture, Royal Aeronautical Society.

Fearon, Peter. 1978. The Vicissitudes of a British Aircraft Company: Handley Page Between the Wars. Business History 20 (1): 63-86.

FH., Page. 1911. The Pressures on Plane and Curved Surfaces Moving Through the Air. Aeronautical journal 15 (58): 47-64.

Fildes, David W. 2012. The Avro Type 698 Vulcan: The Secrets Behind its Design and Development. Pen and Sword.

Floyd, J. C. 1958. The Fourteenth British Commonwealth Lecture: The Canadian Approach to All-Weather Interceptor Development. The Journal of the Royal Aeronautical Society 62 (576): 845-866.

Fozard, J W. 1969. The Harrier—An Engineering Commentary. The Aeronautical Journal 73 (705): 769-788.

Fozard, J. W. 1988. 9th Sir Sydney Camm Memorial Lecture: Jubilees in Design and Development—Some Comments on Change Over the Period of Camm's Work and Influence. The Aeronautical Journal 127-144.

Francis, R.H. 1964. The Development of Blue Steel. The Journal of the Royal Aeronautical Society 68 (641): 303-320.

Gamble, H. E. and Clarke, D. A. April 1948. High Speed Windtunnel Tests on a High Altitude Bomber, Handley Page B35/46 (H.P.80); RAE Aero R.2264."

Gates, S. B., and A. D. Young. 1962. Herbert Brian Squire, 1909-1961. Biographical Memoirs of the Fellows of the Royal Society 8.

Green, J. E. 2015. Obituary—Dr Johanna Weber. 12 January. https://www.aerosociety.com/news/obituary-dr-johanna-weber/.

Greenwood, G. H. 1964. Free-Flight Model Drag Measurements on a Transonic Fighter (Gloster Javelin); ARC-CP-678. Aeronautical Research Council.

Haines, A. B. 1987. 27th Lanchester Memorial Lecture Scale Effect in Transonic Flow. The Aeronautical Journal 91 (907): 291-313.

Haines, A. B. 1957. Wing Section Design for Swept-Back Wings at Transonic Speeds. The Journal of the Royal Aeronautical Society 61 (556): 238-244.

Haines, A.B, and D.S. Capps. November 1948. High Speed Wind Tunnel Test on a 1/12th Scale Model of the Supermarine Swift (Single-Jet Experimental Sweptback Fighter); RAE Aero R.2303. Farnborough: Royal Aircraft Establishment.

Hamel, Peter G. 2005. Birth of Sweepback: Related Research at Luftfahrtforschungsanstalt, Germany. Journal of Aircraft (AIAA) 42 (4): 801-813.

Hansen, James R. 1986. Engineer in Charge: A History of the Langley Aeronautical Laboratory 1917-1958 (NASA SP-4305).

Hansen, James R., ed. 2007. The Wind and Beyond: A Documentary Journey Into the History of Aerodynamics in America (NASA SP-2007-4409). Vol. II. NASA.

Harris, Charles D. 1990. NASA Supercritical Airfoils: A Matrix of Family Related Airfoils (NASA-TP-2969). NASA.

Hartley, Keith. 2020. Costs and Prices of UK Military Aircraft in War and Peace. In Selected Topics on Defence Economics and Terrorism, 1-14. Ekin.

Haywood, K. 2018. "Government and British Civil Aerospace—1945-64. Journal of Aeronautical History (Royal Aeronautical Society) 100-136.

n.d. HC Deb 01 February 1966 vol 723 cc895.

n.d. HC Deb 19 July 1954 vol 530 c957.

n.d. HC Deb 27 Jan 1958 vol 581 col 52.

Hills, R., and Küchemann. 1947. Note on Cranked Sweptback Wings. TN Aero 1911, Royal Aircraft Establishments.

Holder, D. W. 1964. The Transonic Flow Past Two-Dimensional Aerofoils. The Journal of the Royal Aeronautical Society 68 (644): 501-516.

Holder, D.W. December 1946. The High Speed laboratory of the Aerodynamics Division, NPL; ARC R&M 2560.

Hooper, Ralph, interview by Thomas Lean. 2010. National Life Stories, An Oral History of British Science The British Library Board. http://sounds.bl.uk.

Howell, A.R. 1948. The Aerodynamics of the Gas Turbine. Journal of the Royal Aeronautical Society 52 (450): 329-356.

Howell, A.R. 1942. The Present Basis of Axial Flow Compressor Design. Part 1. Cascade Theory and Performance (R&M 2095). Aeronautical Research Council.

Jacobs, E, and K, Pinkerton, R Ward. 1933. The Characteristics of 78 Related Airfoil Sections from Tests in the Variable-Density Wind Tunnel (NACA-TR-460).

Jones, R. T. 1953. Theory of Wing-Body Drag at Supersonic Speeds, NACA Report 1284. NACA.

Kármán, Theodore von. August 1945. Where We Stand; A Report Prepared for the AAF Scientific Advisory Group. Wright Field, Dayton, Ohio.: Headquarters, Air Material Command.

Küchemann, D, and J. Weber. 1947. Calculation of the Velocity Distribution at Zero Lift on the Handley Page Crescent Wing, with Suggested Modifications to Improve This. (RAE TN Aero No.1). Farnborough: Royal Aircraft Establishment.

Küchemann, Dietrich. 1978. The Aerodynamic Design of Aircraft. Oxford: Pergamon Press.

Lachmann, G. V. 1937. Aerodynamic and Structural Features of Tapered Wings. The Journal of the Royal Aeronautical Society 41 (315): 161—237.

—. 1961. Boundary Layer and Flow Control. Pergamon Press.

Lachmann, G. V. 1964. Sir Frederick Handley Page: The Man and his Work. Journal of the Royal Aeronautical Society 68 (643): 433-452.

Land, Norman S, and Annie G. Fox. 1957. An Experimental Investigation of the Effects of Mach Number, Stabilzer Dihedral and Fin Torsional Stiffness on the Transonic Flutter Characteristics of a T-Tail; NACA-RM-L57A24.

Lee, and K. W. Newby. May 1955. Tests in the RAE 10ft x 7ft High Speed Tunnel of a Winged Bomb Mounted Under the Fuselage of the Avro Vulcan; RAE Tech Aero TN 2365.

Lee, G. H. 1947. Tailless Aircraft Design Problems. The Journal of the Royal Aeronautical Society 51 (434): 109-131.

Lee, G. H. 1955. The Aerodynamic and Aeroelastic Characteristics of the Crescent Wing. The Journal of the Royal Aeronautical Society 59 (529): 37-44.

Lee, G. H. 1946. The Case for the Tailless Aircraft. The Journal of the Royal Aeronautical Society 50 (431): 872-887.

Lee, P. 1963. 8ft x 6ft Transonic Tunnel Tests on a Skybolt Missile Installation on the Avro Vulcan; RAE TN Aero 2907. Royal Aircraft Establishment, Farnborough: Ministry of Aviation.

Lloyd, Peter. 1968. Hayne Constant, CB, CBE, MA, FRS, Fellow. The Aeronautical Journal (Royal Aeronautical Society) 72 (688): 285-286.

Lock, R. C, and J. L. Fulker. 1974. Design of Supercritical Aerofoils. Aeronautical Quarterly 25 (4): 245-265.

Lorell, Mark A. 1980. Multinational Development of Large Aircraft: The European Experience. Rand Corporation.

Lyons, D J. 1952. Investigations on Stalling Behaviour, Rudder Oscillations, Take-off Swing and Flow Round Nacelles on the Tudor 1 Aircraft (R&M 2789). Aeronautical Research Council.

Mabey, D G. 1965. Comparison of Seven Wing Buffet Boundaries Measured in Wind Tunnels and in Flight (CP-840). Aeronautical Research Council.

Mair, W. A. 1950. High Speed Aerodynamics at the Royal Aircraft Establishment from 1942 to 45, R&M 2222. Aeronautical Research Council.

May, Ernest R., John D. Steinbruner, and Thomas W. Wolfe. 1981. History of the Strategic Arms Competition 1945-72 Part II. By Alfred Goldberg. Office of the Secretary of Defense Historical Office.

McNaughton, I. I., and D. A. Perfect. 1965. Bird Impact Tests on Vulcan Aircraft Windscreens; RAE TR 65148. Royal Aircraft Establishment.

McRae, D. M. 1973. The Aerodynamic Development of the Wing of the A 300B. The Aeronautical Journal 77 (751): 367—379.

Moore, Richard. 2021. Peierls's Outline of the Development of the British Tube Alloy Project: A 1945 Account of the Earliest UK Work on Atomic Energy. Nuclear Technology 374-379.

Murden, Michael. 2001. Bomber Command Aircraft Structural Defects and the Use of NDT in the Early 1960s. Royal Air Force Historical Society 123-129.

Murrow, Rhyne and. 1957. Effects of Airplane Flexibility on Wing Strains in Rough Air at 5,000ft as Determined by Flight Tests of a Large Swept-wing Airplane; NACA-TN-4107.

Negaard, Gordon R. 1980. The History of the Aircraft Structural Integrity Program (680.1B). Aerospace Structures Information and Analysis Centre.

Neumark, S, J Collingbourne, and E.J. York. 1955.

Velocity Distribution on Thin Tapered Arrowhead and Delta Wings with Spanwise Constant Thickness Ratio at Zero Incidence; ARC-R&M-3008. Aeronautical Research Council.

Neustadt, Prof. Robert E. 1963. Report to the President, Skybolt and Nassau: American Policy-making and Anglo-American relations.

Newby, K W. 1957. R.A.E. High-Speed Wind-Tunnel Tests of the Trailing-edge Controls on a Delta Wing with 52-deg Sweepback, ARC R&M 2999. Aeronautical Research Council.

Newby, K W, A B Haines, and D S Capps. 1952. RAE 10ft x 7ft High Speed Tunnel Tests on 1/30 Scale Models Representative of a Delta Wing Bomber (Avro 698—B35/46)—RAE TN Aero 2204. Farnborough: Royal Aircraft Establishment.

Newby, K. W. 1955. 10ft x 7ft High Speed Tunnel Tests on a 1/30th Scale Model of a Four-Jet Delta Wing Bomber (Avro Vulcan); Report Aero 2558. Royal Aircraft Establishment.

Newby, K. W. 1955. The Effect of Leading-Edge Modifications on the Buffet Boundary of the Avro Vulcan. Royal Aircraft establishment, Farnborough.

Owner, F. M. 1963. Bristol Gas Turbines —The First Decade. Journal of the Royal Aeronautical Society 67 (631): 427-436.

Peacey, H. H. 1955. Some Effects of Shock-Induced Separation in Turbulent Boundary Layers in Transonic Flow Past Aerofoils (ARC R&M 3108). Aeronautical Research Council.

Pearcey, H. H. 2015. Private Communciation to Dr R. W. Pleming.

—. 1960. The Aerodynamic Design of Section Shapes for Swept Wings. Edited by T. von Kármán. Proceedings of the Second International Congress in the Aeronautical Sciences. Zurich. 277-322.

Pearcey, H. H., and D.W Holder. 1967. Examples of the Effects of Shock-Induced Boundary Layer Separation in Transonic Flight, ARC R&M 3510. Aeronautical Research Council.

Perring, Thom &. 1948. The Design and Work of the Farnborough High Speed Tunnel. The Journal of the Royal Aeronautical Society 52 (448): 205-250.

Perry, D. H., J. C. Morrall, and W. G. A. Port. 1970. Low Speed Flight Tests on a Tailless Delta Wing Aircraft (Avro 707B) Part 3 Lateral Stability and Control; ARC-CP-1106. Aeronautical Research Council.

Port, W. G. A., and J. C. Morrall. 1967. Low Speed Flight Tests on a Tailless Delta Wing aircraft (Avro 707B) Part 2. Longitudinal Stability and Control; ARC-CP-1105. Aeronautical Research Council.

Ramaswamy, M.A. July 1978. Supercritical Aerofoils: a Survey. Proceedings of the Indian Academy of Sciences vol 1 C 27-55.

Raney, D. J. 1950. Low Speed Wind Tunnel Tests of Brake Flaps on a Model of a Delta Wing Bomber (Avro B35/46); R. Aero 2366. Royal Aircraft Establishment.

Raney, D. J., and W. J. G. Trebble. 1951. Low Speed Wind Tunnel Tests on a Model of a Revised Version of a Delta Wing Bomber (Avro B35/46); Report Aero 2416."

Richards, M. J., and D. J. Harper. 1963. Tests to High Subsonic Speeds in the 10ft x 7ft Tunnel, of Several Wing-Mounted Air-Brakes on a Half-Model of a Four-Jet Bomber (Vickers Valiant), ARC/CP-621. Aeronautical Research Council.

Rogerson, H. 1956. The Life and Work of Roy Chadwick. Journal of the Royal Aeronautical Society 60 (548): 501—514.

Ross, J.G., R. Hills, and R.C. Lock. 1949. Wind Tunnel Tests on a 90° Apex Delta Wing of Variable Aspect Ratio (Sweepback 36.8°); Aero 2533. Farnborough, UK: Royal Aircraft Establishment.

Rubbra, A.A. 1964. Alan Arnold Griffith, 1893-1963. Biogr. Mems Fell. R. Soc. 10: 117-136.

Schairer, G. S. 1969. The Role of Competition in Aeronautics. The Aeronautical Journal 73 (699): 195—207.

Seal, Diana M. 1962. A Survey of Buffeting Loads ARC/CP-0584. Aeronautical Research Council, Aeronautical Research Council.

Sedden, J., and D. J. Kettle. 1950. Low-Speed Wind-Tunnel Tests on the Characteristics of Leading-Edge Air Intakes in Swept Wings (ARC R&M 3353).

Seddon, J. 1952. Air Intakes For Aircraft Gas Turbines. The Journal of the Royal Aeronautical Society 56 (502): 747-781.

Shields, R. T., J. Stephenson, and I. E. Utting. 1956. An Investigation of High Altitude Cruising Conditions for Turbo-jet Aircraft; ARC-CP-0215. Aeronautical Research Council.

Smelt, R. 1946. A Critical Review of German Research on High-Speed Airflow. The Journal of the Royal Aeronautical Society 50 (432): 899—934.

Smith, J. P. 1969. The Development of the Trident Series. The Aeronautical Journal 73 (707): 935-940.

Smith, Joan. n.d. Early Computing in the Aircraft Industry: Avro's at Chadderton. Archivesit.org.uk.

Staff of RAE Aerodynamics and Structures groups. 1948. Report of the R.A.E. Advanced Bomber Project Group; R. Aero. 2246, Structures 16. Royal Aircraft Establishment.

Staples, K. J. 1965. Speed Stability and the Landing Approach with an Appendix of the Avro 707A Longitudinal Characteristics (ARC/R&M-3476). Aeronautical Research Council.

Summerfield, P.H. 1992. 36th Roy Chadwick Lecture—Manufacturing Breakout 1941-1991. Development in Aerospace Industry Manufacturing Techniques. The Aeronautical Journal (RAeS) 35-46.

Sutton, E. P., and A. Stanbrook. 1959. A Wind-Tunnel Investigation of the Directional and Longitudinal Stability of the Javelin Aircraft at Transonic Speeds, Including Comparison with Flight Test Results; ARC-R&M-3403. Aeronautical Research Council.

Swann, Maj. Ralph. 1986. A Unit History of the 315th Bomb Wing: 1944-1946, 86-2460. Maxwell AFB, Alabama: Air Command and Staff College.

Taylor, A.S., and D.J. Eckford. 1966. Aircraft Loading Actions Problems—Proceedings of a Symposium Held at Farnborough on 28th October 1966; ARC-CP-1003. Aeronautical Research Council.

Taylor, C. R., J. R. Hall, and R. W. Hayward. 1970. Super VC10 Cruise Drag—A Wind Tunnel Investigation Part 1 Experimental Techniques. Aeronautical Research Council.

Termena, Bernard J. 1959. History of the Rascal Weapon System 1952-1958. Wright-Patterson Air Force Base: Historical Division, Office of Information Services, Air Material Command (USAF).

The NACA. 1934. The De Havilland Comet Long-Range Airplane (British): A Low-Wing Cantilever Monoplane; NACA-AC-197.

Thom, A., and W.G.A. Perring. 1948. The Design and Work of the Farnborough High Speed Tunnel. The Journal of the Royal Aeronautical Society 52 (448): 205-250.

Thomas, H. H. B. M. 1955. The Calculations of the Derivatives Involved in the Damping of the Longitudinal Short Period Oscillations of an Aircraft and Correlation with Experiment; R. Aero 2561. Royal Aircraft Establishment.

Vann, F. W. 1968. Loading actions from the designer's viewpoint. Aircraft Loading Actions Problems—Proceedings of a Symposium Held at Farnborough on 28th October 1966 (ARC CP 1003). Farnborough: HMSO.

Wallace, Richard E. 1968. Recent Transonic Airfoil Developments and Some Business Aircraft Implications. SAE Transactions 77: 689-703.

Wallis, R.A. January 1965. Wind Tunnel Studies of Leading Edge Separation Phenomena on a Quarter-Scale Model of the Outer Panel of the Handley Page Victor Wing, With and Without Nose Droop; ARC/R&M-3455. Aeronautical Reseach Council.

Wells, E. W. 1958. Fatigue Loadings in Flight Loads in the Nose Undercarriage and Wing of a Valiant; ARC-CP-0521. Aeronautical Research Council.

Whitcomb, R T. 1974. Review of NASA Supercritical Airfoils. ICAS 74-10, August.

Whitcomb, R. T. 1958. Special Bodies Added on a Wing to Reduce Shock Induced Boundary-layer Separation at High Subsonic Speeds; NACA-TN-4293.

Whitfield, J. 2013. Metropolitan Vickers, the Gas Turbine, and the State: A Socio-Technical History, 1935-1960. Student Thesis, The University of Manchester, UK.

Widgery, W. M. 1947. Pressurisation and Cabin Air Control. The Aeronautical Journal 51 (444): 949-980.

Wilby. 1967. The Pressure Drag of an Aerofoil with Six Different Round Leading Edges, at Transonic and Low Supersonic Speeds (ARC-CP-0921). Aeronautical Research Council.

Wimpenny, J. C. 1954. Stability and Control in Aircraft Design. Journal of the Royal Aeronautical Society 58 (521): 329-360.

Woodward-Nutt, A. E. 1964. Aeronautical Research and Development in the Commonwealth. The Journal of the Royal Aeronautical Society 68 (638): 75-105.

Worman, Charles G. 1967. History of the GAM-87 Skybolt Air-to-Surface Ballistic Missile. Historical Division, Information Office, Aeronautical Systems Division, AFSC, USAF.

Young, A.D., and B.S. Squire. 1942. A Review of Some Stalling Research With an Appendix on Wing Sections and their Stalling Characteristics; ARC-R&M-2609.

Younger, John E. 1938. Engineering Aspects of Commercial High Altitude Flying. Journal of the Royal Aeronautical Society 1055-1083.

ENDNOTES

1. Private communication, H.H. Pearcey to R. W. Pleming, 2015
2. There are many volumes that would fill these criteria, but perhaps a few should be recommended. For an understanding of the Royal Air Force's use of the V-Bombers, the Victor/Vulcan/Valiant Boys series by Tony Blackman abounds with stories. For an impressive collection of technical background on the Vulcan, the works of David Fildes should be referred to, while the definitive account of Black Buck 1 is Vulcan 607 by Rowland White.
3. (Hartley 2020)
4. (Lachmann 1964)
5. Stanley Baldwin delivered this speech as Lord President of the Council, in the House of Commons, November 10, 1932.
6. HC Deb November 10, 1932 vol 270 cc525
7. A copy of the Frisch-Peierls memorandum is held for the National Archives in the Bodleian Library, NCUACS 57.6.95, Supplementary catalogue of the papers and correspondence of SIR RUDOLF (ERNST) PEIERLS FRS (1907-1995). Available online from, for example, https://www.atomicarchive.com/resources/documents/beginnings/frisch-peierls-2.html (retrieved 4/2/24)
8. Report by M.A.U.D. Committee on the use of Uranium for a Bomb, Held by Trinity College Cambridge as CSAC 75.5.80/D.11, also available online, for example https://fissilematerials.org/library/maud.pdf (retrieved 4/2/24)
9. See for example, the commentary on Peierls's writing in (Moore 2021)
10. The New World 1939/46, A History of the United States Atomic Energy Commission, Volume 1, pp263
11. Bush to FDR; Memorandum for The President: Tubealloy – Interchange with the British; August 23, 1943 – Franklin D Roosevelt library.
12. https://www.thebritishacademy.ac.uk/documents/801/certificate.pdf
13. This was described by Truman in his autobiography, *Harry S. Truman, Memoirs: Volume 1, Year of Decisions* (New York: Doubleday & Company, 1955).
14. Truman's announcement speech is published online by the Harry S. Truman Presidential library, https://www.trumanlibrary.gov/library/public-papers/93/statement-president-announcing-use-bomb-hiroshima.
15. GEN 75/1, 'The Atomic Bomb' memorandum by the PM (Attlee). CAB 130/3, taken from Matthew Jones 2017.
16. https://www.belfercenter.org/sites/default/files/files/publication/Andrew_Brown_Chapter3.pdf
17. Oppenheimer's letter to Stimson can be found, for example, online at https://teachingamericanhistory.org/document/letter-to-secretary-of-war-henry-stimson/
18. https://www.jewishvirtuallibrary.org/president-truman-message-to-congress-on-the-atomic-bomb-october-1945
19. https://www.presidency.ucsb.edu/documents/address-foreign-policy-the-navy-day-celebration-new-york-city
20. C.P (45) 272 page 76, TNA CAB-129-4-CP-272
21. HC Deb November 7, 1945, vol 415, col 1292
22. HC Deb November 7, 1945, vol 415, col 1334
23. Published online by the Harry S. Truman Presidential library, https://www.trumanlibrary.gov/library/public-papers/191/presidents-news-conference-following-signing-joint-declaration-atomic
24. (http://filestore.nationalarchives.gov.uk/pdfs/small/cab-128-2-cm-45-55-8.pdf)
25. Attley: https://hansard.parliament.uk/commons/1945-11-22/debates/12ac1527-edfb-4b2e-abf3-1dbc51eaf618/ForeignAffairs#
26. Wilson Harris: https://hansard.parliament.uk/Commons/1945-11-22/debates/12ac1527-edfb-4b2e-abf3-1dbc51eaf618/ForeignAffairs#
27. Nunn dictated these statements to a family member, shortly before his death. They were published in The *Guardian* Newspaper in 2003. Online: https://www.theguardian.com/uk/2003/jan/27/science.internationaleducationnews
28. TNA, FO 371/51624.
29. https://www.osti.gov/atomicenergyact.pdf
30. https://www.parliament.uk/globalassets/documents/lords-information-office/2011-Lord-Hennessy-Robing-Room-Lecture-.pdf
31. D M Hallowes; An Examination of the Interception problem; RAE Aero 2035; April 1945, quoted in Andrew Nahum; World War to Cold War: Formative episodes in the development of the British aircraft industry, 1943-1965; PhD thesis, LSE, March 2002.
32. Sutcliffe, P. (1955). Effect of Variations in Weight or Normal Acceleration on the Ceiling of Jet-Propelled Aircraft. T*he Aeronautical Journal*, 59(534), 435-437.
33. (Driggs 1950)
34. Ibid.
35. Lissaman, as published in Air and Space magazine (Smithsonian)

ENDNOTES

36 Smith, Joan, 2016 (J. Smith n.d.)
37 (Neumark, Collingbourne and York 1955)
38 (D. Holder December 1946)
39 (Thom and Perring 1948)
40 (Haines and Capps November 1948)
41 This description was included on a report for an aircraft that might have been considered, in the first instance, to have been not so dominated by aerodynamics. (G. H. Lee; The aerodynamic design philosophy of the Handley Page Jetstream; Aircraft Engineering, September 1967)
42 (Jacobs and Ward 1933)
43 (Gates and Young 1962)
44 (Holder 1964)
45 (Pearcey 1960)
46 (Mair 1950)
47 (Becker 1980)
48 (Ramaswamy July 1978)
49 Mair, Bridgland and Rose; Flight tests on a Welkin I to investigate pitching oscillations at high Mach numbers; RAE Aero 2114, February 1946
50 Busemann; Compressible Flow in the thirties; Annual Review of Fluid Mechanics, 1971.3:1-12.
51 (Hamel 2005)
52 R T Jones; Adolf Busemann 1901-1986; Memorial Tribute; National Academy of Sciences.
53 (Lachmann 1957)
54 Hans-Ulrich Meier. Historic Review of the Development of High-Speed Aerodynamics, German Development of the Swept Wing 1935—1945. August 2010. 1-68.
55 (Smelt 1946)
56 Ibid.
57 (Jones 1953)
58 (G. H. Lee 1946)
59 From Avro's tender document to, A.M Specification P.13/36 Medium Bomber, dated January 1937 (Via Frank Pleszak).
60 (Lachmann 1964)
61 This would have been the contemporary explanation of the flow physics. For all of his brilliance, Lachmann's explanation of the workings of the slotted wing described it as fundamentally a boundary layer control device. It would be into the 1970s before the more sophisticated explanations of A.M.O Smith, which are regarded as the accepted model today, would be articulated.
62 (Constant 1958)
63 (Armstrong 1976)
64 From Hayne Constant's obituary from the Royal Aeronautical Society, by Peter Lloyd. (Lloyd 1968)
65 (Constant 1958)
66 Howell's 1942 report was ARC R&M 2095 (Howell 1942), while an overall description of the axial flow gas turbine theory in use at the time of B35/46 is given in (Howell 1948)
67 For a comprehensive description of the aeronautical gas turbine activities at Metropolitan-Vickers, the reader is directed to (Whitfield 2013)
68 (Owner 1963)
69 Sir Stanley Hooker; Not Much Of An Engineer: An Autobiography; The Crowood Press; 1984.
70 Sinnette, John T., Jr. & Voss, William J. Extension of Useful Operating Range of Axial-Flow Compressors by Use of Adjustable Stator Blades, report, December 29, 1944
71 (von Kármán August 1945)
72 (Rubbra 1964)
73 (Bridle 1948)
74 (Widgery 1947)
75 Ludlow-Hewitt to HQ Bomber Command, IG/597, June 30, 1941
76 (Younger 1938)
77 This limitation was described by Lieutenant John H. Stickell, and American who had enlisted in the Royal Canadian Air Force and flew as a bomber pilot with 214 sqn and then 7 sqn in the Pathfinder force. He noted wryly that the 18,000ft limitation was not really an issue with the Stirling, but had become one with the Halifax and certainly the Lancaster. This information was imparted to the US Navy Bureau of Aeronautics, in an interview of May 6, 1943.
78 As described in the first annual Chadwick Memorial Lecture. Held under the auspices of the Manchester Branch of the RAeS and given by Mr H. Rogerson, M.B.E., A.M.I.Mech.E., F.R.Ae.S on March 21, 1956, at the College of Technology, Manchester. The lecture series continues to this day.(Rogerson 1956)
79 (Rogerson 1956)
80 (Summerfield 1992)
81 Recollections of Sir William Farren, in response to Rogerson's lecture.
82 Davies made these comments at a 50th Anniversary symposium on the Hurricane, and they were recorded by J. W. Fozard. (J. W. Fozard 1988)
83 These concepts were published by Davies in a paper entitled Aeroplane Design for Production, in Aircraft Engineering, March 1939. (Summerfield 1992)
84 Lachmann's MI5 personal file is extensive and contained in the TNA folders KV-2-2233/34/35.
85 (Davies 1969)
86 (G. H. Lee 1947)
87 (Ewans 1951)
88 (Perring 1948)
89 Avro initial brochure to B35/46, May 1947, as quoted in Fyldes (2012)
90 Rosemary Lapham's account is contained in a collection of Roy Chadwick's documents at the IBCC, Lincoln. (R Lapham, "Turn down an empty glass," IBCC Digital archive, accessed June 29, 2024)
91 (Staff of RAE Aerodynamics and Structures groups 1948)
92 (Mair 1950)
93 (Port and Morrall 1967)
94 Ibid.
95 (Perry, Morrall and Port 1970)
96 (Seddon 1952)
97 (Davies 1969)
98 (A. B. Haines, Wing Section Design for Swept-Back Wings at Transonic Speeds 1957)
99 (Seddon 1952)
100 Described in (Seddon and Kettle 1950)
101 RAE Aero 2416, p8 (Raney and Trebble 1951)
102 (K. W. Newby 1957)
103 RAE Aero R.2366 (Raney 1950)
104 RAE Aero 2561 (Thomas 1955)
105 Elevator-angle-to trim tests conducted as part of high-speed WT tests at RAE Farnborough, sometime in the early 1950s. (K. W. Newby 1955)
106 https://vulcantothesky.org/aircraft/vulcan-vx770/
107 https://www.avweb.com/recent-updates/business-military/guest-blog-an-me-262-personal-connection/
108 https://www.key.aero/article/read-rare-report-vulcan-1952
109 (Davies 1969)
110 (Blackman 2012)
111 Küchemann 1978)
112 R. Hooper, Interview in National Life Stories, An oral history of British science. 2010
113 (Mabey 1965)
114 (Seal 1962)
115 G. Edwards, Looking Ahead with Hindsight, The Aeronautical Journal, vol. 78, no. 760, p. 134—146, 1974.
116 (Richards and Harper 1963)
117 (Seal 1962)
118 (Pearcey and Holder 1967)
119 (Baldock 1958)
120 Parliamentary remarks Beswick
121 (Farren 1956)
122 (K. W. Newby 1955)
123 Private correspondence, H H Pearcey to R W Pleming, 2015.
124 (Pearcey 1955)
125 (Wilby 1967)
126 (K. W. Newby 1955)
127 (Lachmann 1961)
128 Private correspondence, H H Pearcey to R W Pleming, 2015.
129 A&AEE report on VX777 flight tests in May 1955, (3rd Part of Report No. A.A.E.E./910.). This is quoted in (Fildes 2012)
130 HC Deb January 9, 1955, Vol 536 Col 1873
131 Ibid.
132 Ibid.
133 Ibid.
134 This exchange occurred an article called, More On The Origin Of The Crescent Wing, Hawker Association Newsletter, Issue 13, 2006.

135 (G. H. Lee 1947)
136 (Hills and Küchemann 1947)
137 Even More on the Origin of the Crescent Wing, Hawker Association Newsletter, Issue 13, 2006.
138 (Barnett and Lange 1950)
139 (Brebner 1953)
140 (Küchemann 1978)
141 (Küchemann and Weber 1947)
142 These tests were reported in (Gamble April 1948)
143 Reported in (Collingbourne 1951)
144 (Collingbourne and J. 1951)
145 Maj Legge Bourke MP, HC Deb July 10, 1960, Vol 619 Col 687
146 Flight Magazine profiled the Victor B.Mk.2 in their issue of Oct 30th, 1959.
147 Flight magazine included coverage of Godfrey Lee's lecture, Aerodynamics of the Crescent Wing, in its issue of May 14, 1954
148 Johnny Allen's description is found in Higgs and Vigar, Test Pilots of the Jet Age; Air World 2001.
149 (Wallis January 1965)
150 (Land and Fox 1957)
151 (Courtney and Somerville 1950)
152 Javelin Rumpus; *Flight International* Magazine, July 8, 1955.
153 (Sutton and Stanbrook 1959)
154 Described in the talk, Aerodynamic Data for Loading Studies, presented by H.H.B.M Thomas and authored by a group including Johanna Weber, included at a symposium held at Farnborough in 1966. (Taylor and Eckford 1966)
155 HC Deb, June 17, 1963
156 (Greenwood 1964)
157 Described in RAE Aero 2533, August 1949.
158 Results and discussion in (K. W. Newby 1955)
159 Private Correspondence, Ed Jarron to author, July 2021.
160 This work was described in "A feasibility study on a 200 Volt, Direct current, Aircraft electrical power system", issued as ARC CP No 1186 in 1971.
161 V. Madonna, P. Giangrande and M. Galea, Electrical Power Generation in Aircraft: Review, Challenges, and Opportunities, in IEEE Transactions on Transportation Electrification, vol. 4, no. 3, pp. 646-659, Sept. 2018, doi: 10.1109/TTE.2018.2834142.
162 Nosley, Charles C., and David J. Hucker. Constant Frequency AC Electrical System for Business Aircraft. SAE Transactions, vol. 77, 1968, pp. 728-35. JSTOR.
163 (R. T. Whitcomb 1958)
164 (Francis 1964)
165 DEFE 7/2333 Air Staff requirement No OR1132A: Propelled Air-to-Surface Missile for the V-Class Bombers, September 3, 1954, as quoted in Finch, Guy; Replacing the V-Bombers: RAF Strategic Nuclear Systems Prcurement and the Bureaucratic Politics of Threat; Ph.D Thesis, University of Wales at Aberystwyth, December 2001
166 Private communication, Ed Jarron to author, July 2021.
167 Private communication, ACM Sir John Allison to author, August 2020.
168 Crocco, Joseph P.; Analysis of Inertial Guidance position errors caused by errors in reference velocity; Masters Thesis, Institute of Technology, Air University, USAF, August 1961.
169 R. H. Francis, RaeS paper (Francis 1964)
170 This comment came from a contribution to a book on the story of the Blue Steel trials in Australia, complied and edited by David Booth and published in 1999. Accessible at www.jsaxon.org/bluesteel/books/1999book.pdf
171 Lee and Newby; Tests in the R.A.E. 10 ft x 7 ft High Speed Tunnel of a winged bomb mounted under the fuselage of the Avro Vulcan; RAE Tech Aero TN 2365, May 1955. (Lee and Newby May 1955)
172 (Lee and Newby May 1955)
173 Described in the official history of the Rascal project (Termena 1959), pg12.
174 Ibid. p18
175 TNA. Conclusions of Cabinet Meeting June 20, 1960, Cabinet Office, CAB-128-34-35, Nuclear Weapons: Skybolt, p3.
176 Solly Zuckerman, recorded interview by Joseph E. O'Connor, August 5, 1966, (page 27), John F. Kennedy Oral History Program.
177 Recorded in, Krohn Peter Leslie (1995) Solly Zuckerman, Baron Zuckerman, of Burnham Thorpe, O. M., K. C. B., May 30, 1904-April 1, 1993; Biogr. Mems Fell. R. Soc.41576—598
178 Ibid.
179 TNA. Conclusions of Cabinet Meeting June 20, 1960, Cabinet Office, CAB-128-34-35, Nuclear Weapons: Polaris, p4.
180 TNA. Conclusions of Cabinet Meeting June 20, 1960, Cabinet Office, CAB-128-34-35, Nuclear Weapons: Skybolt, p2.
181 ACM Sir John W Baker to George Carter, June 16, 1954.
182 The work on understanding Mach 2 fighter requirements was reported in RAE Aero 2513, in three parts. The specific aspects of the aircraft specification are described in part B, Appendix IV Collingbourne and Cool 1954.
183 Described in the Handley Page brochure of 1956, 'Victor B Phase 4 Supersonic High-Altitude bomber: 4 Rolls-Royce Conway 31 engines with Reheat (Secret).' With thanks to Barry Hinchliffe.
184 NACA report 1284, R.T. Jones (Jones 1953)
185 Carlson and Smith large strore M 1.61
186 NACA RM S56H07 — Relation of turbojet propulsion system development to the strategic bomber mission.
187 TNA. Conclusions of Cabinet Meeting June 20, 1960, Cabinet Office, CAB-128-34-35, Nuclear Weapons: Skybolt, p1.
188 This report was prepared by Capt. Ralph C. Graves Jr., of the Avionics Branch, GAM-87A System Program Office, and issued as GAM-87 Engineering Memorandum No.11. To underline its significance, it was included as Appendix H of the USAF's official history of the system, (Worman 1967)
189 Described in Goldberg (ed), 1981 p513 (May, Steinbruner and Wolfe 1981)
190 John F. Kennedy Presidential Library; Department of Defense: General: McNamara testimony before senate armed services committee, April 4, 1961; Collected in: Papers of John F. Kennedy, Presidential papers, National Security Files; Digital ID JFKNSF-273-007.
191 Ibid.
192 Ibid.
193 Ibid.
194 Ibid.
195 (P. Lee 1963)
196 These tests were documented in Lee's report for the RAE; it references concurrent Avro work at low speed.
197 (Francis 1964)
198 (Worman 1967)
199 (Worman 1967), p71
200 Neustadt report, page 5. (Neustadt 1963)
201 Neustadt report, page 9.
202 Thornycroft interview, JFK library.
203 PM to Ambassador, FO Telegram 7395 of October 22, 1962 In TNA FO 598/29.
204 From the transcript of a lecture entitled, "Summary of the previous RAFHS seminar on the origin and development of the British nuclear deterrent 1945-60", presented by AVM Michael Robinson, and published in the Journal of the RAF Historical Society Vol 26, 2001.
205 Macmillan to Ormsby-Gore, October 30, 1962. In TNA FO 598/29.
206 Macmillan to Ormsby-Gore, October 31, 1962. In TNA FO 598/29.
207 Ormsby-Gore to Macmillan, November 6, 1962. In TNA FO 598/29.
208 Neustadt, p13.
209 John H. Rubel, recorded interview by William m W. Moss, August 12, 1970, John F. Kennedy Library oral history program.
210 Ibid.
211 These quotes are taken from Rubel's "verbatim" notes, captured in Neustadt, page 64. The official notes communicated by telegram to the Secretary of State, and thence to the President, used different but consistent words. "Secretary replied that he was not prepared to say that Skybolt was technically impossible, but it was impossible to achieve objectives which had been set for the program in time period planned." (London to SoS, December 11 1962, from JFKPOF-127a-008a-p0065).
212 Neustadt, p74.
213 Neustadt, p78.
214 Transcript of telephone call recorded December 17, 1962, from White House to Eisenhower. In archives of JFK Presi-

dential library, JFKPOF-TPH-06C1-TR.
215 (George Edward) Lord Peter Thorneycroft, recorded interview by Roebert Leiman, June 19, 1964, John F. Kennedy Library oral history program.
216 (Neustadt 1963), p.91
217 This text is taken from the report by Neustadt, but is corroborated in the the State Department memoranda of the meeting: Department of State, Conference Files: Lot 65 D 533, CF 2209, Line 1100.
218 Prime Minister to The President, December 24 1962. In archives of JFK Presidential library: Papers of John F. Kennedy. Presidential Papers. President's Office Files. Countries. United Kingdom: Security, 1962: October-December, JFKPOF-127a-008-p0080.
219 Private communication, H.H. Pearcey to R. W. Pleming, 2015
220 (Floyd 1958)
221 (Bore 1993)
222 (Fozard 1969)
223 (K. W. Newby 1955)
224 (Haywood 2018)
225 (Edwards 1974)
226 (Pearcey 1955)
227 (Bowes 1974)
228 Parliamentary remarks by Mr Watkinson on Monday, January 27, 1958, https://hansard.parliament.uk/Commons/1958-01-27/debates/00cf0434-8064-46df-913ac0aae46b4bd7/CommonsChamber#contribution-9876a2d2-e1b1-4e4f-a4bad26941431517, accessed July 6, 2021.
229 (Green 2015)
230 (Taylor, Hall and Hayward 1970)
231 (Browne, et al. 1972)
232 (J. P. Smith 1969)
233 (Vann 1968)
234 (Dykins, Jupp and McRae 1988)
235 (Dykins, Jupp and McRae 1988)
236 (Haywood 2018)
237 (Wallace 1968)
238 Parliamentary remarks by Mr Mulley on February 1, 1966 https://api.parliament.uk/historic-hansard/commons/1966/feb/01/aircraft-industryplowden-report#column_895, accessed July 7, 2021.
239 (Lorell 1980)
240 (Brown 1980)
241 (Lock and Fulker 1974)
242 (Harris 1990)
243 (Whitcomb 1958)
244 (Wells 1958)
245 This story is related by the pilot, Milt Cottee, in Tony Blackman and Anthony Wright's book, Valiant Boys.
246 (Murrow 1957)
247 (Negaard 1980)
248 Ibid.
249 (Curran 1973)
250 Private communication, Martin Withers to author, 2021.
251 Private communication, Ed Jarron to author, 2021.
252 Norman Bonnor's account of his time on the Victor B.2 is featured in Tony Blackman's book Victor Boys. (Blackman and O'Keefe, Victor Boys 2012)
253 (Murden 2001)
254 The functional importance of windscreen heating on the Vulcan at low-level can be gleaned from the findings in (McNaughton and Perfect 1965)
255 Lee's lecture was published by the RAeS (G. H. Lee 1955) and, as frequently occurred in the period, covered by Flight magazine.
256 (Shields, Stephenson and Utting 1956)

INDEX

10 Squadron 273
XV Squadron 273
35 Squadron 177, 203
49 Squadron 265
55 Squadron 273
57 Squadron 273
74 Squadron 203
100 Squadron 241, 268
148 Squadron 267
230 Operational Conversion Unit 136, 169
232 Operational Conversion Unit 188
509th Composite Group 23

Aachen 31, 86
Abyssinia 116
Acheson, Dean 232
Ackroyd, John 13
Admiralty 42, 85
Advanced Bomber Project Group (ABPG) 96, 98
Advanced Research Projects Agency (ARPA) 215
Aerodynamische Versuchsanstalt (AVA) 59, 62, 107, 140
Aeroplane and Armament Experimental Establishment (A&AEE) 52, 60, 100, 134, 136, 178, 179, 184, 187, 222, 267
Aeroplane magazine 118
Airbus A300 258-260
Airbus A300B 13, 259, 260
Airbus A310 261
Airbus Industrie 13, 260, 261
Aircraft and Weapons Board 214

Aircraft Research Association (ARA) 188, 191, 208, 234, 235, 255, 259
Airfix 10
Air Force Materiel Command (AMC) 212, 213
Air Force One 240, 246, 248
Air Ministry 68, 74, 75, 77, 84, 92, 141, 199, 217
Air Registration Board 116
Air Staff 35, 40-42, 80, 220
Akers, Wallace 20
Albatros Flugzeugwerke 86
Allam, Johnny 153, 186, 222
Allison J71 158
Allison, John 203
Anderson, John 20
Appleton, Edward 21
ARA *Belgrano* 290
Arado Ar 234 141, 142
Arado E-395 141
Armed Services Committee 232
Armstrong Siddeley 76, 78
Armstrong Siddeley Sapphire 76, 79, 119, 148, 182, 183, 279
Armstrong Siddeley Stentor 207, 214
Armstrong Whitworth Siskin 84
Ascension Island 9, 287, 288, 290
Atkin, Alec 121
Atomic Energy Act 199, 210
Atomic Energy Commission 217
Attlee, Clement 23, 26, 29-31, 33, 35, 135, 210
Avro 500 82

Avro 504 82
Avro 652 82
Avro 698 49, 89, 94, 96, 98, 102-104, 111-113, 116, 118, 128, 132, 175, 250
Avro 707 51, 98, 100, 102, 103, 111, 112, 176, 250
Avro 707A 98, 100, 102, 106, 111, 119, 128, 129, 131
Avro 707B 102, 103, 111
Avro 710 98, 103
Avro 722 Atlantic 251
Avro Anson 82, 83
Avro Ashton 286
Avro Athena 100
Avro Blue Steel (W100) 170, 172, 189, 192, 204, 207-210, 212, 214, 225, 226, 229, 235, 236, 237, 249, 264, 268, 275, 287, 288, 289
Avro Blue Steel Phase II (W114) 214, 230
Avro Canada 250
Avro Canada CF-105 Arrow 164, 250
Avro Flight Research Unit 116
Avro Lancaster 9, 15, 42, 44, 56, 60, 70, 81-83, 85, 115
Avro Lincoln 42, 43, 70, 135
Avro Manchester 38, 56, 71, 82, 83, 94
Avro Tudor 83, 94, 95, 124
Avro Tudor II 94
Avro Vulcan 9, 10, 13, 15, 42, 43, 51, 60, 79, 87, 91, 102, 105, 109, 110, 112, 113, 116, 118, 119, 121, 124, 127, 129, 131, 133, 134, 136, 150, 161, 163-165, 168-170, 172, 173, 175-178, 182, 183, 187, 189, 191, 203, 205, 208, 209, 213, 214, 220-222, 229, 230, 234, 235-237, 248-251, 256-258, 260, 264, 265, 267, 269, 273, 275, 277, 279, 280, 283, 286-290
Avro Vulcan B.Mk.1 136, 169, 170, 177, 178, 179, 183, 234, 235, 275, 277, 286
Avro Vulcan B.Mk.2 177, 183, 187, 203, 229, 248-250, 265, 273, 275, 286-288
Avro Vulcan Phase 2C 170, 172
Avro Vulcan Phase 3 169, 170, 173, 286
Avro Vulcan Phase 6 249, 250, 257
Avro Vulcan VX770 113, 115, 116, 118, 119
Avro Vulcan VX777 115, 119, 134, 136, 183
Avro Vulcan XA889 136
Avro Vulcan XA895 136, 178
Avro Vulcan XA897 136, 169
Avro Vulcan XH532 169
Avro Vulcan XH533 183
Avro Vulcan XH539 267
Avro Vulcan XH558 9, 10, 113, 229, 264
Avro Vulcan XM598 288
Avro Vulcan XM607 288
Avro Weapons Research Division 202

Avro York 83

B9/32 74
B25/37 80
B35/46 40, 42-44, 51, 52, 54, 61, 64, 66, 67, 74, 79-81, 83-85, 87, 90, 95, 97, 98, 111, 115, 119, 128, 131, 136, 139, 159, 169, 170, 183, 199, 202, 220, 221, 225, 277, 286
Baker, John W. 220
Ball, George 246
Bangalore 60
Barking 85
Barnett, U. Reed 142, 143
Becker, John 60
Baldwin, Stanley 17, 26
Beetham, Michael 289
Beijing 262
Belgrade 34
Bell Aircraft 212
Bell Rascal 212-214, 226
Bell X-1 40
Berkeley 19
Berlin 17, 26, 34, 73, 210, 289
Bermuda 242
Beswick, Frank 127
Betz, Albert 62, 140
Bevin, Ernest 23, 30, 35
Biglands, Steve 288
Blackburn Aircraft 84, 121
Blackman, Tony 111, 119, 186
Board of Trade 35
Boeing 63, 80, 96, 226, 253, 254, 257
Boeing 247 80
Boeing 307 80
Boeing 707 251-255, 257
Boeing 747 257, 260
Boeing 757 261
Boeing B-17 Flying Fortress 36, 44
Boeing B-29 Superfortress 16, 23, 42, 44, 60, 61, 65, 81, 94, 135, 212
Boeing B-47 Stratojet 54, 65-67, 94, 212, 213, 233, 266, 267, 275, 279, 280, 283, 287
Boeing B-50 Superfortress 212
Boeing B-52 Stratofortress 66, 199, 212, 213, 215, 217, 226, 227, 229, 232-234, 237, 238, 249, 264
Boeing LGM-30 Minuteman 230, 232-234, 238, 246
Boeing Wichita 238
Bohr, Niels 18
Bomber Command 16, 17, 26, 43, 54, 80, 241, 242, 264, 268, 287

Bonnor, Norman 268
Boot, Harry 42
Booth, David 207
Bore, C. L. 251
Boscombe Down 100, 136, 184, 187, 222, 269
Boulton Paul P.111 176
Brabazon Committee 81
Braun, Wernher von 199
Braunschweig 62
Brebner, G. G. 143, 144
Breguet Aviation 258, 283
Bristol Aeroplane Company 47, 77
Bristol Beaufighter 42
Bristol Britannia
Bristol Engines 77, 214
Bristol Olympus 118, 141, 169, 183
Bristol Olympus 6, 79, 118, 170
Bristol Olympus 100 119
Bristol Olympus 101 134, 178
Bristol Olympus 102 169
Bristol Olympus 104 178
Bristol Olympus B016 172
Bristol Olympus B0121 172
Bristol Proteus 78
Bristol Siddeley 141, 249
Bristol TE1/46 79, 92
Bristol Theseus 77, 78
British Aircraft Corporation BAC 3-11 260
British Aircraft Corporation BAC 111 257
British Aircraft Corporation Concorde 226, 258, 260
British Aircraft Corporation TSR2 202, 250, 251
British Airways 289
British Army 141, 216
British European Airways (BEA) 255, 256, 259, 260
British Overseas Airways Corporation (BOAC) 44, 251, 252, 254, 256, 257
British Thompson Houston 83
British Westinghouse 82
Brixton 269
Brown, Andrew 26
Brown, Eric 'Winkle' 84
Brown, Harold 217
Brussels 191
Bucharest 34
Budapest 34
Bundy, McGeorge 247
Bureau of Aeronautics 44
Busemann, Adolf 61, 62, 64
Bush, Vannevar 19, 20
Byrnes 35

Cabinet Committee on Atomic Energy 26
California Institute of Technology (Caltech) 47, 48, 62, 223
Cambridge University 31
Camm, Sydney 83
Camp David 216, 238
Cape Canaveral 219, 240
Carter, George 220
Castle AFB 238
Central Flying School 84
Central Intelligence Agency (CIA) 262, 263
Central Park 28
Chadwick, Roy 82-84, 87, 91, 94, 95, 220
Cheshire 82
Christmas Island 210, 242
Chrysler PGM-19 Jupiter 201
Church Fenton 267
Churchill, Winston 19
Civil Aviation Authority 113
Cohn, Ben 63
Collingbourne, Bob 149
Columbia University 248
Congress of the United States 23, 27, 28, 30, 31, 35, 210, 227, 229, 230
Consolidated B-24 Liberator 36, 44, 116
Consolidated B-32 Dominator 16
Constant, Hayne 75, 76
Convair B-36 Peacemaker 57, 212
Convair B-58 Hustler 215, 226
Convair F-102 Delta Dagger 223
Convair/General Dynamics SM-65 Atlas 232, 233, 210
Cook, James 43
Cornell University 250
Coventry 26, 31
Cranfield 127
Cricklewood 47, 85, 141
Cripps, Stafford 35, 77
Cuba 240
Cuban Missile Crisis 263
Curtiss-Wright J67 250

Daily Express 162
Dale, Henry 20
Dallas 248
Dalton, Hugh 35
Dassault 258
Dassault Mirage 290
Davies, Stuart Duncan 83, 87-89, 94, 95, 116, 119, 170, 289
Dedelsdorf 141

Defense Research and Engineering 215, 238
De Gaulle, Charles 244, 246, 248
De Havilland Blue Streak 200-202, 207, 216, 217, 229, 230, 234, 246, 248, 264, 269
De Havilland Comet 150, 256, 260, 265
De Havilland Comet 1 250, 286
De Havilland Comet 3 250, 254
De Havilland Comet 4 251
De Havilland DH.108 Swallow 118
De Havilland DH.110 Sea Vixen 118, 161, 277
De Havilland DH.121 Trident 254-260
De Havilland Mosquito 16, 38, 80, 81, 107, 161, 262
De Havilland Propellers 200
De Havilland Vampire 96
Denning, Ralph 141, 142
Department of Defense 214, 216, 226, 227, 232, 239, 245, 248
Derry, John 118
Dobson, Roy 82, 84
Doncaster 9
Doomsday Clock 263
Douglas DC-2 80
Douglas DC-4 80
Douglas DC-8 254, 257
Douglas DC-9 257
Douglas DC-10 260
Douglas GAM-87 Skybolt 215-219, 226, 229, 230, 232-234, 236-250, 257, 264, 267, 275, 287, 288
Douglas PGM-17 Thor 199
Douglas Skymaster 29
Dowding, Hugh 161
Dresden 86
Driggs, Ivan 44, 48, 49
DTD683 268, 269
Dulles, John Foster 210
Dunlop 81
Dunne, John William 67
Dykes, C. 44
Dykins 256, 260

East Anglia 42
East Falkland 289
Ecclestone, Ronald 127
Eden, Anthony 23
Edwards, George 63, 125, 128, 251, 256
Edwards AFB 238
Eglin AFB 236
Eisenhower, Dwight D. 210, 215-217, 219, 227, 229, 244, 246
Elliott Brothers 204, 207

English Electric 61
English Electric A.1 61, 81
English Electric Canberra 132, 135, 262, 267
English Electric Canberra WD952 169
English Electric Lightning 121, 203, 220, 262
English Electric P.1 112
Ernst Heinkel Flugzeugwerke 62
Esler, Eric 100, 102
European Economic Community 244-247, 258, 261
Ewans, Roy 89

F4/48 94
Falk, Roly 116, 118
Falkland Islands 9
Farren, William 63, 64, 75, 82-84, 128, 129, 142
Federal Bureau of Investigation (FBI) 33
Ferranti Mark 1* 49
Festival of Britain 1951 223
Flight magazine 161, 267, 280
Florida 236, 240
Fluid Dynamics Group 256
Focke-Wulf Fw 190 56
Fokker 100
Fort Halstead 49
Francis, R. H. 202, 205
Fraser, Charles 77
Fraser-Mitchell, Harry 141, 142, 286
Frisch, Otto 18, 19, 42
Fulton 33

Gadget 21
Gates, Thomas 218, 244
GEC 42
GEN 75 26, 35
General Aircraft 84
Gilpatric, Roswell 245, 248
Gilze-Rijen 116
Gloster E28/39 139, 220
Gloster Javelin 106, 134, 161-165, 167, 177, 178, 182, 221, 277
Gloster Meteor 76, 100, 102, 161, 220, 265
Gloster Thin Wing Javelin 164, 165, 167, 168, 170, 221
Goose Bay
GOR.148 212
GOR.177 215
Göttingen 59, 62, 73, 85, 140, 142, 143
Gouzenko, Igor 31-33
Graham, Gordon 289
Greatrix, F. B. 103
Green 254

303

Green Granite 202
Green Satin 43
Griffith, Alan Arnold 74-76, 79
Griswald, Roger 62
Guernica 26

H2S 43, 81, 92, 289
Haack, Wolfgang 223
Haines, A. B. 106, 165, 191
Hallows, Don 36
Hamburg 17
Hamble 82
Hampshire 118
Hampton, Virginia 55
Handley Page, Frederick 71, 73, 74, 80, 84, 95, 139, 141, 156
Handley Page Limited 15, 38, 41, 47-49, 52, 54, 60, 63, 68, 70, 71, 77, 84-86, 92, 96, 104, 113, 119, 124, 139, 142, 145, 146, 150, 153, 159, 169, 170, 178, 182, 184, 186, 189, 191, 208, 221, 229, 249, 279, 280, 287
Handley Page Halifax 15, 16, 38, 43, 44, 70, 74, 81, 85, 86
Handley Page Hampden 15, 74, 85, 86
Handley Page Hastings 70
Handley Page Heyford 15
Handley Page HP.42 85
Handley Page HP.65 Super Halifax 60, 71
Handley Page HP.75 Manx 70, 85, 86, 156
Handley Page HP.80 48, 49, 60, 85, 92, 96, 98, 144-146, 149, 156, 178
Handley Page HP.88 51
Handley Page Type O/400 85
Handley Page Rider-Plane 68, 70, 71, 85
Handley Page V/1500 15
Handley Page Victor 9, 13, 15, 48, 51, 52, 54, 89, 119, 121, 127, 134, 135, 139, 141, 142, 144, 150, 153, 158, 159, 168, 169, 175, 181-183, 187-192, 208, 213, 214, 220-223, 225, 226, 235, 241, 250, 256, 258, 264, 267, 268, 273, 275, 277, 279, 280, 283, 286, 287-290
Handley Page Victor II 182, 183
Handley Page Victor III 183
Handley Page Victor Phase 4 221, 223, 224, 226
Handley Page Victor B.Mk.1 79, 153, 182, 223, 273-275, 279, 280
Handley Page Victor B.Mk.2 48, 169, 182, 187, 189, 229, 234, 241, 264, 265, 268, 275, 279, 286, 287
Handley Page Victor K.Mk.2 288
Handley Page Victor WB771 127
Handley Page Victor XA930 189
Handley Page Victor XH667 183
Handley Page Victor XH668 183, 184, 187, 288

Handley Page Victor XH669 187, 188, 286, 288
Handley Page Victor XH670 187
Handley Page Victor XL164 192
Handley Page Victor XL189 288, 289
Hartley, Christopher 217
Harvey, Air Commodore 135
Hatfield 255, 256, 259
Hawker P.1052 102
Hawker Hunter 118, 161, 178, 203, 251, 277
Hawker Siddeley 102, 161, 220, 254, 258, 259, 260, 261, 288
Hawker Siddeley Harrier 121, 251
Hawker Siddeley HS.125 259
Hawker Siddeley HS.681 259
Hawker Siddeley Kestrel 251
Hawker Siddeley P.1127 251
Hawker Siddeley P.1154 251
Hawker Siddeley Sea Harrier 251
Hawker Siddeley Trident 2 259
Hawker Siddeley Trident 3 259
HBN-100 259
Heathrow 257
Heinkel He 177 140
Heinkel-Hirth HeS 109-011 64, 141
Hennesey, Peter 35
High Wycombe 199
Hill, G. T. 67
Hiroshima 16, 23
Hitch, Charles 238, 239
Hitler, Adolf 247
HMS *Beagle* 289
HMS *Conqueror* 290
Holder, D. W. 59
Holy Loch 242, 244, 246, 248
Honington 273
Hood, Samuel 210
Hooker, Stanley 78, 79, 289
Hooper, Ralph 121
Hoover, J. Edgar 33
House of Commons 29, 31, 134
Howell, Alun Raymond 76

IBM AN/ASQ-38(v) 237
Imperial Airways 85
Imperial Chemical Industries (ICI) 19
Imperial College 75, 85
Institute of Mechanical Engineers 80
Ishikawajima 86
Isle of Wight 85

Jacobs, Eastman 55-57, 60
Jamaica 242
Jarron, Ed 177, 203, 268
Joint Chiefs of Staff 215
Joint Committee on Atomic Energy 210
Johnson, Kelly 263
Jones, Melvill 82
Jones, Robert T. 61-63, 66, 139, 140, 223, 224
Journal of Aeronautical History 13
Junkers Ju 287 140, 141
Junkers Jumo 004 141

Kármán, Theodore von 63, 223
Kennedy, John F. 227, 229, 236, 240, 242, 244, 246-248
Kennet, Susan 116
Kent 18, 49, 54
KGB 33
King George VI 20
King, Mackenzie 29, 33, 210
Korean War 135
Kosin, Rüdiger 141
Küchemann, Dietrich 48, 107, 119, 121, 140, 141, 144-146, 156, 189, 192, 251, 254, 255, 257, 288

Lacelles, Alan (Tommy) 20
Lachmann, Gustav Victor 15, 18, 48, 62, 68, 70, 71, 73, 74, 85, 86, 88, 113, 133, 139
Lange, Roy H. 143
Langley Memorial laboratory 55
Langston, Paul 222
Lapham, Rosemary 94
Las Vegas 47
League of Nations 31
Lee, Godfrey 48, 49, 54, 67, 71, 85, 88, 141, 156, 280, 283, 288
Leeds Bradford Airport 83
Lehman, Walter 141
LeMay, Curtis E. 212
Lewis, George W. 63
Lincolnshire 42, 267, 286
Lindemann, Frederick 19
Lindley, Bob 87, 91, 289
Lippisch, Alexander 74, 223
Lippisch DM-1 223
Lissaman, Peter 47
Little Boy 16
Livermore Laboratory 217
Lloyd, Peter 75
Lockheed C-141 Starlifter 52, 132
Lockheed Electra 80

Lockheed F-104 Starfighter 220
Lockheed Hudson 116
Lockheed SR-71 Blackbird 226
Lockheed TriStar 259, 260
Lockheed U-2 229, 240, 262-264
Lockheed UGM-27 Polaris 201, 213, 216-219, 226, 232-234, 243-247, 261, 264
Lockheed XC-35 80
Lockspeiser, Ben 63
London 26, 29, 31, 47, 83, 85, 216, 243, 245
Lovell, Bernard 289
Ludlow-Hewitt, Edgar 16, 80
Ludwieg, Hubert 62, 140, 142
Luftfahrtforshungsanstalt (LFA) 62, 64
Lufthansa 259

Mabey, D. G. 121
Macmillan, Harold 210, 216, 219, 240, 242, 243, 244, 246-248
Magdeburg 26
Manhattan Project 21, 23, 27-29
Marbury, William 28
Marianas 23
Martin-Baker 100, 289
Martin B-57/RB-57 Canberra 262
Martin Bold Orion/WS.199 213, 214
Martin LGM-25 Titan 232, 233
Martin P6M Seamaster 158
Maud committee 18, 19
Maudling, Reginald 164
May-Johnson Bill 28, 31
McDonnell Douglas F-4 Phantom II 203
McDonnell Douglas F/A-18 Hornet 210
McElroy, Neil H. 215
McGraw Hill 107
McMahon, Brien 31
McMahon Act 210
McNamara, Robert 229, 232, 233, 236-240, 243, 244, 246-248
McRae, D. M. 259
Messerschmitt, Willy 62
Messerschmitt AG 62
Messerschmitt Bf 109 141
Messerschmitt Me 163 61
Messerschmitt Me 262 116
Messerschmitt P 1108 64
Metropolitan-Vickers (Metrovick) F.2 70
MI5 32, 86
Mikoyan-Gurevich MiG-15 212
Miles M.52 79, 163

Military Construction Act 1959 232
Ministry of Aircraft Production (MAP) 18, 70, 76, 84, 85
Ministry of Defence 243
Ministry of Supply 15, 35, 40, 49, 79, 86, 98, 115, 277
Missouri 33
Moore, Norton B. 223
Morrison, Herbert 35
Moscow 31, 32, 34, 36, 199, 230, 232, 246, 286
Mount Usborne 289
Multhopp, Hans 158
Multi-Lateral Force (MLF) 245, 246
Munk, Max 55, 62, 98
Murden, Michael 269
Mutual Defence Agreement (MDA) 210, 212

NACA/NASA Langley 55, 62, 63, 261
Nagasaki 23, 26
Napier Sabre 70
National Advisory Committee for Aeronautics (NACA) 55, 56, 59, 60, 62, 63, 71, 78, 84, 98, 103-106, 131, 133, 139, 142-144, 150, 158, 164, 175, 183, 189, 190, 225, 256, 265
National Aeronautics and Space Administration (NASA) 261
National Aerospace Laboratories 60
National Gas Turbine Establishment (NGTE) 79, 80
National Physical Laboratory (NPL) 13, 50, 51, 56, 59, 102, 129, 131, 133, 134, 251, 257, 259, 260, 261
NATO 232, 244-248
Neustadt, Richard E. 245, 247, 248
Newby, Ken 10, 13, 119, 121, 129, 132, 133, 136, 173, 175, 176, 256
New Mexico 21
New York 28, 242
Nord Aviation 258
Normalair 80
North American Aviation 210, 226
North American F-108 Rapier 227
North American GAM-77/AGM-28 Hound Dog 212, 213, 214, 215, 233, 238, 244, 247, 248
North American P-51 Mustang 60, 152
North American RS-70 238
North American SM-64 Navaho 210, 212
North American X-10 210
North American XB-70 Valkyrie 226, 227, 229, 230, 232-234, 238
North Field 23
Northampton Institute 85
Northampton Polytechnic 95
Northrop B-35 57, 113

Northrop YB-49 113
North Weald 267
Nunn May, Alan 32, 35

Oliphant, Mark 42
Operation Black Buck 13, 287, 290
Operation Gomorrah 17
Operation Grapple 210, 265, 269
Operation Meetinghouse 16
Operation Sturgeon 141
Oppenheimer, J. Robert 27
OR.229 35
OR.314 202
OR.339 202
OR.1001 35
OR.1132 202, 204, 212
OR.1139 199
OR.1159 214
Ormsby-Gore, David 240, 246, 247
Owner, Frank 77
Oxford University 18

P13/36 38, 68, 71, 82, 83
Palm Beach 248
Pan American 254
Parsons 74, 75
Pearcey, Herbert 10, 13, 59, 129, 131-134, 136, 251, 252, 257, 261
Pearl Harbor 19, 23
Peierls, Rudolf 18, 19, 42
People's Liberation Army (PLA) 262
Perring, William 51, 88, 90
Petter, W. E. W. 61
Pleming, Robert 10, 59, 261
Plymouth 31
Portal, Charles 35
Potomac 30
Potsdam conference 21, 23
Power Jets 79
Power Jets LR.1 79
Power Jets W.2/700 79
Powers, Francis Gary 229, 263, 264
Prague 34
Prandtl, Ludwig 55, 73
Pratt & Whitney J57 223, 262
Project Milk Bottle 266

Quebec Agreement 19, 21, 30, 31
Queen Elizabeth II 247

Radar Research Establishment 43
Radlett 187, 222
RAE Bedford 162, 164, 188, 208, 214
RAF Central Servicing Development Establishment (CSDE) 269
RAF Coltishall 203
RAF Coningsby 275
RAF Cottesmore 187, 273, 275
RAF Cranwell 199
RAF Gaydon 269
RAF Marham 269
RAF Scampton 275
RAF Waddington 136, 169, 203, 229, 273, 275, 289
RAF Wittering 241, 275
Ramaswamy, M. A. 60, 131
Randall, John 42
Red Arrows 9
Red Cat 202, 203
Red Flag 1982 249
Reeve, John 288
Richards, Tony 118
Robinson, Michael 241
Rocketdyne S-3D 200
Rolls-Royce AJ.65 71, 76, 79
Rolls-Royce Avon 79, 98, 103, 119, 169, 250, 262
Rolls-Royce Conway 169, 183, 187, 192, 221, 225, 252, 257, 279
Rolls-Royce Derwent 100, 111
Rolls-Royce Griffon 116, 279
Rolls-Royce Hucknall 103, 267
Rolls-Royce Merlin 60, 78
Rolls-Royce Nene 38, 286
Rolls-Royce Olympus 9, 277, 289
Rolls-Royce Olympus 21 249, 279
Rolls-Royce Olympus 22 250
Rolls-Royce Olympus 23 250
Rolls-Royce Olympus 301 250, 279
Rolls-Royce RB.207 258-260
Rolls-Royce RB.211 259, 260
Rolls-Royce RZ.2 200
Rolls-Royce Vulture 82
Roosevelt, Theodore 19, 21, 23
Rowbotham, Norman 77
Royal Aeronautical Society 10, 44, 48, 62, 74, 80, 85, 141, 280
Royal Aircraft Establishment (RAE) 10, 38, 44, 51, 52, 59, 60, 64, 74-76, 79, 80-82, 84, 88, 90, 94-98, 102, 104-107, 109-113, 116, 118, 119, 121, 124, 125, 129, 131, 132, 139, 140, 143, 145, 149, 152, 153, 161-165, 168, 173, 175, 176, 178, 181, 183, 184, 189, 202, 208-210, 220, 234, 251, 254, 256, 259, 261, 264, 267, 277, 287
Royal Aircraft Factory BE.2A 85
Royal College of Science 85
Royale, Kenneth 28
Royal Flying Corps 84
Royal Radar Establishment 220
Royal Navy 251
Rubel, John H. 242
Russell, Archibald 77
Russell, Dick 289, 290

SA-2 Guideline 229, 263, 264
SA-5 Gammon 230, 232
Sandys, Duncan 127, 199, 247
Saunders-Roe 85
Schairer, George S. 63, 65, 253
Schleswig 116
Sears, William 223
Seddon, John 107
Sequoia 30, 31
Senate Special Committee on Atomic Energy 31
Shoreham Airshow crash 118
Short Stirling 44, 81
Sikorsky XR-4 116
Skunk Works 263
Skylon 223
Smelt 64
Smith, D. M. 76
Smith, Joan 50
Society of British Aircraft Constructors (SBAC) 85
Society of British Aircraft Constructors Airshow 100, 118
Sofia 34
Southampton Water 82
South Kensington 85
Squire, Brian 59
Spanish Civil War 116
Sputnik 199, 210, 212, 214, 215, 227, 232
Stafford, Reginald 63, 71, 142
Stalin 21, 23, 31, 34
Stalingrad 31
Stanley 9, 289, 290
Stettin 33
Stimson, Henry L. 27
Stranraer 235
Sud Aviation 258
Sud Aviation Caravelle 156
Suez Crisis 210, 269
Sunday Express 161
Sunstrand 181

Supermarine Spitfire 13, 54, 89, 98, 116, 161, 262, 279
Supermarine Swift 52, 162, 165, 277
Supersonic Air Transport Committee 251
Sutcliffe, P. L. 38
Systems Directive 138A 214

Talbot, Eddie 95
Taylor, Peter 289
Tedder, Arthur 35
Theodorsen, Theodore 57
Thirlby, David 116, 118
Thorn, Bill 94, 95
Thorneycroft, Peter 239, 240, 243, 244, 246, 247
Thwaites, Brian 59
Thynne, Una 85
Tinian Island 23
Tizard, Henry 18, 20
Tokyo 16, 86
Tongzhou 262
Trafford Park 75
Trenchard, Hugh 199
Trieste 33
Trinity College 83
Trinity Test 21, 23
Truman, Harry S. 21, 23, 26-30, 33, 210
Tsien, Hsue-shen 62, 63
Tube Alloys 19, 20
Tuxford, Bob 288-290

United Airlines 254
United Nations 29
United Nations Charter 34
United Nations Organisation (UNO) 28, 29
University College Hull 121
University of Birmingham 18, 42, 161
University of Liverpool 18, 19
Urmston Church Choir 82
USAAF 36, 62, 63, 65, 210
USAF 135, 158, 201, 210, 212-215, 225-227, 229, 232, 236-238, 246, 266
USAF Strategic Air Command 199
US Navy 158, 201, 213, 219, 248
USS *George Washington* 219

V-1 16, 36
V-2 16, 199
VE Day 116, 140
Vickers-Armstrong 63
Vickers V-1000 127, 251, 252-255
Vickers Valiant 79, 125, 127, 135, 158, 159, 199, 212, 251, 265, 267-269, 275, 287
Vickers Valiant WB215 265
Vickers Valiant WP217 268
Vickers Valiant XD816 269
Vickers Valiant XD818 265, 269
Vickers Vanguard 252-254
Vickers VC7 254
Vickers VC10 250, 254, 255, 257, 258, 260
Vickers Viscount 269
Vickers Wellington 80, 81
Vickers Windsor 16, 81
Vienna 34
Vinson, Carl 227
Virginia 55
VJ-Day 26
Völkenrode 62, 63, 85, 141, 208
Volkert, George 16, 38, 68, 71, 85, 95, 220
Volta Congress on High Speed Flight 61, 62, 116
Vulcan to the Sky Trust (VTST) 10, 113

Wagner, Wolfgang 141
Wallace 257
Wallis, R. A. 153, 155
Wang, Ying-Chin 262, 263
Ward, George 134, 135
Warnemünde 141
War Office 85
Warsaw 31, 34
Warsaw Pact 264
Washington, D.C. 30, 214, 217, 234, 239, 242, 245, 248
Wass, Geoffrey 222
Waterton, Bill 162
Watkinson, Harold 216, 217, 229, 243, 254
Weber, Johanna 48, 107, 140, 144-146, 156, 254, 255, 257, 288
Webster, John 95
West Freugh 235
Westland Aircraft 60, 61, 67, 80
Westland Pterodactyl 67
Westland Welkin 60, 61, 81
Westminster College 33
Whitcomb, Richard 178, 190, 191, 223-225, 261
White, Peter 191
White, Thomas H. 214
Whitehall 85, 199
White House 240, 248
Whitley, Mrs 47, 48
Whittle, Frank 74-76, 78-80, 84
Wideawake Air Force Base 287, 289
Widgery, W. M. 80

Wigg, George 135
Wilbur Wright Memorial Lecture 129
Wilson, Charles 199
Wilson, David 95
Wilson, Woodrow 55
Wilson-Harris, Henry 31
Window 189, 191, 192
Withers, Martin 9, 267, 288-290
Woodford 9, 49, 82, 100, 116, 118, 119, 136, 234, 268, 288, 289
Woomera 236
Wright, Anthony 267
Wright, Bob 289
Wright, Orville 73
Wright Brothers 85, 246
Wright Field 80
Wright Flyer 85

WS.199 (see also Martin Bold Orion) 213, 214
WS.110A 226
Wyatt, Woodrow 134

Yalta 21
Yeadon 83
Yeager, Chuck 40
Yellow Sun 275, 289
York, E. 50
York, Herbert 215-217
Young, Alec David 59
Younger, John E. 80, 81

Zabotin, Nicholai 32
Zip fuel 226
Zobel, Theodor 140
Zuckerman, Solly 217, 243